## Characteristic ¹H NMR chemical shifts

| Compound | Chemical shift |
|---|---|
| TMS | 0.0 |
| Alkanes (CH—C) | 0.9–1.5 |
| Ketones, aldehydes, carboxylic acids, esters, and other acid derivatives (CH—C=O) | 1.9–3.0 |
| Halides (CH—X) | 2.1–4.52 |
| Alcohols, esters, ethers (CH—O) | 3.0–4.5 |
| Alkenes (CH=C) | 4.0–6.8 |
| Aromatic (Ar—H) | 6.5–8.5 |
| Aldehydes [(C=O)H] | 9.0–10.0 |
| Amines (—NH) | 1.0–3.0 |
| Alcohols (—OH) | 1.0–5.5 |
| Phenols (ArOH) | 4.0–8.0 |
| Amides [(C=O)NH] | 5.0–8.0 |
| Carboxylic acids [(C=O)OH] | 10.0–13.0 |

## Substituent-shift parameters for alkyl protons compiled from a variety of sources*

**Base values**
Methyl (R = R′ = H)    0.9 ppm
Methylene (R′ = H)    1.2 ppm
Methine    1.5 ppm

| Group (Y) | Alpha substituent RR′C$\underline{H}$Y | Beta substituent RR′C$\underline{H}$CH$_2$Y |
|---|---|---|
| —C=C | 0.8 | 0.2 |
| —C≡C—H, —C≡C—R | 1.0 | 0.2 |
| —C≡C—C$_6$H$_5$ | 1.2 | 0.2 |
| —C$_6$H$_5$ | 1.3 | 0.4 |
| —(C=O)OH, —(C=O)OR | 1.0 | 0.3 |
| —(C=O)R, —(C=O)H | 1.1 | 0.4 |
| —(C=O)C$_6$H$_5$ | 1.6 | 0.6 |
| —I | 2.0 | 0.6 |
| —Br | 2.2 | 0.5 |
| —Cl | 2.3 | 0.4 |
| —F | 3.2 | 0.2 |
| —NH$_2$ | 1.3 | 0.3 |
| —NH(C=O)R | 1.8 | 0.3 |
| —NH(C=O)C$_6$H$_5$ | 2.3 | 0.3 |
| —OR | 1.9 | 0.3 |
| —OH | 2.1 | 0.2 |
| —OC$_6$H$_5$ | 2.5 | 0.6 |
| —O(C=O)R | 2.6 | 0.4 |

*   There may be small differences in the chemical-shift values calculated from this table and those measured from individual spectra.

2.0 mL
1.5 mL
1.0 mL
0.75 mL
0.50 mL
0.25 mL
0.10 mL
0.05 mL

# Techniques
# in Organic Chemistry

## Miniscale, Standard Taper Microscale, and Williamson Microscale

**JERRY R. MOHRIG**
Carleton College

**CHRISTINA NORING HAMMOND**
Vassar College

**PAUL F. SCHATZ**
University of Wisconsin, Madison

**TERENCE C. MORRILL**
Rochester Institute of Technology

**W. H. Freeman and Company**
New York

*Acquisitions Editor:* Jessica Fiorillo/Yolanda Cossio
*Development Editor:* Robert Jordan
*Marketing Manager:* Mark Santee
*Project Editor:* Penelope Hull
*Media and Supplements Editor:* Charlie Van Wagner
*Assistant Editor:* Guy Copes
*Text designer:* Marcia Cohen
*Cover designer:* Michael Jung
*Cover photo:* Juniper Pierce
*Illustrator:* Fine Line Illustrations
*Illustration Coordinator:* Bill Page
*Production Coordinator:* Julia DeRosa
*Compositor:* Progressive Information Technologies
*Printer and Binder:* R R Donnelley & Sons

**Library of Congress Cataloging-in-Publication Data**
Techniques in organic chemistry : miniscale, standard taper microscale,
    and Williamson microscale / Jerry R. Mohrig ... [et al.].
        p. cm.
    Includes bibliographical references and index.
    ISBN 0-7167-6638-8 (paper)
    1. Chemistry, Organic—Laboratory manuals. 2. Microchemistry—Laboratory manuals.
    I. Mohrig, Jerry R.
QD261 .T44 2003
543′.0858—dc21
2002003464

Printed in the United States of America

First printing 2002

**W. H. Freeman and Company**
41 Madison Avenue, New York, NY 10010
Houndmills, Basingstoke RG21 6XS, England

# Contents

**PART 3     SPECTROSCOPIC METHODS / 195**

# Preface

Organic chemistry is an experimental science, and students learn its process in the laboratory. Our primary goal in teaching laboratories should be to involve students in the process, to take them beyond the "cookbook" approach to address questions whose answers come from their experimental data. Teaching students how to interpret experimental results and draw reasonable conclusions from them is at the heart of the process of science. Synthesizing organic compounds and answering questions that concern their reactions require the mastery of good laboratory technique and build scientific skills. *Techniques in Organic Chemistry* presents the most important techniques used in modern organic chemistry.

## Who Should Use This Book?

This book is intended to serve as course material and future reference material for all students of organic chemistry. It can be used in conjunction with any laboratory experiments to provide the skills and understanding necessary to master the organic chemistry laboratory. It also serves as a useful reference for laboratory practitioners and instructors.

This book is also designed to accompany *Modern Projects and Experiments in Organic Chemistry,* a significant revision of the Projects and Experiments parts of *Experimental Organic Chemistry: A Balanced Approach.* As was done in *Experimental Organic Chemistry,* all projects and experiments contain cross-references to relevant techniques.

## Flexibility

*Techniques in Organic Chemistry* offers a great deal of flexibility. It can be used for any organic laboratory with any glassware. The basic techniques for using standard taper miniscale glassware as well as 14/10 standard taper microscale and Williamson microscale glassware are all covered in this book. The miniscale glassware that is described is appropriate with virtually any 14/20 or 19/22 standard taper glassware kit.

## Modern Instrumentation

Modern instrumental methods play a crucial role in supporting the investigative experiments, which provide the active learning opportunities that many instructors seek for their students. We feature instrumental methods that provide quick, reliable, quantitative data. Nuclear magnetic resonance (NMR) and infrared (IR) spectroscopy, as well as gas-liquid chromatography (GC), are particularly important. The section on spectroscopic methods breaks away from the emphasis on theory, which is already present in lecture textbooks, to provide the kinds of information that students need to use spectroscopy effectively in the organic laboratory. The chapters on NMR, IR, and mass spectrometry have been totally rewritten with a focus on the practical interpretation of spectra and how they can be used to answer questions posed in an experimental context. They describe how to deal with real laboratory samples. Subsections on sample preparation and sources of confusion walk students through the pitfalls that could easily discourage them if they did not have practical support.

## Organization

The book is divided into three parts. Part 1 covers the basic techniques for synthesis and purification of organic compounds, and it begins with a chapter on laboratory safety, of paramount importance from the very beginning of any laboratory experience. Part 2, Chromatography, discusses the three chromatographic techniques widely used in the organic laboratory: thin-layer, gas-liquid, and liquid chromatography. Modern spectroscopy has revolutionized structure determination in organic chemistry, and Part 3, Spectroscopic Methods, discusses them in detail. An appendix discusses the sources for retrieval of chemical information. Additionally, a special section, "Organic Quatitative Analysis," is available on our Web site, www.whfreeman.com/mohrig.

## Modern Projects and Experiments in Organic Chemistry

The accompanying laboratory manual, Modern Projects and Experiments in Organic Chemistry, comes in two complete versions:

- *Modern Projects and Experiments in Organic Chemistry: Miniscale and Standard Taper Microscale* (ISBN 0-7167-9779-8)
- *Modern Projects and Experiments in Organic Chemistry: Miniscale and Williamson Microscale* (ISBN 0-7167-3921-6)

A unique combination of standard and inquiry-based experiments, plus multiweek projects, Modern Projects and Experiments in Organic Chemistry is designed to provide the utmost in quality content, student accessibility, and instructor flexibility. This laboratory manual introduces students to the way the contemporary organic lab actually functions.

## Custom Publishing

All experiments and projects are available through W. H. Freeman's custom publishing service. Instructors can use this service to create their own customized lab manual, selecting specific experiments and projects from Modern Projects and Experiments in Organic Chemistry, even incorporating their own material, so that the manual is organized to suit their course.

Visit http://custompub.whfreeman.com to learn more.

We hope that teachers and students of organic chemistry find our text useful in their laboratory work. We would be pleased to hear from users of the book regarding ways in which it might be improved. Please write to us in care of the Chemistry Acquisitions Editor at W. H. Freeman and Company, 41 Madison Avenue, New York, NY 10010, or e-mail us at chemistry@whfreeman.com.

# Acknowledgments

We have benefited greatly from the insights and thoughtful critiques of the following reviewers:

Jamie L. Anderson, Santa Monica College
Luis Avila, Columbia University
Christine DiMeglio, Yale University
Mark Forman, St. Joseph's University
Alvan C. Hengge, Utah State University
Susan L. Knock, Texas A&M University, Galveston
Barbara Mayer, California State University, Fresno
Lynne O'Connell, Boston College
Dan Steffek, North Carolina State University

We wish to thank Jessica Fiorillo, our editor at W. H. Freeman and Company, for her vision and direction of this revision, Robert Jordan, Development Editor, for his helpful discussions in preparing the manuscript, Marsha Cohen, Designer, for the elegant book design, and Penelope Hull, Project Editor, for her masterly orchestration of the production stages. JRM and CNH express heartfelt thanks for the patience and support of our spouses, the late Jean Mohrig and Bill Hammond, during the years that we have worked together on these books.

# PART

# 1

# Basic Techniques

As you begin your study of experimental organic chemistry, you need to learn how to work safely with organic chemicals. Technique 1 discusses laboratory safety, and we urge you to read it carefully before you begin laboratory work. When you work in the laboratory, always keep in mind the safe ways of performing laboratory operations and the safe handling practices for the chemicals you use.

Chemistry is an experimental science. The laboratory is where you learn its processes. To learn experimental organic chemistry, you need to master an array of techniques for carrying out chemical reactions, separating products from their reaction mixtures, purifying them, and analyzing the results. Part 1 covers the basic techniques of the organic laboratory, including the following:

- Heating and cooling methods, and how to assemble glassware for organic reactions
- Extraction, recrystallization, distillation, and sublimation for separating and purifying organic compounds
- Melting points, polarimetry, and refractometry for analyzing products

The three fundamental chromatographic methods used for separation and analysis in organic chemistry—thin-layer chromatography, gas-liquid chromatography, and liquid chromatography—are discussed in Part 2, Techniques 15–17. The important modern analytical tools of infrared spectroscopy, nuclear magnetic resonance spectroscopy, and mass spectrometry are discussing in Part 3, Techniques 18–20.

# 1

# SAFETY

As you begin your study of experimental organic chemistry, you need a basic understanding of general laboratory safety principles. The organic chemistry laboratory is a place where accidents can occur and where safety is everyone's business. As a consumer, you are protected by laws. In the laboratory, you are protected by the instructions in an experiment and by the design of the laboratory. Your laboratory is designed to safeguard you from most routine hazards, but it cannot protect you from the worst hazard—your own or your neighbors' carelessness. In this chapter we will familiarize you with basic laboratory safety information and common problems relating to safety.

In addition to a knowledge of basic laboratory safety, you need to learn how to work safely with organic chemicals. Many organic compounds are flammable, toxic, or can be absorbed through the skin; others are volatile and vaporize easily into the air within the laboratory. Despite the hazards, organic compounds can be handled with a minimum of risk if you are adequately informed about the hazards and necessary safe-handling procedures and if you use common sense while you are in the laboratory.

At the first meeting of your lab class, local safety issues will be discussed—the chemistry department's policies on safety goggles, the location of safety showers, eye wash stations, protective gloves, and the procedures to be followed in emergency situations. The information in this chapter is intended to complement the safety rules and instructions given by your instructor.

Part of learning to do organic chemistry in the laboratory is learning how to do it safely. Consider this chapter to be required reading before you perform any experiments in the organic laboratory.

## 1.1 Causes of Laboratory Accidents

Laboratory accidents are of three general types: accidents involving fires and explosions; those producing cuts or burns; and those occurring from inhalation, ingestion, or absorption through the skin of hazardous materials.

*Fires and Explosions* Fire is the chemical union of a fuel with an oxidizing agent, usually oxygen, and is accompanied by the evolution of heat and a flame. Most fires are those of ordinary combustible materials, which are hydrocarbons or their derivatives. Such fires are *extinguished* by removing oxygen or the combustible material or by decreasing the heat of the burning fire. Fires are *prevented* by keeping flammable materials away from a flame source or from oxygen (obviously, the former is easier).

There are four sources of ignition in the organic laboratory: *open flames, hot surfaces* such as hot plates or hot heating mantles, *faulty electrical connections,* and *chemical fires.* The most obvious way to prevent a fire is to prevent ignition.

Open-flame ignition is easily prevented: **Never bring a lighted Bunsen burner or a match near a low-boiling-point flammable liquid.** Furthermore, because vapors from organic liquids travel over long distances toward the floor (they are heavier than air), an open flame within 10 ft of diethyl ether, pentane, or other low-boiling-point (and therefore volatile) organic solvents is an unsafe practice. In fact, the use of a Bunsen burner or any other flame in an organic laboratory should be a rare occurrence and should be done only with the permission of your instructor.

Hot surfaces, such as a hot plate or heating mantle, present a trickier problem (Figure 1.1). An organic solvent spilled or heated recklessly on a hot plate surface may burst into flames (Figure 1.2). The thermostat on most hot plates is not sealed and sparks when it cycles on and off. The spark can ignite flammable vapors from an open container such as a beaker. It is also possible for the vapors from a volatile organic solvent to be ignited by the hot surface of a hot plate or a heating mantle, even if you are not actually heating a container of the substance with the heating device. Remove any hot heating mantle or hot plate from the vicinity before pouring a volatile organic liquid.

Electrical fires pose the same danger in the laboratory as they do in the home; do not use appliances with frayed or damaged electrical cords.

Chemical reactions sometimes produce enough heat to cause a fire and explosion. Consider, for example, the reaction of metallic sodium with water. The hydrogen gas that forms as a product of this reaction can explode and then ignite a volatile solvent that happens to be nearby.

*Cuts and Injuries*

Cuts and mechanical injuries are hazards anywhere, including in the laboratory.

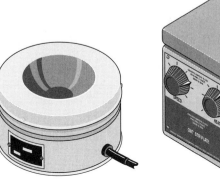

**FIGURE 1.1**
Heating devices.

Ceramic heating mantle                Hot plate

**FIGURE 1.2**
Fire hazards with a hot plate.

Flammable liquid spilled
on a hot plate

Flammable vapors emitted when
heating a volatile solvent
in an open container

When you break a glass rod or a glass tube, do it correctly. Score (scratch) a small line on one side of the tube with a file. Wet the scored line with a drop of water. Then, holding the tube on both sides with a towel and with the scored part away from you, quickly snap it by pulling the ends toward you (Figure 1.3).

Insert glass tubes or thermometers into cork or rubber stoppers carefully and correctly. First, lubricate the end of the tube with a drop of water or glycerol; then, while holding it with a towel close to the lubricated end, insert the tube slowly by firmly rotating it into the stopper. Never hold the glass tube by the end far away from the stopper—it may break and the shattered end be driven into your hand.

Check the upper rim of beakers, flasks, and other glassware for chips. Discard any piece of glassware that is chipped, because you could be cut very easily by the sharp edge.

**Inhalation, Ingestion, and Skin Absorption**

The hoods in your laboratory protect you from inhalation of noxious fumes, hazardous vapors, or dust from finely powdered materials. A hood is an enclosed space with a continuous flow of air that sweeps

**FIGURE 1.3**
Breaking a glass rod properly.

over the bench top, removing vapors or fumes from the area. Because a great many compounds are at least potentially dangerous, the best practice is to run every experiment in a hood, if possible. Your instructor will tell you when an experiment must be carried out in a hood. Make sure that the hood is turned on before you use it. Position the sash for the optimal air flow through the hood. If the optimum sash position is not indicated on the hoods in your laboratory, consult your instructor about how far to open the sash.

Ingestion of chemicals by mouth is easily prevented. **Never taste any substance or pipet any liquid by mouth.** Wash your hands with soap and water before you leave the laboratory. Food or drink of any sort should neither be brought into a laboratory nor eaten there.

Many organic compounds are absorbed through the skin. Wear gloves while handling these reagents and reaction mixtures. If you spill any substance on your skin, wash the affected area thoroughly with water for 10–15 min.

---

## 1.2  Safety Features in Your Laboratory

Your organic laboratory contains many safety features for your protection and comfort. It is unlikely that you will have to use them, but in the event that you do, you must know what and where they are and how they operate.

*Fire Extinguishers*

Colleges and universities all have standard policies regarding the handling of fires. Your instructor will inform you as to whether evacuation of the lab or the use of a fire extinguisher takes priority at your institution. **Learn where the exits from your laboratory are located.**

Fire extinguishers are strategically located in your laboratory. There may be several types and your instructor should demonstrate their use.

Water as a fire extinguisher in an organic laboratory has a disadvantage. Because many flammable liquids are lighter than water, they rise to the surface. When water is splashed on this type of fire, it will actually spread it. Your lab, therefore, is equipped with either class BC or class ABC dry chemical fire extinguishers suitable for solvent or electrical fires.

*Fire Blankets*

Fire blankets are used for one thing and one thing only—to smother a fire involving a person's clothing. Fire blankets are available in most labs. If a person's clothing catches fire, **drop the person to the floor** and **roll** him/her to wrap the blanket tightly around the body. To wrap the blanket around a person who is standing may direct the flames toward the person's face.

*Safety Showers*

Safety showers are for acid burns and any other spill of corrosive, irritating, or toxic chemicals on the skin or clothing. If a shower is nearby, it can also be used when a person's clothing or hair is ablaze. The typical safety shower dumps a huge volume of water in a short period of time and thus is also effective in the case of acid spills, when speed is of the essence. **Do not use the safety shower routinely, but do not hesitate to use it in an emergency.**

*Eye Wash Stations*

**You should always wear safety goggles while working in a laboratory,** but if you accidentally splash something in your eyes, immediately rinse them with copious quantities of tepid water for at least 15 min at an eye wash station. Learn the location of the eye wash stations in your laboratory and examine the instructions on them during the first (check-in) lab session. You need to keep your eyes open while using the eyewash. Because this is difficult, assistance may be required. Do not hesitate to call for help.

If you are wearing contact lenses, they must be removed for the use of an eye wash station to be effective, an operation that is extremely difficult if a chemical is causing severe discomfort to your eyes. Therefore, **it is prudent not to wear contact lenses in the laboratory.**

*First Aid Kits*

Your laboratory contains a basic first aid kit consisting of such items as adhesive bandages, sterile pads, and adhesive tape for treating a small cut or burn. All injuries, no matter how slight, should be reported to your instructor immediately. Your instructor will send you to the college health service, accompanied by another person, after any accident except for the most trivial. However, some immediate first aid will almost always be required. Your instructor will indicate the location of the first aid station and instruct you in its use.

---

**1.3**     **Preventing Accidents**

Accidents can largely be prevented by common sense and a few simple safety rules:

1. Think about what you are doing while you are in the laboratory. Read your experiment ahead of the laboratory time and perform laboratory operations with careful forethought.
2. It is a state law in many states and common sense in the remainder to **wear safety glasses or goggles at all times in the laboratory.** Your institution may have a policy regarding wearing contact lenses in the laboratory; learn what it is.
3. Be aware of what your neighbors are doing. Many accidents and injuries in the laboratory are caused by other people. Often the person hurt worst in an accident is the one standing next to the place where the accident occurred.
4. **Never work alone in the laboratory.** Being alone in a situation in which you may be helpless is dangerous.
5. Make yourself aware of the procedures that should be followed in case of any accident. (See Technique 1.4, What to Do If an Accident Occurs.)

6. **Never taste, ingest, or sniff directly any chemical.** Always use the hood when working with toxic or noxious materials. Handle all chemicals carefully, and remember that many chemicals can enter the body through the skin and eyes, as well as through the mouth or lungs.

7. Disposable gloves are available in all laboratories. **Wear gloves** to prevent chemicals from coming into contact with your skin unnecessarily. Wear a lab coat or apron when working with hazardous chemicals; cotton is the preferred fabric because synthetic fabrics could melt in a fire or undergo a reaction that causes the fabric to adhere to the skin. If a chemical is spilled on the skin, wash it off immediately with soap and water. If you are handling a particularly hazardous compound, find out how to remove it from your skin before you begin the experiment. Consult your instructor if in doubt about the toxicity of the chemical. Always wash your hands at the end of the laboratory period.

8. Wear clothing that covers and protects your body. Shorts, tank tops, and sandals (or bare feet) are not suitable attire for the laboratory. Avoid loose clothing and long hair, which are a fire hazard or could become entangled in an apparatus. Laboratory aprons or lab coats may be required by your instructor.

9. Keep your laboratory space clean. In addition to your own bench area, the balance and chemical dispensing areas should be left clean and orderly. **If you spill anything while measuring out your materials, notify your instructor and clean it up immediately.** Do not leave a mess that other students must work around. After weighing your chemicals, replace the caps on the containers.

10. **Never heat a closed system! Never completely close off an assembly of apparatus in which a gas is being evolved.** Always provide a vent in order to prevent an explosion.

11. Flammable solvents with boiling points of less than 100°C, such as diethyl ether, methanol, pentane, hexane, ethanol, and acetone, should be distilled, heated, or evaporated on a steam bath or heating mantle, **never** with a Bunsen burner. Keep flammable solvents in flasks and not in open beakers.

12. **All laboratories have explicit directions for waste disposal,** and you should obtain details from your lab instructor. Local regulations take precedence over waste disposal methods suggested by any experiment.

13. **Remember that both glass and hot plates look the same when hot as when cold.** When heating glass, do not touch the hot spot. Do not lay hot glass on a bench where someone else might pick it up.

14. Keep gas and water valves closed whenever they are not in use.

15. **Steam and boiling water cause severe burns.** Turn off the steam source before removing containers from the top of a steam bath or steam cone. Handle containers of boiling water very carefully.

16. **Never eat, chew gum, drink beverages, or apply cosmetics in the lab.**

17. Floors can become very slippery if water is spilled; wipe up any spill immediately.

18. Women who are pregnant or who become pregnant should discuss working in the organic laboratory with the appropriate medical professionals.

| 1.4 | **What to Do If an Accident Occurs** |

Act quickly, but think first. The first few seconds after an accident may be crucial. Acquaint yourself with the following instructions so that you can be of immediate assistance.

**Fire**

In case of fire, get out of danger and then immediately notify your instructor. If possible, remove any containers of flammable solvents from the fire area. You should consult your laboratory instructor about the proper response to a fire. **It is important to know the policy of your institution concerning when to evacuate the building and when to use a fire extinguisher.**

Know the location of the fire extinguishers and how they operate. A fire extinguisher will always be available. Aim low and direct its nozzle first toward the edge of the fire and then toward the middle. Tap water is not always useful for extinguishing chemical fires and can actually make some fires worse, so always use the fire extinguisher.

If a person's clothes catch fire, smother the flames with either a lab coat, a fire blanket, or a water shower. Be sure you know where the fire blanket and safety shower are located. If your clothing is on fire, do not run. Rapid movement fans flames.

**Cuts and Burns**

Be aware of the location of the first aid kit and of the materials it contains for the treatment of burns and cuts. The first thing to do for any chemical wound, unless you have been specifically told otherwise, is to wash it well with water. This treatment will rinse away the excess chemical reagent. For acids, alkalis, and toxic chemicals, quick dousing with water will save pain later. Specific treatments for chemical burns are available in *The Merck Index*. **Notify your instructor immediately if you are cut or burned.**

If chemicals get into your eye, use the eye wash fountain and wash your eye immediately with a large amount of water. Hold your head in the shower or under the tap. **Hold your eyes open** to allow the water to flush the eyeballs. Do not use very cold water, however, because it can also damage the eyeball. Then seek medical treatment immediately.

**General Policy**

Always inform the instructor of any accident that has happened to you or your neighbors. **Let your instructor decide whether a physician's attention is needed.** If a physician's attention is necessary, the injured person should always be accompanied to the medical facility, even when he/she protests that he/she is fine. The injury may be more serious than it initially appears.

## 1.5    Chemical Toxicology

The toxicity of a compound generally refers to its ability to produce injury once it reaches a susceptible site in the body. The toxicity of a substance is related to its probability of causing injury and is a species-dependent term. What is toxic for people may not be toxic for other animals and vice versa.

A substance is *acutely toxic* if it causes a toxic effect in a short time; it is *chronically toxic* if it causes toxic effects with repeated small exposures over a long duration. A major concern in chemical toxicology is quantity or dosage. A *poison* is a substance that causes harm to living tissues when applied in small doses. Poison is a word that comes to us through many centuries of usage; it has meaning but is not a scientifically useful term. Remember, the dose makes the poison.

Most compounds in their pure form are toxic or poisonous in one way or another. For example, oxygen is strongly implicated in many forms of cell aging and cell destruction and is also important in cancer cell growth. Oxygen at a partial pressure of 2 atm is toxic; for this reason, scuba tanks are filled not with $O_2$ but with compressed air. Thus oxygen, although essential for life, is toxic to people if the concentration is sufficiently high. In the organic laboratory, most compounds you use are more toxic than oxygen, however, and you must be knowledgeable about how to handle them safety.

***How Toxic Substances Enter the Body***

Three main routes exist for any toxic substance to enter the body: through the lungs, through the mouth and gastrointestinal tract, and through the skin. Through any of these routes, a toxic substance can reach the bloodstream, where it can be carried to other body tissues.

Many organic solvents are absorbed through the skin. Wearing gloves while working in the organic laboratory protects your hands. Table 1.1 lists commonly available types of gloves and their chemical resistance. The barrier properties of each glove may be affected by differences in glove thickness, chemical concentration, temperature, and length of exposure to the chemical.

Many organic solvents are also absorbed through the lungs, because most of them have an appreciable vapor pressure at room temperature. Working in a properly ventilated hood can prevent exposure to chemical vapors.

Fortunately, not all poisons that accidentally enter the body get to a site where they can be deleterious. Even though a toxic substance is absorbed, it is often excreted rapidly. Our body protects us with various devices: the nose, scavenger cells, metabolism, cell replacement, and rapid exchange of good air for bad. Most toxic substances are detectably toxic only because they are present in our environment over long periods of time. Most foreign substances are discharged from the body immediately.

The action of toxic substances varies from individual to individual. Although many substances are toxic to the entire system (arsenic, for example), many others act on specific sites; that is,

| TABLE 1.1 | Chemical resistance of common types of gloves to various compounds[a] | | |
|---|---|---|---|

| | GLOVE TYPE | | |
|---|---|---|---|
| Compound | Neoprene | Nitrile | Latex |
| Acetone | good | fair | good |
| Chloroform | good | poor | poor |
| Dichloromethane | fair | poor | poor |
| Diethyl ether | very good | good | poor |
| Ethanol | very good | excellent | excellent |
| Ethyl acetate | good | poor | fair |
| Hexane | excellent | excellent | poor |
| Methanol | very good | fair | fair |
| Nitric acid (conc.) | good | poor | poor |
| Sodium hydroxide | very good | excellent | excellent |
| Sulfuric acid (conc.) | good | poor | poor |
| Toluene | fair | fair | poor |

a. The information in this table is compiled from the Web site http://www.inform. umd.edu/CampusInfo/Departments/EnvirSafety/Is/gloves.html and from "Chemical Resistance and Barrier Guide for Nitrile and Natural Rubber Latex Gloves," Safeskin Corporation, San Diego, CA, 1996.

they are site specific. Carbon monoxide, for example, forms a complex with blood hemoglobin, destroying the blood's ability to absorb and release oxygen. Acute benzene toxicity seems to affect only the platelet-producing bone marrow.

In certain instances (e.g., methanol poisoning), the metabolites (in this case, formaldehyde and formic acid) are more deleterious than the original compound; formic acid affects the optic nerve, causing blindness. Impaired health can be the result of a substance acting at some body site to cause a structural change; as far as the body is concerned, it does not matter whether the poison is the original substance or a metabolic product of it.

**Toxicity Testing and Reporting**

Consumers are protected by a series of laws that define toxicity, the legal limits and dosages of toxic materials, and the procedures for measuring these toxicities.

Acute oral toxicity is measured in terms of $LD_{50}$ ("LD" stands for "lethal dose"). The $LD_{50}$ represents the dose, in milligrams per kilogram of body weight, that will be fatal to 50% of a certain population of experimental animals (mice, rats, and so on). Other tests include dermal toxicity (skin sensitization) and irritation of mucous membranes (eyes, nose).

The toxicity of almost every chemical compound commercially available has been reported, and every year the toxicity of many more compounds becomes known. A wall chart of toxicities for many common organic compounds may be hanging in your laboratory or near your stockroom. A useful and handy reference for toxicity information is *The Merck Index*, and a page from the twelfth edition (1996) is reproduced in Figure 1.4.

Anilinephthalein                                                698

White to tan crystals, mp 159-160°. Insol in water. Soly at 30° (g/100 ml): toluene, 5; xylene, 4; acetone, 10. Subject to hydrolysis; not compatible with oils and alkaline materials. $LD_{50}$ orally in rats: > 5000 mg/kg, Mobay Technical Information Sheet, Jan. 1979.
USE: Fungicide.

**695. Anileridine.** *1-[2-(4-Aminophenyl)ethyl]-4-phenyl-4-piperidinecarboxylic acid ethyl ester; 1-(p-aminopheneth-yl)-4-phenylisonipecotic acid ethyl ester; ethyl 1-(4-amino-phenethyl)-4-phenylisonipecotate; N-[β-(p-aminophenyl)eth-yl]-4-phenyl-4-carbethoxypiperidine; N-β-(p-aminophenyl)-ethylnormeperidine; Leritine; Nipecotan; Alidine; Apodol.* $C_{22}H_{28}N_2O_2$; mol wt 352.48. C 74.97%, H 8.01%, N 7.95%, O 9.08%. Synthesis: Weijlard *et al., J. Am. Chem. Soc.* **78,** 2342 (1956); U.S. pat. 2,966,490 (1960 to Merck & Co.).

mp 83°.
Dihydrochloride, $C_{22}H_{28}N_2O_2.2HCl$, crystals from methanol + ether, mp 280-287° (dec). Freely soluble in water, methanol. Solubility in ethanol: 8 mg/g. pH of aq solns 2.0 to 2.5. Solns are stable at pH 3.5 and below. At pH 4 and higher the insol free base is precipitated. uv max (pH 7 in 90% methanol contg phosphate buffer): 235, 289 nm ($A_{1cm}^{1\%}$ 293, 34.5). Distribution coefficient (water, pH 3.6/n-butanol): 0.9.
*Note:* This is a controlled substance (opiate) listed in the U.S. Code of Federal Regulations, Title 21 Part 1308.12 (1995).
THERAP CAT: Analgesic (narcotic).

**696. Aniline.** *Benzenamine; aniline oil; phenylamine; aminobenzene; aminophen; kyanol.* $C_6H_7N$; mol wt 93.13. C 77.38%, H 7.58%, N 15.04%. First obtained in 1826 by Unverdorben from dry distillation of indigo. Runge found it in coal tar in 1834. Fritzsche, in 1841, prepared it from indigo and potash and gave it the name aniline. Manuf from nitrobenzene or chlorobenzene: Faith, Keyes & Clark's *Industrial Chemicals,* F. A. Lowenheim, M. K. Moran, Eds. (Wiley-Interscience, New York, 4th ed., 1975) pp 109-116. Procedures: A. I. Vogel, *Practical Organic Chemistry* (Longmans, London, 3rd ed.; 1959) p 564; Gattermann-Wieland, *Praxis des organischen Chemikers* (de Gruyter, Berlin, 40th ed., 1961) p 148. Brochure *"Aniline"* by Allied Chemical's National Aniline Division (New York, 1964) 109 pp, gives reactions and uses of aniline (877 references). Toxicity study: K. H. Jacobson, *Toxicol. Appl. Pharmacol.* **22,** 153 (1972).

Oily liquid; colorless when freshly distilled, darkens on exposure to air and light. *Poisonous!* Characteristic odor and burning taste; combustible; volatile with steam. $d_4^{20}$ 1.022. bp 184-186°. Solidif −6°. Flash pt, closed cup: 169°F (76°C). $n_D^{20}$ 1.5863. pKb 9.30. pH of 0.2 molar aq

soln 8.1. One gram dissolves in 28.6 ml water, 15.7 ml boil. water; misc with alcohol, benzene, chloroform, and most other organic solvents. Combines with acids to form salts. It dissolves alkali or alkaline earth metals with evolution of hydrogen and formation of anilides, e.g., $C_6H_5NHNa$. Keep well closed and protected from light. Incompat. Oxidizers, albumin, solns of Fe, Zn, Al, acids, and alkalies. $LD_{50}$ orally in rats: 0.44 g/kg (Jacobson).
Hydrobromide, $C_6H_7N.HBr$, white to slightly reddish, crystalline powder, mp 286°. Darkens in air and light. Sol in water, alc. *Protect from light.*
Hydrochloride, $C_6H_7N.HCl$, crystals, mp 198°. d 1.222. Darkens in air and light. Sol in about 1 part water; freely sol in alc. *Protect from light.*
Hydrofluoride, $C_6H_7N.HF$, crystalline powder. Turns gray on standing. Freely sol in water; slightly sol in cold, freely in hot alc.
Nitrate, $C_6H_7N.HNO_3$, crystals, dec about 190°. d 1.36. Discolors in air and light. Sol in water, alc. *Protect from light.*
Hemisulfate, $C_6H_7N.\frac{1}{2}H_2SO_4$, crystalline powder. d 1.38. Darkens on exposure to air and light. One gram dissolves in about 15 ml water; slightly sol in alc. Practically insol in ether. *Protect from light.*
Acetate, $C_6H_5NH_2.HOOCCH_3$. Prepd from aniline and acetic acid: Vignon, Evieux, *Bull. Soc. Chim. France* [4] 3, 1012 (1908). Colorless liquid. d 1.070-1.072. Darkens with age; gradually converted to acetanilide on standing. Misc with water, alc.
Oxalate, $C_6H_5NH_2.HOOCCOOH.H_2NC_6H_5$. Prepd from aniline and oxalic acid in alc soln: Hofmann, *Ann.* 47, 37 (1843). Triclinic rods from water, mp 174-175°. Readily sol in water; sparingly sol in abs alc. Practically insol in ether.
*Caution:* Intoxication may occur from inhalation, ingestion, or cutaneous absorption. *Acute Toxicity:* cyanosis, methemoglobinemia, vertigo, headache, mental confusion. *Chronic Toxicity:* anemia, anorexia; wt loss, cutaneous lesions, *Clinical Toxicology of Commercial Products,* R. E. Gosselin *et al.,* Eds. (Williams & Wilkins, Baltimore, 4th ed., 1976) Section III, pp 29-35.
USE: Manuf dyes, medicinals, resins, varnishes, perfumes, shoe blacks; vulcanizing rubber; as solvent. Hydrochloride used in manuf of intermediates, aniline black and other dyes, in dyeing fabrics or wood black.

**697. Aniline Mustard.** *N,N-Bis(2-chloroethyl)benzen-amine; N,N-bis(2-chloroethyl)aniline; phenylbis[2-chloroeth-ylamine]; β,β'-dichlorodiethylaniline; Lymphochin; Lymphocin; Lymphoquin.* $C_{10}H_{13}Cl_2N$; mol wt 218.13. C 55.06%, H 6.01%, Cl 32.51%, N 6.42%. Prepd by the action of phosphorus pentachloride on N,N-bis-[2-hydroxyethyl]-aniline (phenyldiethanolamine): Robinson, Watt, *J. Chem. Soc.* 1934, 1538; Korshak, Strepikheev, *J. Gen. Chem. USSR* 14, 312 (1944).

Stout prisms from methanol, mp 45°. $bp_{14}$ 164°; $bp_{0.5}$ 110°. Sol in hot methanol, ethanol. Very slightly sol in ether.
Hydrochloride, $C_{10}H_{14}Cl_3N$, crystals. *Vesicant.* Freely sol in water. Sol in alcohol.
USE: In cancer research.

**698. Anilinephthalein.** *3,3-Bis(4-aminophenyl)-1(3H)-isobenzofuranone; 3,3-bis(p-aminophenyl)phthalide.* $C_{20}H_{16}N_2O_3$; mol wt 316.36. C 75.93%, H 5.10%, N 8.85%, O 10.11%. Prepn: Hubacher, *J. Am. Chem. Soc.* 73, 5885 (1951).

*Consult the Name Index before using this section.*                    Page 111

---

## 1.6    Where to Find Chemical Safety Information

All laboratories must make available Material Safety Data Sheets (MSDSs) for every chemical used. Figure 1.5 shows the MSDS for dichloromethane (methylene chloride), detailing the hazards, safe handling practices, first aid information, storage, and disposal of dichloromethane.

MSDS information for thousands of compounds can also be obtained on the Internet. The number of Web sites changes almost daily; the following list of sites provides access to MSDS and is current at the time of publication.

**FIGURE 1.5**

Material Safety Data Sheet (MSDS) for dichloromethane. (Reprinted with permission from Aldrich Chemical Co., Inc., Milwaukee, WI.)

```
Sigma Chemical Co.        Aldrich Chemical Co., Inc.    Fluka Chemical Corp.
P.O. Box 14508            1001 West St. Paul            980 South Second St.
St. Louis, MO 63178       Milwaukee, WI 53233           Ronkonkoma, NY 11779
Phone: 314-771-5765       Phone: 414-273-3850           Phone: 516-467-0980
                                             Emergency Phone: 516-467-3535


SECTION 1. - - - - - - - - - CHEMICAL IDENTIFICATION- - - - - - - - - -
    CATALOG #:          D65100
    NAME:               DICHLOROMETHANE, 99.6%, A.C.S. REAGENT
SECTION 2. - - - - - COMPOSITION/INFORMATION ON INGREDIENTS - - - - - -
    CAS #:      75-09-2
    MF: CH2CL2
    EC NO:  200-838-9
    SYNONYMS
    AEROTHENE MM * CHLORURE DE METHYLENE (FRENCH) * DICHLOROMETHANE (DOT:
    OSHA) * METHANE DICHLORIDE * METHYLENE BICHLORIDE * METHYLENE
    CHLORIDE (ACGIH:DOT:OSHA) * METHYLENE DICHLORIDE * METYLENU CHLOREK
    (POLISH) * NARKOTIL * NCI-C50102 * R 30 * R30 (REFRIGERANT) * RCRA
    WASTE NUMBER U080 * SOLAESTHIN * SOLMETHINE * UN1593 (DOT) *
SECTION 3. - - - - - - - - - - HAZARDS IDENTIFICATION - - - - - - - - -
  LABEL PRECAUTIONARY STATEMENTS
    TOXIC (USA)
    HARMFUL (EU)
    HARMFUL BY INHALATION, IN CONTACT WITH SKIN AND IF SWALLOWED.
    IRRITATING TO EYES, RESPIRATORY SYSTEM AND SKIN.
    POSSIBLE RISK OF IRREVERSIBLE EFFECTS.
    CALIF. PROP. 65 CARCINOGEN.
    POSSIBLE CARCINOGEN/MUTAGEN.
    NEUROLOGICAL HAZARD.
    READILY ABSORBED THROUGH SKIN.
    TARGET ORGAN(S):
    LIVER, PANCREAS
    IN CASE OF ACCIDENT OR IF YOU FEEL UNWELL, SEEK MEDICAL ADVICE
    IMMEDIATELY (SHOW THE LABEL WHERE POSSIBLE).
    IN CASE OF CONTACT WITH EYES, RINSE IMMEDIATELY WITH PLENTY OF
    WATER AND SEEK MEDICAL ADVICE.
    AFTER CONTACT WITH SKIN, WASH IMMEDIATELY WITH PLENTY OF WATER.
    WEAR SUITABLE PROTECTIVE CLOTHING, GLOVES AND EYE/FACE
    PROTECTION.
    HANDLE AND STORE UNDER NITROGEN.
SECTION 4. - - - - - - - - - FIRST-AID MEASURES- - - - - - - - - - -
    IN CASE OF CONTACT, IMMEDIATELY WASH SKIN WITH SOAP AND COPIOUS
    AMOUNTS OF WATER.
    CONTAMINATION OF THE EYES SHOULD BE TREATED BY IMMEDIATE AND PROLONGED
    IRRIGATION WITH COPIOUS AMOUNTS OF WATER.
    ASSURE ADEQUATE FLUSHING OF THE EYES BY SEPARATING THE EYELIDS
    WITH FINGERS.
    IF INHALED, REMOVE TO FRESH AIR. IF NOT BREATHING GIVE ARTIFICIAL
    RESPIRATION. IF BREATHING IS DIFFICULT, GIVE OXYGEN.
    IF SWALLOWED, WASH OUT MOUTH WITH WATER PROVIDED PERSON IS CONSCIOUS.
    CALL A PHYSICIAN.
    WASH CONTAMINATED CLOTHING BEFORE REUSE.
                                 Page 1
```

```
SECTION 5. - - - - - - - - FIRE FIGHTING MEASURES - - - - - - - - - -
  EXTINGUISHING MEDIA
    NONCOMBUSTIBLE.
    USE EXTINGUISHING MEDIA APPROPRIATE TO SURROUNDING FIRE CONDITIONS.
  SPECIAL FIREFIGHTING PROCEDURES
    WEAR SELF-CONTAINED BREATHING APPARATUS AND PROTECTIVE CLOTHING TO
    PREVENT CONTACT WITH SKIN AND EYES.
  UNUSUAL FIRE AND EXPLOSIONS HAZARDS
    EMITS TOXIC FUMES UNDER FIRE CONDITIONS.
SECTION 6. - - - - - - - - ACCIDENTAL RELEASE MEASURES- - - - - - - -
    EVACUATE AREA.
    WEAR SELF-CONTAINED BREATHING APPARATUS, RUBBER BOOTS AND HEAVY
    RUBBER GLOVES.
    ABSORB ON SAND OR VERMICULITE AND PLACE IN CLOSED CONTAINERS FOR
    DISPOSAL.
    VENTILATE AREA AND WASH SPILL SITE AFTER MATERIAL PICKUP IS COMPLETE.
SECTION 7. - - - - - - - - - HANDLING AND STORAGE- - - - - - - - - -
    REFER TO SECTION 8.
SECTION 8. - - - - - EXPOSURE CONTROLS/PERSONAL PROTECTION- - - - - -
    WEAR APPROPRIATE NIOSH/MSHA-APPROVED RESPIRATOR, CHEMICAL-RESISTANT
    GLOVES, SAFETY GOGGLES, OTHER PROTECTIVE CLOTHING.
    USE ONLY IN A CHEMICAL FUME HOOD.
    SAFETY SHOWER AND EYE BATH.
    DO NOT BREATHE VAPOR.
    AVOID CONTACT WITH EYES, SKIN AND CLOTHING.
    AVOID PROLONGED OR REPEATED EXPOSURE.
    READILY ABSORBED THROUGH SKIN.
    WASH THOROUGHLY AFTER HANDLING.
    TOXIC.
    POSSIBLE CARCINOGEN.
    IRRITANT.
    NEUROLOGICAL HAZARD.
    POSSIBLE MUTAGEN.
    KEEP TIGHTLY CLOSED.
    KEEP AWAY FROM HEAT AND OPEN FLAME.
    STORE IN A COOL DRY PLACE.
SECTION 9. - - - - - - - - PHYSICAL AND CHEMICAL PROPERTIES - - - - - -
  APPEARANCE AND ODOR
    COLORLESS LIQUID
  PHYSICAL PROPERTIES
    BOILING POINT:        39.8 C TO 40 C
    MELTING POINT:        -97 C
    FLASHPOINT            NONE
    EXPLOSION LIMITS IN AIR:
      UPPER                              22%
      LOWER                              14%
    AUTOIGNITION TEMPERATURE:       1223 F         661C
    VAPOR PRESSURE:      6.83PSI 20 C    24.48PSI 55 C
    SPECIFIC GRAVITY:    1.325
    VAPOR DENSITY:       2.9
SECTION 10. - - - - - - - - -STABILITY AND REACTIVITY - - - - - - - - -
  INCOMPATIBILITIES
    ALKALI METALS
    ALUMINUM
                                 Page 2
```

```
    HEAT
  HAZARDOUS COMBUSTION OR DECOMPOSITION PRODUCTS
    CARBON MONOXIDE, CARBON DIOXIDE
    HYDROGEN CHLORIDE GAS
    PHOSGENE GAS
SECTION 11. - - - - - - - - - TOXICOLOGICAL INFORMATION - - - - - - -
  ACUTE EFFECTS
    HARMFUL IF SWALLOWED, INHALED, OR ABSORBED THROUGH SKIN.
    VAPOR OR MIST IS IRRITATING TO THE EYES, MUCOUS MEMBRANES AND UPPER
    RESPIRATORY TRACT.
    CAUSES SKIN IRRITATION.
    DICHLOROMETHANE IS METABOLIZED IN THE BODY PRODUCING CARBON MONOXIDE
    WHICH INCREASES AND SUSTAINS CARBOXYHEMOGLOBIN LEVELS IN THE BLOOD,
    REDUCING THE OXYGEN-CARRYING CAPACITY OF THE BLOOD.
    EXPOSURE CAN CAUSE:
    NAUSEA, DIZZINESS AND HEADACHE
    MAY CAUSE NERVOUS SYSTEM DISTURBANCES.
  CHRONIC EFFECTS
    POSSIBLE CARCINOGEN.
    LABORATORY EXPERIMENTS HAVE SHOWN MUTAGENIC EFFECTS.
    TARGET ORGAN(S):
    LIVER
    PANCREAS
    NERVES
    CARDIOVASCULAR SYSTEM
    TO THE BEST OF OUR KNOWLEDGE, THE CHEMICAL, PHYSICAL, AND
    TOXICOLOGICAL PROPERTIES HAVE NOT BEEN THOROUGHLY INVESTIGATED.
    RTECS #: PA8050000
    METHANE, DICHLORO-
  IRRITATION DATA
    SKN-RBT 810 MG/24H SEV              EJTXAZ 9,171,76
    SKN-RBT 100 MG/24H MOD             85JCAE -,88,86
    EYE-RBT 162 MG MOD                  EJTXAZ 9,171,76
    EYE-RBT 10 MG MLD                   TXCYAC 6,173,76
    EYE-RBT 500 MG/24H MLD             85JCAE -,88,86
  TOXICITY DATA
    ORI-HMN LDLO:357 MG/KG              34ZIAG -,390,69
    ORI-RAT LD50:1600 MG/KG            FAONAU 48A,94,70
    IHL-RAT LC50:52 GM/M3               TPKVAL 15,64,79
    IPR-RAT LD50:916 MG/KG              ENVRAL 40,411,86
    UNR-RAT LD50:5350 MG/KG            GISAAA 53(6),78,88
    IHL-MUS LC50:14400 PPM/7H          NIHBAZ 191,1,49
    IPR-MUS LD50:437 MG/KG             AGGHAR 18,109,60
    SCU-MUS LD50:6460 MG/KG            TXAPA9 4,354,62
    UNR-MUS LD50:4770 MG/KG            ESKGA2 28,P31,82
    UNR-RBT LD50:1225 MG/KG            GISAAA 53(6),78,88
  TARGET ORGAN DATA
    PERIPHERAL NERVE AND SENSATION (PARESTHESIA)
    BEHAVIORAL (ALTERED SLEEP TIME)
    BEHAVIORAL (EUPHORIA)
    BEHAVIORAL (SOMNOLENCE)
    BEHAVIORAL (CONVULSIONS OR EFFECT ON SEIZURE THRESHOLD)
    BEHAVIORAL (ATAXIA)
    CARDIAC (CHANGE IN RATE)
                                 Page 3
```

```
PRODUCT #: D65100     NAME: DICHLOROMETHANE, 99.6%, A.C.S. REAGENT
         MATERIAL SAFETY DATA SHEET, Valid 11/96 - 1/97
              Printed Thursday, November 21, 1996  4:38PM
```

```
  LUNGS, THORAX OR RESPIRATION (TUMORS)
  LIVER (LIVER FUNCTION TESTS IMPAIRED)
  SPECIFIC DEVELOPMENTAL ABNORMALITIES (MUSCULOSKELETAL SYSTEM)
  SPECIFIC DEVELOPMENTAL ABNORMALITIES (UROGENITAL SYSTEM)
  TUMORIGENIC (CARCINOGENIC BY RTECS CRITERIA)
  ONLY SELECTED REGISTRY OF TOXIC EFFECTS OF CHEMICAL SUBSTANCES
  (RTECS) DATA IS PRESENTED HERE. SEE ACTUAL ENTRY IN RTECS FOR
  COMPLETE INFORMATION.
SECTION 12. - - - - - - - - - ECOLOGICAL INFORMATION - - - - - - - - -
     DATA NOT YET AVAILABLE.
SECTION 13. - - - - - - - - - DISPOSAL CONSIDERATIONS - - - - - - - - -
     DISSOLVE OR MIX THE MATERIAL WITH A COMBUSTIBLE SOLVENT AND BURN IN A
     CHEMICAL INCINERATOR EQUIPPED WITH AN AFTERBURNER AND SCRUBBER.
     OBSERVE ALL FEDERAL, STATE AND LOCAL ENVIRONMENTAL REGULATIONS.
SECTION 14. - - - - - - - - - TRANSPORT INFORMATION - - - - - - - - -
     CONTACT ALDRICH CHEMICAL COMPANY FOR TRANSPORTATION INFORMATION.
SECTION 15. - - - - - - - - - REGULATORY INFORMATION - - - - - - - - -
  EUROPEAN INFORMATION
     EC INDEX NO:     602-004-00-3
     HARMFUL
     R 40
     POSSIBLE RISK OF IRREVERSIBLE EFFECTS.
     S 23
     DO NOT BREATHE VAPOR.
     S 24/25
     AVOID CONTACT WITH SKIN AND EYES.
     S 36/37
     WEAR SUITABLE PROTECTIVE CLOTHING AND GLOVES.
  REVIEWS, STANDARDS, AND REGULATIONS
     OEL=MAK
     ACGIH TLV-TWA 50 PPM                           85INA8 6,981,91
     ACGIH TLV-SUSPECTED HUMAN CARCINOGEN           85INA8 6,981,91
     IARC CANCER REVIEW:ANIMAL SUFFICIENT EVIDENCE IMEMDT 41,43,86
     IARC CANCER REVIEW:HUMAN INADEQUATE EVIDENCE  IMEMDT 41,43,86
     IARC CANCER REVIEW:GROUP 2B                   IMSUDL 7,194,87
     EPA FIFRA 1988 PESTICIDE SUBJECT TO REGISTRATION OR RE-REGISTRATION
        FEREAC 54,7740,89
     MSHA STANDARD-AIR:TWA 500 PPM (1750 MG/M3)
        DTLVS* 3,171,71
     OSHA PEL (GEN INDU):8H TWA 500 PPM;CL 1000 PPM;PK 2000 PPM/5M/2H
        CFRGBR 29,1910.1000,94
     OSHA PEL (CONSTRUC):SEE 56 FR 57036
        CFRGBR 29,1926.55,94
     OSHA PEL (SHIPYARD):8H TWA 500 PPM (1740 MG/M3
        CFRGBR 29,1915.1000,93
     OSHA PEL (FED CONT):8H TWA 500 PPM (1740 MG/M3)
        CFRGBR 41,50-204.50,94
     OEL-AUSTRALIA:TWA 100 PPM (350 MG/M3);CARCINOGEN JAN93
     OEL-AUSTRIA:TWA 100 PPM (360 MG/M3) JAN93
     OEL-BELGIUM:TWA 50 PPM (174 MG/M3);CARCINOGEN JAN93
     OEL-DENMARK:TWA 100 PPM (175 MG/M3);SKIN;CARCINOGEN JAN93
     OEL-FINLAND:TWA 100 PPM (350 MG/M3);STEL 250 PPM (870 MG/M3) JAN93
     OEL-FRANCE:TWA 100 PPM (360 MG/M3);STEL 500 PPM (1800 MG/M3) JAN93
     OEL-GERMANY:TWA 100 PPM (360 MG/M3);CARCINOGEN JAN93
```

```
PRODUCT #: D65100     NAME: DICHLOROMETHANE, 99.6%, A.C.S. REAGENT
         MATERIAL SAFETY DATA SHEET, Valid 11/96 - 1/97
              Printed Thursday, November 21, 1996  4:38PM
```

```
  OEL-HUNGARY:STEL 10 MG/M3;CARCINOGEN JAN93
  OEL-JAPAN:TWA 100 PPM (350 MG/M3) JAN93
  OEL-THE NETHERLANDS:TWA 100 PPM (350 MG/M3);STEL 500 PPM JAN93
  OEL-THE PHILIPINES:TWA 500 PPM (1740 MG/M3) JAN93
  OEL-POLAND:TWA 50 MG/M3 JAN93
  OEL-RUSSIA:TWA 100 PPM;STEL 50 MG/M3 JAN93
  OEL-SWEDEN:TWA 35 PPM (120 MG/M3);STEL 70 PPM (250 MG/M3);SKIN JAN93
  OEL-SWITZERLAND:TWA 100 PPM (360 MG/M3);STEL 500 PPM JAN93
  OEL-THAILAND:TWA 500 MG/M3;STEL 1000 MG/M3 JAN93
  OEL-TURKEY:TWA 500 PPM (1740 MG/M3) JAN93
  OEL-UNITED KINGDOM:TWA 100 PPM (350 MG/M3);STEL 250 PPM JAN93
  OEL IN BULGARIA, COLOMBIA, JORDAN, KOREA CHECK ACGIH TLV
  OEL IN NEW ZEALAND, SINGAPORE, VIETNAM CHECK ACGIH TLV
  NIOSH REL TO METHYLENE CHLORIDE-AIR:CA LOWEST FEASIBLE CONCENTRATION
     NIOSH* DHHS #92-100,92
  NOHS 1974: HZD 47270; NIS 374; TNF 89025; NOS 192; TNE 975696
  NOES 1983: HZD 47270; NIS 363; TNF 87086; NOS 212; TNE 1438196; TFE
     352536
  ATSDR TOXICOLOGY PROFILE (NTIS** PB/89/194468/AS)
  EPA GENETOX PROGRAM 1988, POSITIVE: CELL TRANSFORM.-RLV F344 RAT EMBRYO
  EPA GENETOX PROGRAM 1988, POSITIVE: HISTIDINE REVERSION-AMES TEST
  EPA GENETOX PROGRAM 1988, POSITIVE: S CEREVISIAE GENE CONVERSION; S
     CEREVISIAE-HOMOZYGOSIS
  EPA GENETOX PROGRAM 1988, POSITIVE: S CEREVISIAE-REVERSION
  EPA GENETOX PROGRAM 1988, NEGATIVE: D MELANOGASTER SEX-LINKED LETHAL
  EPA TSCA SECTION 8(B) CHEMICAL INVENTORY
  EPA TSCA 8(A) PRELIMINARY ASSESSMENT INFORMATION, FINAL RULE
     FEREAC 47,26992,82
  EPA TSCA SECTION 8(D) UNPUBLISHED HEALTH/SAFETY STUDIES
  ON EPA IRIS DATABASE
  EPA TSCA TEST SUBMISSION (TSCATS) DATA BASE, JULY 1996
  NIOSH CURRENT INTELLIGENCE BULLETIN 46, 1986
  NIOSH ANALYTICAL METHOD, 1994: METHYLENE CHLORIDE, 1005
  NTP CARCINOGENESIS STUDIES (INHALATION);CLEAR EVIDENCE:MOUSE,RAT
     NTPTR* NTP-TR-306,86
  NTP 7TH ANNUAL REPORT ON CARCINOGENS, 1992 : ANTICIPATED TO BE
     CARCINOGEN
  OSHA ANALYTICAL METHOD #ID-59
U.S. INFORMATION
     THIS PRODUCT IS SUBJECT TO SARA SECTION 313 REPORTING REQUIREMENTS.
SECTION 16. - - - - - - - - - OTHER INFORMATION - - - - - - - - - - - -
     THE ABOVE INFORMATION IS BELIEVED TO BE CORRECT BUT DOES NOT PURPORT TO
     BE ALL INCLUSIVE AND SHALL BE USED ONLY AS A GUIDE. SIGMA, ALDRICH,
     FLUKA SHALL NOT BE HELD LIABLE FOR ANY DAMAGE RESULTING FROM HANDLING
     OR FROM CONTACT WITH THE ABOVE PRODUCT. SEE REVERSE SIDE OF INVOICE OR
     PACKING SLIP FOR ADDITIONAL TERMS AND CONDITIONS OF SALE.
     COPYRIGHT 1996 SIGMA CHEMICAL CO., ALDRICH CHEMICAL CO., INC.,
     FLUKA CHEMIE AG
     LICENSE GRANTED TO MAKE UNLIMITED PAPER COPIES FOR INTERNAL USE ONLY
```

http://www.ilpi.com/msds/index.html
http://www.chem.plu.edu/cirrus.html
http://physchem.ox.ac.uk/MSDS/
http://www.sigma-aldrich.com

An extensive collection of chemical safety information is found in the three-volume set *The Sigma-Aldrich Library of Regulatory and Safety Data* (Aldrich Chemical Co., 1993).

Chemical suppliers put a four-diamond symbol—developed by the National Fire Protection Association—on the container label, describing the hazards associated with the compounds they sell (Figure 1.6). The four diamonds provide information about *fire or flammability* (top, red diamond), *instability* (right, yellow diamond), *special information* such a oxidizing power or reactivity with water (bottom, white diamond), and *health hazards* (left, blue diamond). The numerical values in the diamonds range from 0 to 4 with 0 indicating no chemical hazard and 4 indicating an extreme chemical hazard. For example, the four-diamond symbol on acetone shows a 3 (severe) rating for flammability, a 0 (none) rating for instability, a blank in the special information diamond, and a 1 (slight) for health hazards. In comparison, nitric acid has a 0 (none) rating for flammability, a 0 (none) rating for instability, "oxy" in the special information diamond indicating that it is an oxidizing agent, and a 3 (severe) rating for health hazards. Many chemical suppliers also print the pictograms approved by the European Union (EU) on the labels of their chemicals to indicate the hazards (Figure 1.7).

**FIGURE 1.6**
A four-diamond label for chemical containers indicating health, fire, instability, and special hazards. The symbol ⛌ in the special information diamond indicates that the compound is reactive with water and should not come into contact with it.

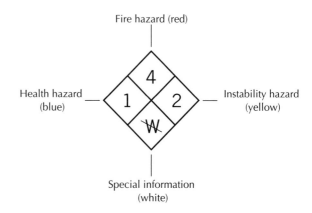

Fire hazard (red)

Health hazard (blue)

Instability hazard (yellow)

Special information (white)

**FIGURE 1.7**
Pictograms indicating chemical hazards.

Explosive          Oxidizing          Highly flammable or          Toxic or
                                      extremely flammable          very toxic

Harmful or          Corrosive          Biohazard          Dangerous for
irritant                                                  the environment

---

### 1.7        Protecting the Environment

Protection in the laboratory does not stop with the persons who work there. Every student and industrial worker must also be aware of the impact of what he or she does on others outside the immediate sphere of the laboratory.

Before disposing of anything in the lab, you must be conscious of how this disposal will affect the environment. Although zero waste is impossible, minimum waste is essential. Industries are now required to account for almost every gas, liquid, or solid waste they put into the environment. In the college laboratory, we must do likewise.

You may be required to produce an environmental profile for some laboratory experiments you do. An environmental profile provides a quantitative accounting, as nearly as possible, for every material used: What goes into a reaction must be accounted for in what comes out. You cannot simply pour things down the drain or throw things into a wastebasket, because that isn't disposing of them — it is just putting them in another place.

To construct an environmental profile, consider a typical organic laboratory synthesis—the preparation of methyl benzoate from methanol and benzoic acid. The equation for the reaction is

$$C_6H_5COOH + CH_3OH \xrightarrow{\text{H}_2\text{SO}_4} C_6H_5COOCH_3 + H_2O$$

$$\underset{MW = 122}{\phantom{C_6H_5COOH}} \quad \underset{MW = 32}{\phantom{CH_3OH}} \quad \underset{MW = 136}{\phantom{C_6H_5COOCH_3}} \quad \underset{MW = 18}{\phantom{H_2O}}$$

The experiment is conducted by refluxing a mixture of 12 g of benzoic acid and 25 mL of methanol with a sulfuric acid catalyst. The density of methanol is $0.79 \text{ g} \cdot \text{mL}^{-1}$, so we can calculate that 0.10 mol of benzoic acid and 0.62 mol of methanol are present. Methanol is in substantial excess, because benzoic acid and methanol react in a 1:1 mole ratio according to the balanced chemical equation.

Ignoring heat and water, the input statement is

| Input | Amount | Moles |
|---|---|---|
| Benzoic acid | 12 g | 0.10 |
| Methanol (25 mL) | 20 g | 0.62 |
| Sulfuric acid | (approx. 1 mL) | |
| Dichloromethane | 25 mL | |

After heating the mixture under reflux for 1 h, the solution is cooled, poured into water, and extracted with dichloromethane (25 mL).

After these steps, the sulfuric acid is neutralized, and we isolate not only the product, methyl benzoate, but also any unused benzoic acid and all the dichloromethane, the solvent used for extraction. Because it is water soluble, the excess methanol is lost from the product solution when the reaction mixture is poured into water.

The output statement might look like this:

| Output | Amount | Moles |
|---|---|---|
| Benzoic acid (recovered) | 3.7 g | 0.033 |
| Methyl benzoate (formed) | 8.3 g | 0.061 |
| Dichloromethane (recovered) | 20 mL | |

Lost in the synthesis were substantial amounts of methanol, 0.006 mole of benzoic acid, 5 mL of dichloromethane, and about 1 mL of sulfuric acid. Where did it go? Into waste waters, into waste containers, or into the air. And in that "going," it contributed to environmental pollution.

We can prevent some of this waste by being more careful, although much of the loss in a small-scale experiment cannot be prevented. On an industrial scale, the percentage loss observed here would be much too large, for both economic and environmental reasons, and the experimental procedures would be changed to minimize it.

## References

1. Lewis, Sr., R. J. *Sax's Dangerous Properties of Industrial Materials*; 10th ed.; Wiley-Interscience: New York, 2000.

2. The Manufacturing Chemists Association, *Guide for Safety in the Chemistry Laboratory*; Van Nostrand-Reinhold: New York, 1972.

3. Furr, A. K. (Ed.) *CRC Handbook of Laboratory Safety*; 5th ed.; CRC Press: Boca Raton, FL, 2000.

4. The Manufacturing Chemists Association, Chemical Safety Data Sheets; Washington, DC.

5. U.S. Department of Labor, *Occupational Exposure to Hazardous Chemicals in Laboratories*; no. OSHA 95-33; Government Printing Office: Washington, DC, 1995.

6. U.S. Department of Health, Education, and Welfare, NIOSH, *Suspected Carcinogens*; 2nd ed.; Government Printing Office: Washington, DC, 1976.

7. Budavari, S.; O'Neil, M. J.; Smith, A.; Heckelman, P. E.; Kinneary, J. F. (Eds.) *The Merck Index: An Encyclopedia of Chemicals, Drugs, and Biologicals*; 12th ed.; Merck & Co., Inc.: Whitehouse Station, NJ, 1996.

8. Lewis, Sr., R. J. *Rapid Guide to Hazardous Chemicals in the Workplace*; 4th ed.; Wiley-Interscience: New York, 2000.

9. Lenga, R. E.; Votoupal, K. L. (Eds.) *The Sigma-Aldrich Library of Regulatory and Safety Data*; Aldrich Chemical Co.: Milwaukee, WI, 1993; 3 volumes.

10. Hajian, H. G. *Working Safely in the Chemical Laboratory*; American Chemical Society: Washington, DC, 1994.

11. National Research Council *Prudent Practices in the Laboratory, Handling and Disposal of Chemicals*; National Academy Press: Washington, DC, 1995.

12. American Chemical Society *Less Is Better: Laboratory Chemical Management for Waste Reduction*; 2nd ed.; American Chemical Society: Washington, DC, 1993.

13. American Chemical Society *Safety in Academic Chemistry Laboratories*; 6th ed.; American Chemical Society: Washington, DC, 1995.

**TECHNIQUE**

# 2  LABORATORY GLASSWARE

You will find an assortment of glassware and equipment in your laboratory desk, some items that will be familiar to you from your earlier lab experiences and other items that you may not have used previously. If your laboratory is equipped for miniscale experimentation you will find specialized glassware that has carefully constructed ground-glass joints designed to fit together tightly. This glassware is called **standard taper glassware,** and it comes in a variety of sizes. If you will be carrying out microscale experimentation, you will use scaled-down glassware designed for the milligram and milliliter quantities of reagents used in microscale work. There are two types of microscale glassware commonly used in the undergraduate organic laboratory—microscale standard taper glassware with threaded screw cap connectors and the Kontes/ Williamson microscale glassware that fastens together with flexible elastomeric connectors.

## 2.1    Desk Equipment

A typical student desk contains an assortment of beakers, Erlenmeyer flasks, filter flasks, thermometers, graduated cylinders, test tubes, funnels, and a variety of other items. Your desk or locker will

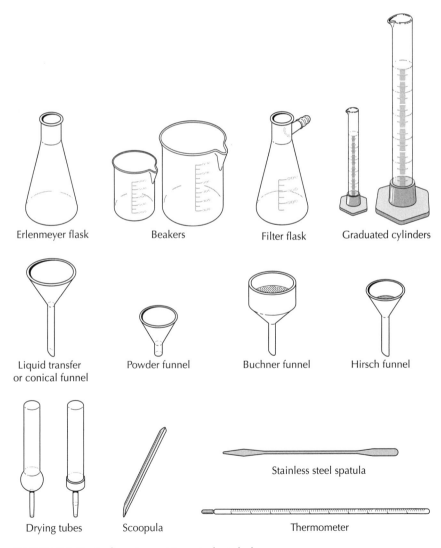

FIGURE 2.1 Typical equipment in a student desk.

probably have most, if not all, of the equipment items shown in
Figure 2.1. Make sure that all glassware is clean and has no chips
or cracks. Replace damaged glassware.

## 2.2          Standard Taper Miniscale Glassware

Standard taper glassware is designated by the symbol ℑ. All the
joints in standard taper glassware have been carefully ground so
that they are exactly the same size, and all the pieces fit together
interchangeably. We recommend the use of ℑ 19/22 or ℑ 14/20
glassware for miniscale experiments. The numbers represent, in mil-
limeters, the diameter and the length of all ground glass surfaces
(Figure 2.2). Figure 2.3 shows a typical set of ℑ 19/22 glassware
found in introductory organic laboratories.

*Greasing Ground
Glass Joints*

Because standard taper joints fit together tightly, they are not
usually put together dry. The ground glass joints of ℑ 19/22 or
ℑ 14/20 glassware are coated with a lubricating grease, as shown
in Figure 2.4. The grease prevents interaction of the ground glass
joints with the chemicals used in the experiment that can cause the
joints to "freeze," or stick together. Taking apart stuck joints, while
not impossible, is often not an easy task, and standard taper glassware (which is expensive) frequently is broken in the process. **Note:**
Microscale glassware with ground glass joints is never greased
unless the reaction involves strong bases such as sodium hydroxide
or sodium methoxide.

Several greases are commercially available. For general purposes in an undergraduate laboratory, a hydrocarbon grease, such
as Lubriseal, is preferred because it can be removed easily. Silicone
greases have a very low vapor pressure and are intended for sealing a system that will be under vacuum. They are nearly impossible
to remove completely because they do not dissolve in detergents or
organic solvents.

Using excess grease is bad practice. Not only is it messy, but
worse, it may contaminate your reacti bon or coat the inside of your
flasks, making them difficult to clean. Just enough grease to coat the

**FIGURE 2.2**
Standard taper
distillation head and
round-bottomed flask
showing the dimensions
of the ground glass
joints.

19/22 distillation head          19/22 flask

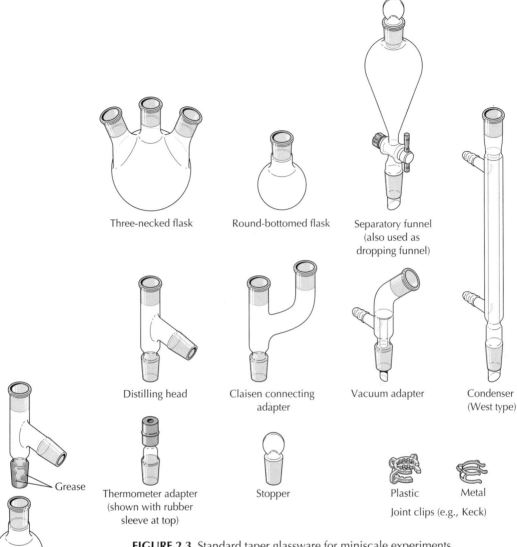

**FIGURE 2.3** Standard taper glassware for miniscale experiments.

Three-necked flask

Round-bottomed flask

Separatory funnel
(also used as
dropping funnel)

Distilling head

Claisen connecting
adapter

Vacuum adapter

Condenser
(West type)

Grease

Thermometer adapter
(shown with rubber
sleeve at top)

Stopper

Plastic    Metal

Joint clips (e.g., Keck)

**FIGURE 2.4**
Sealing a standard taper joint with grease. Apply two *thin* strips of grease that run the entire length of the inner joint and are about 180° apart. Gently insert the inner joint into the outer joint and rotate one of the pieces. The joint should rotate easily and the grease should be uniformly distributed so that the frosted surfaces now appear clear.

entire ground surface thinly is sufficient. If grease oozes above the top or below the bottom of the joint, you have used too much. Take the joint apart, wipe off the excess grease with a towel or tissue, and assemble the pieces again.

When you have finished your experiment, clean the grease from the joints by using a brush, detergent, and hot water. If this scrubbing does not remove all the grease, dry the joint and clean it with a towel (for example, a Kimwipe) moistened with toluene or hexane.

**SAFETY PRECAUTION**

Toluene or hexane are irritants and pose a fire hazard; wear gloves and work in a hood.

## 2.3    Microscale Glassware

When the amounts of reagents used for experiments are in the 100–300-mg or 0.1–2.0-mL range, microscale glassware is used. Recovering any product from an operation at this scale would be almost impossible if you were using 19/22 standard taper glassware; most of the material would be lost on the glass surfaces. Two types of microscale glassware are commonly used in undergraduate organic laboratories—standard taper glassware with threaded screw cap connectors or Kontes/Williamson glassware that fastens together with flexible elastomeric connectors. Your instructor will tell you which type of microscale glassware is used in your laboratory.

***Standard Taper Microscale Glassware***

The pieces of microscale standard taper glassware needed for typical experiments in the introductory organic laboratory are shown in Figure 2.5.

The microscale glassware fits together with 14/20 standard taper joints. **Grease is NOT used with microscale glassware,** except when the reaction mixture contains a strong base, because its presence could cause significant contamination of the reaction mixture. Instead, a threaded cap and O-ring ensure a tight seal and hold the pieces together, thus eliminating the use of clamps or clips. Place the threaded cap over the inner joint, then slip the O-ring

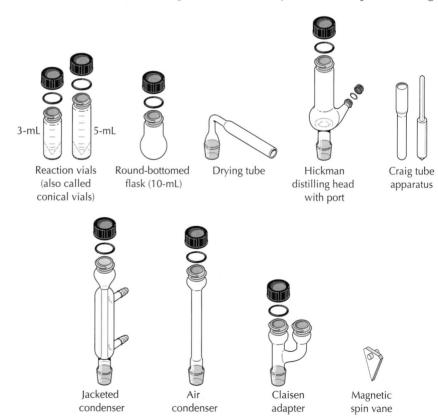

Reaction vials (also called conical vials)  3-mL  5-mL

Round-bottomed flask (10-mL)

Drying tube

Hickman distilling head with port

Craig tube apparatus

Jacketed condenser

Air condenser

Claisen adapter

Magnetic spin vane

**FIGURE 2.5** Standard taper microscale glassware.

**FIGURE 2.6**
Assembling a standard
taper joint on standard
taper microscale
glassware.

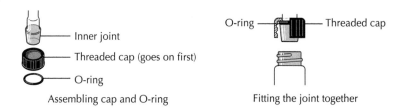

over the tapered portion. Fit the inner joint inside the outer joint and screw the threaded cap tightly onto the outer joint (Figure 2.6). A securely screwed connection effectively prevents the escape of vapors and is also vacuum tight.

***Kontes/Williamson Microscale Glassware***    The various pieces of Kontes/Williamson microscale glassware used in typical experiments in the organic laboratory are shown in Figure 2.7.

*Available from Kontes, Catalog no. 748008-9001.

**FIGURE 2.7** Williamson microscale glassware and other microscale apparatus. (Manufactured by Kontes Glass Co., Vineland, NJ.)

**FIGURE 2.8**
Assembling Williamson microscale glassware with a flexible connector.

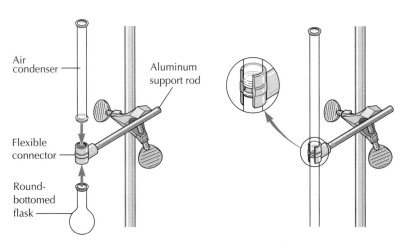

Fitting the glassware into the flexible connector one piece at a time

Cutaway showing the two pieces of glassware fastened in connector

This type of microscale glassware fits together with flexible elastomeric connectors that are heat and solvent resistant. **Grease is NOT used with this type of glassware connector.** A flexible connector with an aluminum support rod serves to fasten two pieces of glassware together and provides a way of attaching the apparatus in a two-way clamp to a ring stand or vertical support rod. One piece of glassware is pushed into the connector, then the second piece is pushed into the other end of the connector, as shown in Figure 2.8. The flexible connector effectively seals the joint and prevents the escape of vapors.

---

## 2.4    Cleaning Laboratory Glassware

Part of careful laboratory technique includes cleaning your glassware before you leave the laboratory. Cleaning up immediately ultimately saves time and reduces everyone's exposure to chemicals. Clean glassware is essential for maximizing the yield in any organic reaction, and in many instances glassware also must be dry. Try not to have to wash something immediately before using it, because then you will waste precious time while it dries in the oven.

Strong detergents and hot water are the ingredients needed to clean most glassware used for organic reactions. Scrubbing with a paste made from water and scouring powder, such as Ajax, will remove many organic residues from glassware. Organic solvents, such as acetone or hexane, will help dissolve the polymeric tars that sometimes coat the inside of a flask after a distillation. You may want to wear gloves when cleaning glassware. A final rinse of clean glassware with distilled water prevents water spots.

SAFETY   PRECAUTION

Solvents such as acetone and hexane are irritants and flammable; wear gloves, use the solvents in a hood, and dispose of them in the flammable (organic waste) container.

A solution of alcoholic sodium hydroxide* is our favorite cleanser for removing grease and organic residues from flasks and other glassware.

SAFETY   PRECAUTION

Strong bases, such as **sodium hydroxide,** cause severe burns and eye damage. Skin contact with alkali solutions starts as a slippery feel to the skin followed by irritation. Wash the affected area with copious amounts of water. Wear gloves and eye protection while cleaning glassware with alcoholic NaOH solution.

*Made by dissolving 120 g of NaOH in 120 mL of water and diluting to 1 L with 95% ethanol.

**TECHNIQUE**

# 3

# THE LABORATORY NOTEBOOK

A few general comments are in order about the laboratory notebook that is the primary record of your experimental work. First, although many college bookstores sell notebooks that are specifically designed as lab notebooks, it is often sufficient to use any notebook with tightly bound pages. Spiral and three-ring binders are inappropriate for lab notebooks because pages can be easily removed. **All entries about your work must be made directly in your laboratory notebook in ink.** Recording data on scraps of paper is an unacceptable practice because the papers may be lost; the practice will probably be strictly forbidden in your laboratory.

## 3.1     Organization of the Laboratory Notebook

The notebook should begin with a table of contents; set aside the first two or three pages for this purpose. The rest of the pages should be numbered sequentially, and **no page should ever be torn**

**out** of a laboratory notebook. Your notebook must be written with accuracy and completeness. It must be organized and legible, but it does not need to be a work of art.

Some flexibility in format and style may be allowed, but proper records of your experimental results must answer certain questions.

- *When* did you do the work?
- *What* are you trying to accomplish in the experiment?
- *How* did you do the experiment?
- *What* did you observe?
- *How* do you explain your observations?

A lab record needs to be written in three steps: **prelab, during lab, and postlab.** It should contain the following sections for each experiment you do.

*To Be Done Before You Come to the Laboratory*

The basic notebook format discussed here is designed to help you prepare for an experiment in an effective and safe fashion. It includes the date and title of the experiment or project, the balanced chemical reaction you are studying, a statement of purpose, a table of reagents and solvents, the way you will calculate your percent yield, an outline of the procedure to be used, and answers to any prelab questions. Your instructor will undoubtedly provide specific guidelines for lab notebook procedures pertaining at your institution.

*Title:* Use a title that clearly identifies what you are doing in this experiment or project.

*Date(s):* Use the date on which an experiment is actually carried out. In some research labs, where patent issues are important, a witnessed signature of the date is required.

*Balanced chemical reaction:* Write balanced chemical equations that show the overall process. Any details of reaction mechanisms go into the summary.

*Purpose statement:* Write a brief statement of purpose for the synthesis or analysis, or state the question you are addressing, with a few words on major analytical or conceptual approaches.

*Table:* Include all reagents and solvents. The table normally lists molecular weights and the number of moles and grams of reagents, as well as the densities of liquids you will be using, boiling points of compounds that are liquids at room temperature and melting points of all organic solids, and pertinent hazard warnings.

*Method of yield calculation:* Outline the computations to be used in a synthesis experiment, including calculation of the theoretical yield [see Technique 3.2].

*Procedure and prelab questions:* Write an experimental outline in sufficient detail so that the experiment could be done without reference to your lab textbook, along with the answers to any assigned prelab questions.

***To Be Done During the Laboratory Session(s)***

Recording observations during the experiment is a crucial part of your laboratory record. If your observations are incomplete, you cannot interpret the results of your experiments; once you have left the laboratory, it is difficult, if not impossible, to reconstruct them.

Observations must be recorded in your notebook in ink while you are doing an experiment. The actual quantities of all reagents must be recorded as they are used, as well as the amounts of crude and purified products that you obtained. Mention which measurements (temperature, time, melting point, and so on) and spectra are taken. Because organic chemistry is primarily an experimental science, your observations are crucial to your success. Things that seem insignificant may be important in understanding and explaining your results later. Typical laboratory observations might be as follows:

> A white precipitate appeared, which dissolved when the sulfuric acid was added; the solution turned cloudy when it was cooled to 10°C; an additional 10 mL of solvent were required to completely dissolve the yellow solid; the reaction was heated at 50°C for 25 min with a warm-water bath; the NMR sample was prepared with [*state amount of compound used*], using $CDCl_3$ added to a height of 50 mm in the NMR tube; a capillary OV-101 GC column was used at a temperature of 137°C; the infrared spectrum was run on a mineral oil mull; a short puff of white smoke appeared when sodium hydroxide was added to the reaction mixture.

Your observations may be recorded in a variety of ways. They may be written on right-hand pages across from the corresponding section of the experimental outline on a left-hand page; the page may be divided into columns with the left column used for procedure and the right column for observations. It is a good idea to cross-index your observations to specific steps in the procedure that you wrote out as part of your prelab preparation. Your instructor will probably provide specific advice on how you should record your observations during the laboratory.

***To Be Done After the Experimental Work Has Been Completed***

In this section of your notebook you evaluate and interpret your experimental results. Entries include a section on interpretation of physical and spectral data, a summary of your conclusions, calculation of the percent yield, and answers to any assigned postlab questions.

> *Percent yield:* The single most important measure of success in a chemical synthesis is the quantity of desired product produced. To be sure, the purity of the product is also crucial, but if a synthetic method produces only small amounts of the needed product, it is not much good. Reactions on the pages of textbooks are often far more difficult to carry out in good yield than the books suggest. Usually the best yields of products come when experienced chemists carry out a reaction a second or third time. Calculation of the percent yield is discussed in Technique 3.2.

*Conclusions and summary:* In an inquiry-based project or experiment, return to the question being addressed and discuss the conclusions you can draw from analysis of your data. For both inquiry-based experiments and those where you learned about laboratory techniques of organic syntheses, discuss how your experimental results support your conclusions. Include a thorough interpretation of NMR and IR spectra and other analytical results, such as TLC and GC analyses. Properly labeled spectra and chromatograms should be stapled into your notebook. Cite any reference sources that you used and include answers to any assigned postlab questions.

## 3.2    Calculation of the Percent Yield

When you report the results of a synthesis reaction, the **percent yield** is always stated, a parameter defined as the ratio of product obtained to the theoretical yield (maximum amount expected) multiplied by 100:

$$\% \text{ yield} = \frac{\text{actual yield of product}}{\text{theoretical yield}} \times 100$$

You calculate the theoretical yield from the balanced chemical equation and the limiting reagent, assuming 100% conversion of the starting materials to product(s). For example, consider the Williamson ether synthesis of 1-ethoxybutane from 1-bromobutane and sodium ethoxide.

$$CH_3(CH_2)_3-Br + CH_3CH_2-O^-Na^+ \xrightarrow{\text{ethanol}} CH_3(CH_2)_3-O-CH_2CH_3 + NaBr$$

| 1-Bromobutane | Sodium ethoxide | 1-Ethoxybutane |
|---|---|---|
| MW 137 | MW 68.1 | MW 102 |
| density 1.27 g·mL⁻¹ | | |

The procedure specified 4.5 mL of 1-bromobutane, 3.7 g of sodium ethoxide, and 20 mL of anhydrous ethanol. To calculate the theoretical yield, it is necessary to ascertain whether 1-bromobutane or sodium ethoxide is the limiting reagent by calculating the moles of each reagent present in the reaction mixture:

$$\text{moles of 1-bromobutane} = \frac{4.5 \text{ mL} \times 1.27 \text{ g·mL}^{-1}}{137 \text{ g·M}^{-1}} = 0.041 \text{ mol}$$

$$\text{moles of sodium ethoxide} = \frac{3.7 \text{ g}}{68.1 \text{ g·M}^{-1}} = 0.054 \text{ mol}$$

Therefore, 1-bromobutane is the limiting reagent.

According to the balanced equation, equimolar amounts of the two reactants are required to produce a similar amount of product. Thus the **theoretical yield,** the maximum amount of product that is possible from the reaction assuming that it goes to completion and that no experimental losses occur, is 0.041 mol or 4.2 g of ethoxybutane:

$$\text{theoretical yield} = 0.041 \text{ mol} \times 102 \text{ g} \cdot \text{M}^{-1} = 4.2 \text{ g of 1-ethoxybutane}$$

The percent yield for a synthesis that produced 2.7 of 1-ethoxy butane is 64%:

$$\% \text{ yield} = \frac{2.7 \text{ g}}{4.2 \text{ g}} \times 100 = 64\%$$

TECHNIQUE

# 4

# USING HANDBOOKS AND ONLINE DATABASES

From time to time, you will need to find physical constants, such as melting and boiling points, densities, and other useful information about organic compounds. Compilations of physical constants are published in a number of handbooks. In addition, databases now available on the Internet provide information about physical constants for many organic compounds.

## 4.1  Handbooks

Three handbooks are particularly useful for physical constants of organic compounds: the *Aldrich Catalog of Fine Chemicals*, the *CRC Handbook of Chemistry and Physics*, and *The Merck Index*.

***Aldrich Catalog of Fine Chemicals***

The *Aldrich Catalog of Fine Chemicals* is published annually by the Aldrich Chemical Company of Milwaukee, Wisconsin. It lists thousands of organic and inorganic compounds and includes the chemical structure for each one, a brief summary of its physical properties, references on IR, UV, and NMR spectra, plus safety and disposal information. There are also references to *Beilstein's Handbuch der Organischen Chemie* and to *Reagents for Organic Synthesis* by Fieser and Fieser (see Appendix B for more information about these reference works). Figure 4.1 shows a page from an *Aldrich Catalog*.

■ **Dimethylbu** ■

$

| 18,310-5 ★ | **3,3-Dimethyl-1-butanol**, 99% [624-95-3] (CH₃)₃CCH₂CH₂OH FW 102.18 mp -60°........... bp 143° n₀ 1.4140 d 0.844 Fp 118°F(47°C) Beil. 1(3),1677 FT-NMR 1(1),172A FT-IR 1(1),115D SI 14,A,9 Safety 2,1359B R&S 1(1),125F | 1g 10g 50g | 8.70 38.90 137.10 |
|---|---|---|---|

| 13,682-4 ★ | **3,3-Dimethyl-2-butanol**, 98% [464-07-3] (pinacolyl alcohol) (CH₃)₃CCH(OH)CH₃........ FW 102.18 mp 4.8° bp 119-121° n₀ 1.4150 d 0.812 Fp 84°F(28°C) [α]²⁵ 0° (neat) Beil. 1,412 FT-NMR 1(1),184B FT-IR 1(1),122D SI 15,C,9 Safety 2,1359C R&S 1(1),133E RTECS# EL2276000 FLAMMABLE LIQUID | 25g 100g | 30.20 84.20 |

**3,3-Dimethyl-2-butanone**, see P4,560-5, Pinacolone page 1184

| 19,040-3 ★ | **2,3-Dimethyl-1-butene**, 97% [563-78-0] (CH₃)₂CHC(CH₃)=CH₂ FW 84.16 mp -158°....... bp 56° n₀ 1.3890 d 0.680 Fp -1°F(-18°C) Beil. 1(3),816 FT-NMR 1(1),31B FT-IR 1(1),22A SI 3,D,5 Safety 2,1359D R&S 1(1),231 FLAMMABLE LIQUID IRRITANT | 5g 25g | 19.35 62.10 |

| 22,015-9 ★ | **2,3-Dimethyl-2-butene**, 99+% [563-79-1] (CH₃)₂C=C(CH₃)₂ FW 84.16 mp -75°........... bp 73° n₀ 1.4120 d 0.708 Fp 2°F(-16°C) Beil. 1,218 FT-NMR 1(1),35B FT-IR 1(1),25A SI 3,B,8 Safety 2,1360A R&S 1(1),25H FLAMMABLE LIQUID IRRITANT Substrate in photoinduced molecular transformations involving 2-hydroxy-1,4-naphthoquinones. J. Org. Chem. 1993, 58, 4614. | 10mL 25mL 100mL | 34.15 52.40 140.80 |

| 12,925-9 ★ | **2,3-Dimethyl-2-butene**, 98% [563-79-1] (CH₃)₂C=C(CH₃)₂............................. | 10mL 25mL 100mL 1L | 18.10 36.15 109.15 521.15 |

| 30,403-4 ★ | **2,3-Dimethyl-2-butene**, 1.0M solution in tetrahydrofuran [563-79-1]....................... (CH₃)₂C=C(CH₃)₂ FW 84.16 d 0.857 Fp -5°F(-20°C) FT-NMR 1(1),35B FT-IR 1(1),25A SI 3,B,8 Safety 2,1360B R&S 1(1),25H FLAMMABLE LIQUID IRRITANT Used for the preparation of thexylborane. See 22,079-5, Thexylborane Preparation Kit. (Packaged under nitrogen in Sure/Seal™ bottles) | 100mL 800mL | 14.65 35.30 |

| 11,905-9 ★ | **3,3-Dimethyl-1-butene**, 95% [558-37-2] (neohexene) (CH₃)₃CCH=CH₂ FW 84.16........ mp -115° bp 41° n₀ 1.3760 d 0.653 Fp -20°F(-28°C) Beil. 1,217 FT-NMR 1(1),29B FT-IR 1(1),25B SI 3,C,4 Safety 2,1360C R&S 1(1),23B FLAMMABLE LIQUID IRRITANT Stabilized with 50-150 ppm BHT | 50mL 250mL | 10.05 41.85 |

| 36,952-7 ★ | **N,N-Dimethylbutylamine**, 99% [927-62-8] CH₃(CH₂)₃N(CH₃)₂ FW 101.19....................... bp 93.3°/750mm n₀ 1.3980 d 0.721 Fp 25°F(-3°C) Beil. 4,1,371 FT-NMR 1(1),483B SI 44,A,3 R&S 1(1),321D FLAMMABLE LIQUID CORROSIVE | 250mL 1L | 13.50 37.45 |

| 12,641-1 ★ | **1,3-Dimethylbutylamine**, 98% [108-09-8] (CH₃)₂CHCH₂CH(CH₃)NH₂ FW 101.19............. bp 108-110° n₀ 1.4085 d 0.717 Fp 55°F(12°C) Beil. 4,191 FT-NMR 1(1),460B FT-IR 1(1),288A SI 42,A,1 Safety 2,1360D R&S 1(1),307D RTECS# EO4460000 FLAMMABLE LIQUID TOXIC | 5g 25g | 22.70 77.80 |

| 18,311-3 | **3,3-Dimethylbutylamine**, 98% [15673-00-4] (CH₃)₃CCH₂CH₂NH₂ FW 101.19.................. bp 114-116° n₀ 1.4135 d 0.752 Fp 42°F(5°C) FT-NMR 1(1),460C FT-IR 1(1),287D SI 42,B,1 Safety 2,1361A R&S 1(1),307E FLAMMABLE LIQUID CORROSIVE | 250mg 1g | 12.65 37.35 |

| 24,439-2 | **3,3-Dimethyl-1-butyne**, 98% [917-92-0] (tert-butylacetylene) (CH₃)₃CC≡CH.............. FW 82.15 mp -78° bp 37-38° n₀ 1.3740 d 0.667 Fp <-30°F(-34°C) Beil. 1,256 FT-NMR 1(3),504C FT-IR 1(2),933B SI 425,A,4 Safety 2,1361B R&S 1(2),2697N FLAMMABLE LIQUID IRRITANT | 5g 25g | 35.00 116.65 |

| 35,990-4 | **3,3-Dimethylbutyraldehyde**, 95% [2987-16-8] (CH₃)₃CCH₂CHO FW 100.16.................. bp 104-106° n₀ 1.3970 d 0.798 Fp 51°F(10°C) Beil. 1,3,2843 FT-NMR 1(1),729B SI 73,B,3 R&S 1(1),509K FLAMMABLE LIQUID IRRITANT | 1mL 5mL | 24.50 81.75 |

| D15,260-9 | **2,2-Dimethylbutyric acid**, 96% [595-37-9] C₂H₅C(CH₃)₂CO₂H FW 116.16.................. bp 94-96°/5mm n₀ 1.4154 d 0.928 Fp 175°F(79°C) Beil. 2,335 FT-NMR 1(1),761A FT-IR 1(3),579A SI 76,A,3 R&S 1(1),539A IRRITANT | 5mL 100mL 500mL | 15.60 17.30 57.60 |

**3,3-Dimethylbutyric acid**, see B8840-3, tert-Butylacetic acid page 264

**3,3-Dimethylbutyryl chloride**, see B8880-2, tert-Butylacetyl chloride page 264

| D15,280-3 ★ | **Dimethylcarbamyl chloride**, 98% [79-44-7] (CH₃)₂NCOCl FW 107.54 mp -33°.............. bp 167-168°/775mm n₀ 1.4530 d 1.168 Fp 155°F(68°C) Beil. 4,73 FT-NMR 1(1),1217A FT-IR 1(1),743C SI 133,A,6 Safety 2,1361C R&S 1(1),871G RTECS# FD4200000 CANCER SUSPECT AGENT CORROSIVE | 5g 100g 500g | 10.45 10.85 35.10 |

| 41,539-1 ↙ | **1-(N,N-Dimethylcarbamoyl)-4-(2-sulfoethyl)pyridinium hydroxide, inner salt**........ [136997-71-2] FW 258.30 mp 185° (dec.) SI 404,C,2 IRRITANT | 10g 50g | 12.50 41.35 |

**41,539-1**

**FOR LABORATORY SUPPLIES SEE THE TECHWARE SECTION** ■ **579** ■

**FIGURE 4.1** Page from the 1996–1997 *Aldrich Catalog*. Listings provide a summary of the physical properties for each compound. (Reprinted with permission from Aldrich Chemical Co., Inc., Milwaukee, WI.)

*CRC Handbook of Chemistry and Physics*

The *CRC Handbook of Chemistry and Physics* is a commonly used handbook and is also published annually. The *CRC Handbook* contains a wealth of information, including extensive tables of physical properties and solubilities, as well as structural formulas, for more than 12,000 organic and 2400 inorganic compounds. Recent editions of the *CRC Handbook* have an arrangement of topics different from that of earlier editions: the tables of organic compounds are now listed before the tables of inorganic compounds.

To locate an organic compound successfully, you must pay close attention to the nomenclature used in the tables. In general, IUPAC nomenclature is followed, but a compound usually known by its common name may be listed under both names or even only under the common name. For example, $CH_3COOH$, ethanoic acid, is listed under its common name of acetic acid with ethanoic acid given as the second name; no ethanoic acid entry occurs. Monosubstituted derivatives of compounds are found under the heading of the parent compound rather than simply in alphabetical order by the first letter of the compound's name. For example, Figure 4.2 shows a page from a recent edition of the *CRC Handbook,* which illustrates how 1-bromobutane (entry 3231) is listed under the parent alkane as "Butane, 1-bromo-"; the second name, "butyl bromide," is another commonly used name for the compound. An explanation of the nomenclature system, plus definitions of abbreviations and symbols, precedes the tables of organic compounds.

*The Merck Index*

*The Merck Index,* currently in its 12th edition, has over 10,000 organic compound entries that give physical properties and solubilities as well as many references to the compounds' syntheses, safety information, and uses. *The Merck Index* is particularly comprehensive for organic compounds of medical and pharmaceutical importance. Figure 4.3 repeats the page from the 12th edition of *The Merck Index* shown in Technique 1.

## 4.2  Online Resources

The World Wide Web provides access to many sites that have information about organic compounds. The number of Web sites changes almost daily; the following sites provided information on physical constants at the time of publication.

http://www.sigma-aldrich.com (Online catalog identical to printed version.)
http://ChemFinder.com (Database)

A Web search using keywords such as "chemical compound databases" or "chemistry databases" will help you locate current Web sites with information about organic compounds.

## PHYSICAL CONSTANTS OF ORGANIC COMPOUNDS (continued)

| No. | Name / Synonym | Mol. Form. / Mol. Wt. | CAS RN / mp/°C | Merck No. / bp/°C | Beil. Ref. / den/g cm$^{-3}$ | Solubility / $n_D$ |
|---|---|---|---|---|---|---|
| 3202 | Butanamide, 2,4-dihydroxy-N-(3-hydroxypropyl)-3,3-dimethyl-, (R)- / Dexpanthenol | $C_9H_{19}NO_4$ / 205.25 | 81-13-0 | 2924 / dec | 4-04-00-01652 / 1.20$^{20}$ | H$_2$O 4; EtOH 4; eth 2; MeOH 4 / 1.497$^{20}$ |
| 3203 | Butanamide, N,N-dimethyl- | $C_6H_{13}NO$ / 115.18 | 760-79-2 / -40 | 186; 125$^{100}$ | 4-04-00-00185 / 0.9064$^{25}$ | ace 4; bz 4; eth 4; EtOH 4 / 1.4391$^{25}$ |
| 3204 | Butanamide, N,N'-(3,3'-dimethyl[1,1'-biphenyl]-4,4'-diyl)bis[3-oxo- / N,N'-Bis(acetoacetyl)-3,3'-dimethylbenzidine | $C_{22}H_{24}N_2O_4$ / 380.44 | 91-96-3 / 212 | | 3-13-00-00490 | DMSO 2 |
| 3205 | Butanamide, N-(4-hydroxyphenyl)- / 4'-Hydroxybutyranilide | $C_{10}H_{13}NO_2$ / 179.22 | 101-91-7 / 139.5 | 4745 | 4-13-00-01109 | H$_2$O 4; EtOH 4 |
| 3206 | Butanamide, 3-methyl- / Isovaleramide | $C_5H_{11}NO$ / 101.15 | 541-46-8 / 137 | 5119 / 226 | 4-02-00-00902 | H$_2$O 3; EtOH 3; eth 3; peth 4 |
| 3207 | Butanamide, 3-oxo-N-phenyl- / Acetoacetanilide | $C_{10}H_{11}NO_2$ / 177.20 | 102-01-2 / 86 | 50 | 4-12-00-00955 | H$_2$O 2; EtOH 3; eth 3; bz 3 |
| 3208 | Butanamide, N-phenyl- | $C_{10}H_{13}NO$ / 163.22 | 1129-50-6 / 97 | 189$^{15}$ | 4-12-00-00387 / 1.134$^{25}$ | H$_2$O 1; EtOH 4; eth 4; chl 2 |
| 3209 | 1-Butanamine / Butylamine | $C_4H_{11}N$ / 73.14 | 109-73-9 / -49.1 | 1543 / 77.0 | 4-04-00-00540 / 0.7414$^{20}$ | H$_2$O 5; EtOH 3; eth 3 / 1 403$^{20}$ |
| 3210 | 2-Butanamine, (±)- / sec-Butylamine, (±)- | $C_4H_{11}N$ / 73.14 | 33966-50-6 / <-72 | 1544 / 63.5 | 4-04-00-00618 / 0.7246$^{20}$ | H$_2$O 3; EtOH 5; eth 5; ace 4 / 1.3932$^{20}$ |
| 3211 | 1-Butanamine, N-butyl- / Dibutylamine | $C_8H_{19}N$ / 129.25 | 111-92-2 / -62 | 3019 / 159.6 | 4-04-00-00550 / 0.7670$^{20}$ | H$_2$O 3; EtOH 4; eth 4; ace 3 / 1.4177$^{20}$ |
| 3212 | 1-Butanamine, N-butyl-N-nitroso- / Dibutylnitrosamine | $C_8H_{18}N_2O$ / 158.24 | 924-16-3 | 105$^8$ | 4-04-00-03389 | |
| 3213 | 1-Butanamine, N,N-dibutyl- / Tributylamine | $C_{12}H_{27}N$ / 185.35 | 102-82-9 / -70 | 9530 / 216.5 | 4-04-00-00554 / 0.7770$^{20}$ | H$_2$O 2; EtOH 4; eth 4; ace 3 / 1.4299$^{20}$ |
| 3214 | 1-Butanamine, 4,4-diethoxy- | $C_8H_{19}NO_2$ / 161.24 | 6346-09-4 | 196 | 4-04-00-01928 / 0.933$^{25}$ | 1.4275$^{20}$ |
| 3215 | 1-Butanamine, N,N-dimethyl- | $C_6H_{15}N$ / 101.19 | 927-62-8 | 95 | 4-04-00-00546 / 0.7206$^{20}$ | H$_2$O 5; EtOH 5; eth 5; ace 5 / 1.3970$^{20}$ |
| 3216 | 2-Butanamine, 2,3-dimethyl- | $C_6H_{15}N$ / 101.19 | 4358-75-2 / -70.5 | 104.5 | 4-04-00-00733 / 0.7683$^0$ | 1 4096$^{17}$ |
| 3217 | 2-Butanamine, 3,3-dimethyl- | $C_6H_{15}N$ / 101.19 | 3850-30-4 / -20 | 102 | 4-04-00-00730 / 0.7668$^{20}$ | H$_2$O 4 / 1.4105$^{25}$ |
| 3218 | 1-Butanamine, N-ethyl- / Butylethylamine | $C_6H_{15}N$ / 101.19 | 13360-63-9 | 107.5 | 4-04-00-00547 / 0.7398$^{20}$ | EtOH 5; eth 5; ace 5; bz 5 / 1 4040$^{20}$ |
| 3219 | 2-Butanamine, N-ethyl-, (±)- / sec-Butyl ethyl amine (DL) | $C_6H_{15}N$ / 101.19 | 116724-10-8 / -104.3 | 98 | 2-04-00-00636 / 0.7358$^{20}$ | ace 4; bz 4; eth 4; EtOH 4 |
| 3220 | 1-Butanamine, 3-methyl- / Isoamylamine | $C_5H_{13}N$ / 87.16 | 107-85-7 / | 4994 / 96 | 4-04-00-00696 / 0.7505$^{20}$ | H$_2$O 5; EtOH 5; eth 5; ace 3 / 1.4083$^{20}$ |
| 3221 | 1-Butanamine, N-methyl- / Butylmethylamine | $C_5H_{13}N$ / 87.16 | 110-68-9 | 91 | 4-04-00-00546 / 0.763$^{15}$ | |
| 3222 | 2-Butanamine, 2-methyl- | $C_5H_{13}N$ / 87.16 | 594-39-8 / -105 | 77 | 4-04-00-00694 / 0.731$^{25}$ | H$_2$O 4; ace 4; eth 4; EtOH 4 / 1.3954$^{25}$ |
| 3223 | 2-Butanamine, 3-methyl- | $C_5H_{13}N$ / 87.16 | 598-74-3 / -50 | 85.5 | 2-04-00-00644 / 0.757$^{419}$ | H$_2$O 4; EtOH 3 / 1.4096$^{18}$ |
| 3224 | 1-Butanamine, 2-methyl-N,N-bis(2-methylbutyl)- | $C_{15}H_{33}N$ / 227.43 | 620-43-9 | 232 | 0-04-00-00179 / 0.913 | ace 4; bz 4; eth 4; EtOH 4 / 1 4330$^{20}$ |
| 3225 | 1-Butanamine, 3-methyl-N,N-bis(3-methylbutyl)- | $C_{15}H_{33}N$ / 227.43 | 645-41-0 | 235 | 4-04-00-00700 / 0.7848$^{20}$ | H$_2$O 1; EtOH 4; eth 5; bz 5 / 1.4331$^{20}$ |
| 3226 | 1-Butanamine, 3-methyl-N-(3-methylbutyl)- / Diisoamylamine | $C_{10}H_{23}N$ / 157.30 | 544-00-3 / -44 | 3178 / 188 | 4-04-00-00699 / 0.7672$^{21}$ | H$_2$O 1; EtOH 3; eth 5 / 1.4235$^{20}$ |
| 3227 | 2-Butanamine, N-(1-methylpropyl)- | $C_8H_{19}N$ / 129.25 | 626-23-3 | 135 | 4-04-00-00620 / 0.7534$^{20}$ | H$_2$O 4; EtOH 3; os 3 / 1.4162$^{20}$ |
| 3228 | 1-Butanamine, 1,1,2,2,3,3,4,4,4-nonafluoro-N,N-bis(nonafluorobutyl)- | $C_{12}F_{27}N$ / 671.10 | 311-89-7 | 178 | 4-02-00-00819 / 1.884$^{25}$ | ace 3 / 1.291$^{25}$ |
| 3229 | Butane | $C_4H_{10}$ / 58.12 | 106-97-8 / -138.2 | 1507 / -0.5 | 4-01-00-00236 / *0.573$^{25}$ | H$_2$O 3; EtOH 4; eth 4; chl 4 / 1.3326$^{20}$ |
| 3230 | Butane, 2,2-bis(ethylsulfonyl)- / Sulfonethylmethane | $C_8H_{18}O_4S_2$ / 242.36 | 76-20-0 / 76 | 8937 / dec | 3-01-00-02790 / 1.199$^{85}$ | chl 3 |
| 3231 | Butane, 1-bromo- / Butyl bromide | $C_4H_9Br$ / 137.02 | 109-65-9 / -112.4 | 1553 / 101.6 | 4-01-00-00258 / 1.2758$^{20}$ | H$_2$O 1; EtOH 5; eth 5; ace 5 / 1.4401$^{20}$ |
| 3232 | Butane, 2-bromo-, (±)- / (±)-sec-Butyl bromide | $C_4H_9Br$ / 137.02 | 5787-31-5 / -111.9 | 1554 / 91.2 | 4-01-00-00261 / 1.2585$^{20}$ | ace 4; eth 4; chl 4 / 1.4366$^{20}$ |
| 3233 | Butane, 1-bromo-4-chloro- | $C_4H_8BrCl$ / 171.46 | 6940-78-9 | 175; 63$^{10}$ | 4-01-00-00264 / 1.488$^{20}$ | H$_2$O 1; EtOH 3; eth 3; ctc 2 / 1.4885$^{20}$ |
| 3234 | Butane, 2-bromo-1-chloro- / 2-Bromo-1-chlorobutane | $C_4H_8BrCl$ / 171.46 | 79504-01-1 | 146.5 | 4-01-00-00264 / 1.468$^{20}$ | bz 4; eth 4; EtOH 4; chl 4 / 1.4880$^{20}$ |
| 3235 | Butane, 1-bromo-3,3-dimethyl- | $C_6H_{13}Br$ / 165.07 | 1647-23-0 | 138 | 3-01-00-00409 / 1.1556$^{20}$ | eth 4; EtOH 4; chl 4 / 1.4440$^{20}$ |
| 3236 | Butane, 2-bromo-2,3-dimethyl- / 2-Bromo-2,3-dimethylbutane | $C_6H_{13}Br$ / 165.07 | 594-52-5 / 24.5 | 133; 87$^{180}$ | 4-01-00-00374 / 1.1772$^{10}$ | eth 4; chl 4 / 1.4517 |
| 3237 | Butane, 1-bromo-4-fluoro- / 1-Bromo-4-fluorobutane | $C_4H_8BrF$ / 155.01 | 462-72-6 | 135 | 4-01-00-00263 | eth 4; EtOH 4 / 1.4370$^{25}$ |
| 3238 | Butane, 1-bromo-2-methyl-, (S)- / d-Amyl bromide | $C_5H_{11}Br$ / 151.05 | 534-00-9 / | 636 / 121.6 | 4-01-00-00327 / 1.2234$^{20}$ | H$_2$O 1; EtOH 3; eth 3; chl 4 / 1.4451$^{20}$ |
| 3239 | Butane, 1-bromo-3-methyl- / Isoamyl bromide | $C_5H_{11}Br$ / 151 05 | 107-82-4 / -112 | 4996 / 120.4 | 4-01-00-00328 / 1.2071$^{20}$ | H$_2$O 1; EtOH 3; eth 3; ctc 2 / 1 4420$^{20}$ |

3-90

**FIGURE 4.2** Page from the 77th Edition of the *CRC Handbook of Chemistry and Physics*. (Reprinted with permission from CRC Press, Inc., Boca Raton, FL.)

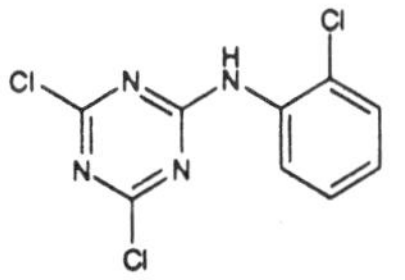

White to tan crystals, mp 159-160°. Insol in water. Soly at 30° (g/100 ml): toluene, 5; xylene, 4; acetone, 10. Subject to hydrolysis; not compatible with oils and alkaline materials. LD$_{50}$ orally in rats: > 5000 mg/kg, Mobay Technical Information Sheet, Jan. 1979.
USE: Fungicide.

**695. Anileridine.** *1-[2-(4-Aminophenyl)ethyl]-4-phenyl-4-piperidinecarboxylic acid ethyl ester; 1-(p-aminophenethyl)-4-phenylisonipecotic acid ethyl ester;* ethyl 1-(4-aminophenethyl)-4-phenylisonipecotate; *N-[β-(p-aminophenyl)ethyl]-4-phenyl-4-carbethoxypiperidine; N-β-(p-aminophenyl)-ethylnormeperidine;* Leritine; Nipecotan; Alidine; Apodol. C$_{22}$H$_{28}$N$_2$O$_2$; mol wt 352.48. C 74.97%, H 8.01%, N 7.95%, O 9.08%. Synthesis: Weijlard *et al., J. Am. Chem. Soc.* **78**, 2342 (1956); U.S. pat. **2,966,490** (1960 to Merck & Co.).

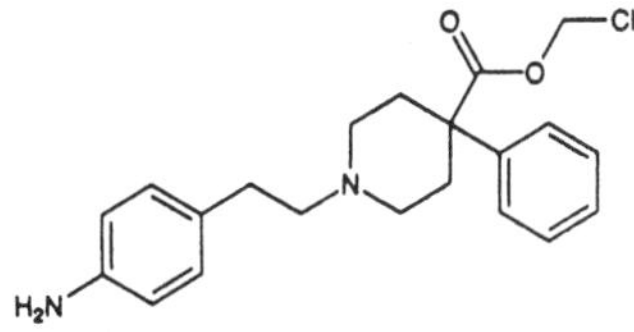

mp 83°.
Dihydrochloride, C$_{22}$H$_{28}$N$_2$O$_2$.2HCl, crystals from methanol + ether, mp 280-287° (dec). Freely soluble in water, methanol. Solubility in ethanol: 8 mg/g. pH of aq solns 2.0 to 2.5. Solns are stable at pH 3.5 and below. At pH 4 and higher the insol free base is precipitated. uv max (pH 7 in 90% methanol contg phosphate buffer): 235, 289 nm (A$^{1\%}_{1cm}$ 293, 34.5). Distribution coefficient (water, pH 3.6/n-butanol): 0.9.
*Note:* This is a controlled substance (opiate) listed in the U.S. Code of Federal Regulations, Title 21 Part 1308.12 (1995).
THERAP CAT: Analgesic (narcotic).

**696. Aniline.** *Benzenamine;* aniline oil; phenylamine; aminobenzene; aminophen; kyanol. C$_6$H$_7$N; mol wt 93.13. C 77.38%, H 7.58%, N 15.04%. First obtained in 1826 by Unverdorben from dry distillation of indigo. Runge found it in coal tar in 1834. Fritzsche, in 1841, prepared it from indigo and potash and gave it the name aniline. Manuf from nitrobenzene or chlorobenzene: Faith, Keyes & Clark's *Industrial Chemicals,* F. A. Lowenheim, M. K. Moran, Eds. (Wiley-Interscience, New York, 4th ed., 1975) pp 109-116. Procedures: A. I. Vogel, *Practical Organic Chemistry* (Longmans, London, 3rd ed., 1959) p 564; Gattermann-Wieland, *Praxis des organischen Chemikers* (de Gruyter, Berlin, 40th ed., 1961) p 148. Brochure *"Aniline"* by Allied Chemical's National Aniline Division (New York, 1964) 109 pp, gives reactions and uses of aniline (877 references). Toxicity study: K. H. Jacobson, *Toxicol. Appl. Pharmacol.* **22**, 153 (1972).

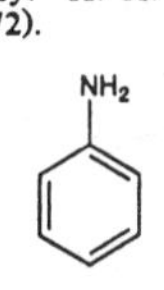

Oily liquid; colorless when freshly distilled, darkens on exposure to air and light. *Poisonous!* Characteristic odor and burning taste; combustible; volatile with steam. d$^{20}_{20}$ 1.022. bp 184-186°. Solidif −6°. Flash pt, closed cup: 169°F (76°C). n$^{20}_D$ 1.5863. pKb 9.30. pH of 0.2 molar aq

soln 8.1. One gram dissolves in 28.6 ml water, 15.7 ml boil. water; misc with alcohol, benzene, chloroform, and most other organic solvents. Combines with acids to form salts. It dissolves alkali or alkaline earth metals with evolution of hydrogen and formation of anilides, e.g., C$_6$H$_5$NHNa. *Keep well closed and protected from light.* Incompat. Oxidizers, albumin, solns of Fe, Zn, Al, acids, and alkalies. LD$_{50}$ orally in rats: 0.44 g/kg (Jacobson).
Hydrobromide, C$_6$H$_7$N.HBr, white to slightly reddish, crystalline powder, mp 286°. Darkens in air and light. Sol in water, alc. *Protect from light.*
Hydrochloride, C$_6$H$_7$N.HCl, crystals, mp 198°. d 1.222. Darkens in air and light. Sol in about 1 part water; freely sol in alc. *Protect from light.*
Hydrofluoride, C$_6$H$_7$N.HF, crystalline powder. Turns gray on standing. Freely sol in water; slightly sol in cold, freely in hot alc.
Nitrate, C$_6$H$_7$N.HNO$_3$, crystals, dec about 190°. d 1.36. Discolors in air and light. Sol in water, alc. *Protect from light.*
Hemisulfate, C$_6$H$_7$N.½H$_2$SO$_4$, crystalline powder. d 1.38. Darkens on exposure to air and light. One gram dissolves in about 15 ml water; slightly sol in alc. Practically insol in ether. *Protect from light.*
Acetate, C$_6$H$_5$NH$_2$.HOOCCH$_3$. Prepd from aniline and acetic acid: Vignon, Evieux, *Bull. Soc. Chim. France* [4] **3**, 1012 (1908). Colorless liquid. d 1.070-1.072. Darkens with age; gradually converted to acetanilide on standing. Misc with water, alc.
Oxalate, C$_6$H$_5$NH$_2$.HOOCCOOH.H$_2$NC$_6$H$_5$. Prepd from aniline and oxalic acid in alc soln: Hofmann, *Ann.* **47**, 37 (1843). Triclinic rods from water, mp 174-175°. Readily sol in water; sparingly sol in abs alc. Practically insol in ether. *Caution:* Intoxication may occur from inhalation, ingestion, or cutaneous absorption. *Acute Toxicity:* cyanosis, methemoglobinemia, vertigo, headache, mental confusion. *Chronic Toxicity:* anemia, anorexia, wt loss, cutaneous lesions, *Clinical Toxicology of Commercial Products,* R. E. Gosselin *et al.,* Eds. (Williams & Wilkins, Baltimore, 4th ed., 1976) Section III, pp 29-35.
USE: Manuf dyes, medicinals, resins, varnishes, perfumes, shoe blacks; vulcanizing rubber; as solvent. Hydrochloride used in manuf of intermediates, aniline black and other dyes, in dyeing fabrics or wood black.

**697. Aniline Mustard.** *N,N-Bis(2-chloroethyl)benzenamine; N,N-bis(2-chloroethyl)aniline; phenylbis[2-chloroethylamine]; β,β'-dichlorodiethylaniline;* Lymphochin; Lymphocin; Lymphoquin. C$_{10}$H$_{13}$Cl$_2$N; mol wt 218.13. C 55.06%, H 6.01%, Cl 32.51%, N 6.42%. Prepd by the action of phosphorus pentachloride on *N,N*-bis-[2-hydroxyethyl]-aniline (phenyldiethanolamine): Robinson, Watt, *J. Chem. Soc.* **1934,** 1538; Korshak, Strepikheev, *J. Gen. Chem. USSR* **14,** 312 (1944).

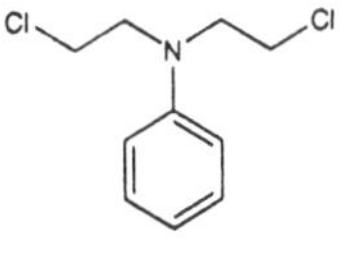

Stout prisms from methanol, mp 45°. bp$_{14}$ 164°; bp$_{0.5}$ 110°. Sol in hot methanol, ethanol. Very slightly sol in ether.
Hydrochloride, C$_{10}$H$_{14}$Cl$_3$N, crystals. *Vesicant.* Freely sol in water. Sol in alcohol.
USE: In cancer research.

**698. Anilinephthalein.** *3,3-Bis(4-aminophenyl)-1(3H)-isobenzofuranone; 3,3-bis(p-aminophenyl)phthalide.* C$_{20}$H$_{16}$N$_2$O$_3$; mol wt 316.36. C 75.93%, H 5.10%, N 8.85%, O 10.11%. Prepn: Hubacher, *J. Am. Chem. Soc.* **73,** 5885 (1951).

*Consult the Name Index before using this section.*                    **Page 111**

---

FIGURE 4.3 Page from the 12th Edition of *The Merck Index.* (Reprinted with permission from Merck and Co., Inc., Whitehouse Station, NJ.)

# 5

# MEASURING MASS AND VOLUME

Whether you are carrying out miniscale or microscale experiments, you need to accurately measure the reagents used in the reaction. The techniques for weighing solids and liquids, methods for measuring liquid volumes, and ways of transferring solids and liquids without loss are discussed in Technique 5.

## 5.1    Weighing

For miniscale reactions using more than 1 g of a substance, a balance that weighs to the nearest centigram (0.01 g) is satisfactory. The small quantities of reagents used in microscale (10–300 mg) and in many miniscale (<1 g) reactions require that their masses be determined on a balance that weighs to the nearest milligram (0.001 g).

Electronic top-loading balances are expensive precision instruments that can be easily rendered inaccurate by corrosion from spilled reagents. If anything spills on the balance or the weighing pan, clean it up immediately. Notify the instructor if the spill is extensive or the substance is corrosive. A milligram balance has a draft shield (Figure 5.1) to prevent air currents from disturbing the weighing pan while a sample is being weighed.

**Weighing Solids**

No solid reagent is weighed directly on the balance pan. Weigh the solid in a glass container (vial or beaker), in an aluminum or plastic weighing boat, or on glazed weighing paper, then transfer it to the reaction vessel. If the mass of the container or weighing paper was not *tared* ("subtracted") by pressing the tare or zero button before the sample was added, the container mass should be determined and recorded *after* the sample is transferred from it. A vial or flask that will be used to hold a product should be weighed with its cap or cork and label **before the product is placed in it.** Be sure to record the mass of the container, called the tare, in your notebook.

To weigh a specific quantity of a solid reagent, place the weighing container or paper on the balance pan and press the zero or tare button. Use a spatula to add small portions of the reagent until the desired mass (within 1–2%) is shown on the digital display. For example, the mass of a sample does not need to be exactly the 0.300 g specified, but it should be within ±0.005 g of that amount. **Record the actual amount that you use** in your notebook. If the reagent is the limiting factor, calculate the theoretical yield based on the actual amount used, not on the amount specified in the procedure.

**FIGURE 5.1**
Milligram top-loading
balance with draft
shield.

Draft shield

*Transferring Solids to the Reaction Vessel*

Once a solid is weighed, it must be transferred to the reaction vessel without mishap. For reactions being run in round-bottomed flasks, a powder funnel aids in transferring the solid reagent from the weighing paper into the small neck of the flask (Figure 5.2).* The stem of a powder funnel has a larger diameter than that of a funnel used for liquid transfers; therefore, solids will not clog it. The powder funnel serves to keep the solid from spilling and prevents any

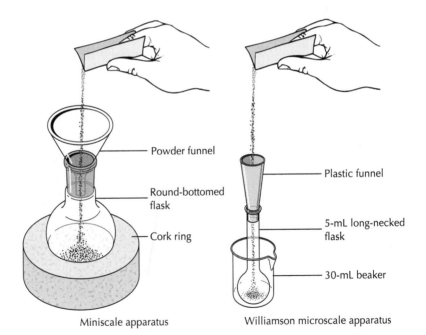

Powder funnel

Round-bottomed
flask

Cork ring

Plastic funnel

5-mL long-necked
flask

30-mL beaker

Miniscale apparatus

Williamson microscale apparatus

**FIGURE 5.2**
Transferring solids with
a powder funnel.

*Use of a powder funnel is essential with Williamson microscale glassware because of the very small opening at the top of the round-bottomed flasks and reaction tubes.

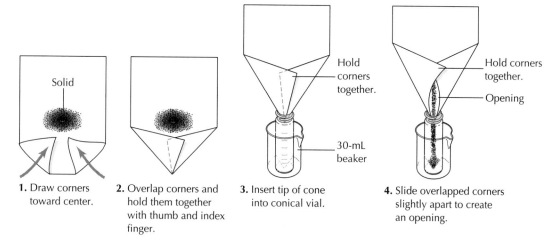

1. Draw corners toward center.

2. Overlap corners and hold them together with thumb and index finger.

3. Insert tip of cone into conical vial.

4. Slide overlapped corners slightly apart to create an opening.

Hold corners together.

30-mL beaker

Hold corners together.

Opening

**FIGURE 5.3** Transferring solids with a weighing paper.

solid from sticking to the inside of the joint at the top of the flask. If a powder funnel is unavailable, roll the weighing paper into a cone by overlapping two corners on the same side of the paper. Place the rolled end well down into the neck of the flask or vial and gently slide one corner of the paper away from the other to create an opening just large enough for the solid to fall into the reaction vessel, as shown in Figure 5.3.

**Weighing Liquids**

To weigh a liquid, the mass of the container (tare) must be ascertained and recorded, or else subtracted by using the zero button on the balance, before the liquid is placed in it. If the liquid is volatile, a cap or cork for the container should be included in the tare so that the sample will not evaporate during the weighing process.

*Be very careful that liquid does not spill on the balance while you are weighing a liquid sample; should a spill occur, clean it up immediately.*

To weigh a specific amount of a liquid compound, determine the volume of the required sample from the density and transfer that volume to a tared container. Ascertain the mass of the tared container plus the liquid to determine the mass of the liquid sample. If the mass of liquid needed is less than 1 g, an alternative to measuring the volume is to add the liquid dropwise to a tared container until the desired mass is obtained.

---

## 5.2      Measuring Volume

Several volume measuring devices are available in the laboratory, including graduated cylinders, assorted types of pipets, burets, dispensing pumps, or even beakers and flasks with volume markings on them. The device used for measuring a specific volume of liquid depends on the accuracy with which the volume needs to be known.

For example, the volume of a reagent that is the limiting factor in a reaction needs to be measured with a graduated pipet or a dispensing pump, then weighed to know the exact amount obtained. If the liquid is a solvent or present in excess of the limiting reagent, volume measurement should be done with a graduated pipet for microscale work and with a graduated pipet or cylinder for miniscale work. The volume markings on beakers and flasks can be used only to estimate an approximate volume and should never be used for measuring a reagent that will go into a reaction.

*Graduated Cylinders*

Graduated cylinders do not provide high accuracy in volume measurement and should be used only to measure quantities of reagents other than the limiting reagent. The volume contained in a graduated cylinder is correctly read from the bottom of the meniscus as shown in Figure 5.4. For microscale extractions, a 5-mL or 10-mL graduated cylinder can be used for measuring volumes of washing solvents greater than 1 mL, but only graduated pipets should be used for measuring reagents used in a microscale reaction.

*Dispensing Pumps*

Dispensing pumps fitted to glass bottles come in a variety of sizes designed to dispense a preset volume of liquid (>0.1 mL). Pumps in the 1-, 2-, and 5-mL range can be used for microscale work; larger-volume pumps may be needed for miniscale work.

   The operation of a dispensing pump consists of slowly pulling the plunger up until it reaches the preset volume stop (Figure 5.5). Hold the receiving container or reaction vessel under the spout, then gently push the plunger down as far as it will go to discharge the preset volume. Be sure to check that the spout is filled with liquid and contains no air bubbles before you begin to draw up the plunger—air bubbles will cause a volume less than the preset one

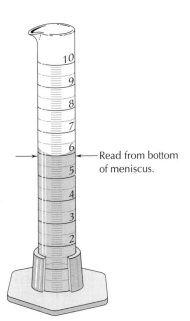

Read from bottom of meniscus.

**FIGURE 5.4**
Graduated cylinder with liquid showing a meniscus.

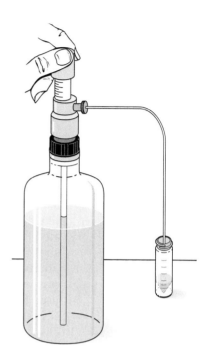

**FIGURE 5.5**
Dispensing pump.

to be delivered. If air bubbles are present in the spout, pull up the plunger and discharge one or two samples into another container until the spout is completely filled with liquid before you dispense a sample into your receiving container. (Place the discarded samples in the appropriate waste container.)

*Graduated Pipets*

The small volumes used in microscale and many miniscale reactions are conveniently and accurately measured with graduated pipets of 1.00-, 2.00-, and 5.00-mL size. A syringe attached to the pipet with a short piece of latex tubing or a pipet pump serves to fill the pipet and expel the requisite volume. The most accurate volumes are obtained by difference measurement, that is, filling the pipet to the zero mark, then discharging the liquid until the required volume has been dispensed. The excess remaining in the pipet should be placed in the appropriate waste container.

Two types of graduated pipets are available: one delivers its total capacity when the last drop is expelled (Figure 5.6a), and the other delivers its total capacity by stopping the delivery when the meniscus reaches the bottom graduation mark (Figure 5.6b). However, both kinds of graduated pipets are more frequently used to deliver a specific volume by stopping the delivery when the meniscus reaches the desired volume.

*Automatic Delivery Pipet*

Small volumes of $10 - 1000$ μL ($0.010 - 1.000$ mL) can be measured very accurately and reproducibly with automatic delivery pipets or pipettors. Automatic pipets have disposable tips that hold the preset volume of liquid; no liquid actually enters the pipet itself and the pipet should never be used without a disposable tip in place (Figure 5.7). Automatic pipets are very expensive, and your

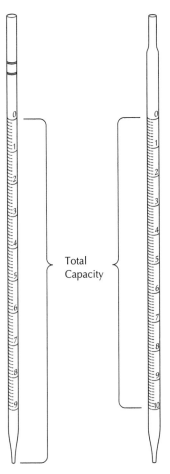

**(a)** Expel entire contents
to deliver total capacity.

**(b)** Deliver total capacity by
draining until the bottom
of the meniscus is at 10.00 mL.

**FIGURE 5.6**
Graduated pipets.

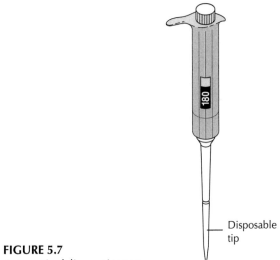

Disposable
tip

**FIGURE 5.7**
Automatic delivery pipettor.

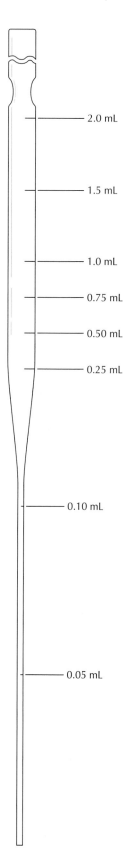

**FIGURE 5.8**
Approximate calibration of a Pasteur pipet.

instructor will demonstrate the specific operating technique for the type in your laboratory. Automatic pipets need to be properly calibrated before use; consult your instructor.

**Pasteur Pipets and
Plastic Transfer Pipets**

As volume measurement devices, Pasteur pipets are suitable for measuring only approximate volumes. Pasteur pipets do not have volume markings; an approximate calibration is shown in Figure 5.8. Pasteur pipets are particularly useful for the transfer of liquids in microscale reactions or extractions and for transferring small volumes of liquid from one container to another. Attaching a syringe with a piece of latex tubing to a Pasteur pipet allows an approximate volume to be estimated from the position of the plunger in the syringe as the liquid is drawn into the pipet (Figure 5.9).

Plastic transfer pipets are also useful for the transfer of liquids (Figure 5.10). Calibrated plastic pipets, available in 1- and 2-mL sizes, are suitable for measuring volumes of aqueous washing solutions used for microscale extractions or for estimating the volume of solvent added in a recrystallization. Most plastic transfer pipets are made from polyethylene and are chemically impervious to aqueous acidic or basic solutions, alcohols such as methanol or ethanol, and diethyl ether; they are not suitable for halogenated hydrocarbons because the plasticizer leaches out of the polyethylene into the liquid being transferred.

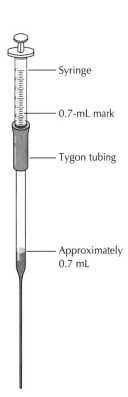

Syringe

0.7-mL mark

Tygon tubing

Approximately 0.7 mL

**FIGURE 5.9**
Using a syringe to estimate the volume of liquid drawn into a Pasteur pipet.

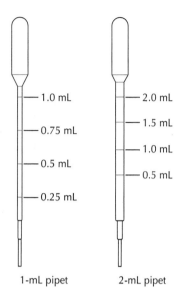

**FIGURE 5.10**
Plastic transfer pipets.

1-mL pipet          2-mL pipet

***Beakers, Erlenmeyer Flasks, Conical Vials, and Reaction Tubes***    The volume markings found on beakers and Erlenmeyer flasks are only approximations and are not suitable for measuring any reagent that will be used in a reaction. The volume markings on conical vials or reaction tubes are also approximations and should be used only to estimate the volume of the contents, such as the final volume of a recrystallization solution, and not for measuring the volume of a reagent used in a reaction.

**TECHNIQUE**

# 6  HEATING AND COOLING METHODS

Technique 6 describes heating and cooling methods used in the organic laboratory. Many organic reactions do not occur spontaneously when the reactants are mixed together but require a period of heating to reach completion. Contrarily, exothermic organic reactions require removal of the heat generated during the reaction period by placing the reaction vessel in a cooling bath. Cooling baths are also used to ensure the maximum recovery of crystallized product from a solution or to cool the contents of a reaction flask before proceeding to the next operation in the procedure.

The use of Bunsen burners in the organic laboratory poses an extreme fire hazard because the volatile vapors of organic compounds can ignite explosively when mixed with air. Utilizing electrical heating devices or steam baths plus working in hoods greatly reduces the chances of an accidental fire. Using a Bunsen burner or other source of open flame should be a very rare event in an organic laboratory.

## 6.1    Boiling Stones

Before describing various heating methods, let's consider the role of boiling stones (also called boiling chips) or boiling sticks. Liquids heated in laboratory glassware tend to boil by forming large bubbles of superheated vapor, a process called "bumping." The inside surface of the glass is so smooth that no tiny crevices exist where air bubbles can be trapped, unlike the surfaces of metal pans used for cooking. The addition of boiling stones supplies a porous surface. The liquid enters the vapor phase at the air-vapor interface of a pore in the boiling stone. As the volume of vapor nucleating at the pore increases, a bubble forms, is released, and continues to grow as it rises through the liquid. The sharp edges on boiling stones also catalyze bubble formation in complex ways not fully understood.

The boiling stones commonly used in the laboratory consist of small pieces of carborundum, a chemically inert compound of carbon and silicon. Their black color makes them easy to identify and remove from your product if you have not removed them earlier by filtration. Boiling sticks are short pieces of wooden applicator sticks. Boiling sticks should not be used in reaction mixtures or with any solvent that might react with wood or in a solution containing an acid.

*You should always add boiling stones or a boiling stick to any unstirred liquid before boiling it—unless instructed otherwise.*

One or two boiling stones suffice for smooth boiling of most liquids. **Boiling stones should always be added before heating the liquid.** Adding boiling stones to a hot liquid may cause the liquid to boil violently and erupt from the flask because superheated vapor trapped in the liquid is released all at once. If you forget to add boiling stones before heating, **the liquid must be cooled well below the boiling point before putting boiling stones into it.**

If a liquid you have boiled requires cooling and reheating, additional boiling stones should be added before reheating commences. Once boiling stones cool, their pores fill with liquid. The liquid does not escape from the pores as readily as air does when the boiling stone is reheated, rendering the boiling stone less effective in promoting smooth boiling.

Mechanical or magnetic stirring agitates a boiling liquid enough to make boiling stones unnecessary, and stirring is frequently used instead of boiling stones.

## 6.2    Heating Devices

*Heating Mantles*

Many reactions and other operations are carried out in round-bottomed flasks, which are heated with electric heating mantles shaped to fit the bottom on the flask. Several types of heating

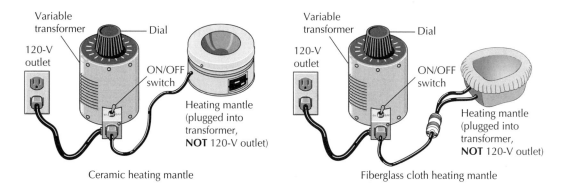

**FIGURE 6.1**
Heating mantle and variable transformer. Note that the transformer dial is calibrated in percentage of line voltage, NOT in degrees.)

mantles are available commercially. One type consists of woven fiberglass, with the heating element embedded between the layers of fabric. Fiberglass heating mantles come in a variety of sizes to fit specific sizes of round-bottomed flasks; one sized for a 100-mL flask will not work well with a flask of any other size. Another type of heating mantle, called a Thermowell, has a metal housing and a ceramic well covering the heating element. Thermowell heating mantles can be used with flasks smaller than the designated size of the mantle because of radiant heating from the surface of the well. Loosely packed glass wool will transfer heat from the mantle more effectively than just air because air is a poor conductor of heat.

Both types of heating mantles have no controls and must be plugged into a variable transformer (or rheostat) or other variable controller to adjust the rate of heating (Figure 6.1). The variable transformer is then plugged into a wall outlet.

Heating mantles are supported underneath a round-bottomed flask by an iron ring or lab jack. Fiberglass heating mantles should not be used on wooden surfaces because the bottom of the heating mantle can become hot enough to char the wood.

*Hot Plates*

Hot plates work well for heating flat-bottomed containers such as beakers, Erlenmeyer flasks, and crystallizing dishes used as water baths or sand baths. Hot plates also serve to heat the aluminum blocks used with microscale glassware. Figure 6.2a shows a microscale setup for heating a standard-taper conical vial fitted with an air condenser; Figure 6.2b shows a microscale setup for heating a Williamson reaction tube and a round-bottomed flask fitted with an air condenser. Several types of aluminum blocks are available commercially. The blocks have holes sized to fit microscale reaction tubes or vials and a depression or hole for 5- or 10-mL microscale round-bottomed flasks. The blocks also have a hole designed to hold a metal probe thermometer, so the temperature of the block can be monitored.

**FIGURE. 6.2**
Heating aluminum
blocks used with
microscale glassware.

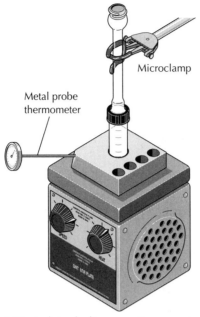

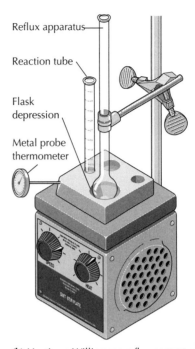

(a) Typical standard taper reaction apparatus
with a conical vial and an air condenser

(b) Heating a Williamson reflux apparatus
and a reaction tube

---

**SAFETY PRECAUTION**

A mercury thermometer should not be used with an aluminum heating block because if it breaks, the heat will vaporize the mercury.

Auxiliary aluminum blocks designed in two sections can be placed on top of the aluminum block around a vial or round-bottomed flask to provide extra radiant heat, as shown in Figure 6.3.

---

**SAFETY PRECAUTION**

These safety precautions pertain to all electrical heating devices.

*Flash point or autoignition temperature is the minimum temperature at which a substance mixed with air ignites in the absence of a flame or spark.*

1. The hot surface of a hot plate or on the inside of a heating mantle is a fire hazard in the presence of volatile, flammable solvents. An organic solvent spilled on the hot surface can ignite if its flash point is exceeded. Remove a hot heating mantle or hot plate from your work area before pouring flammable liquid.
2. Never heat a flammable solvent in an open container on a hot plate; a buildup of flammable vapors around the hot plate could result. The thermostat on most laboratory hot plates is not sealed and it arcs each time it cycles on and off, providing an ignition source for flammable vapors. Steam baths, oil baths, or heating mantles are safer choices (Ref. 1).

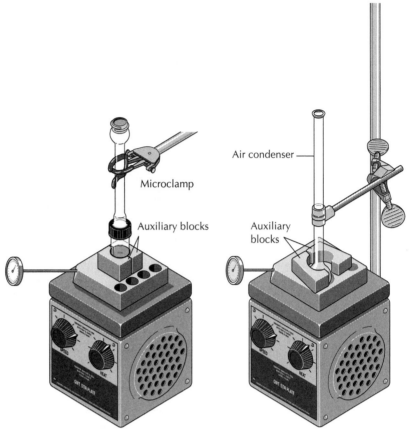

Auxiliary blocks used around a standard taper conical vial and an air condenser

Auxiliary blocks used around a Williamson reflux apparatus for extra radiant heat

**FIGURE 6.3**
Using auxiliary aluminum blocks to provide extra radiant heat with microscale glassware.

***Sand Baths***

A sand bath provides another method for heating microscale reactions. Sand is a poor conductor of heat, so a temperature gradient exists along the various depths of the sand, with the highest temperature occurring at the bottom of the sand and the lowest temperature near the top surface of the sand.

One method of preparing a sand bath uses a ceramic heating mantle, such as a Thermowell, about two-thirds full of washed sand (Figure 6.4a). A second method employs a crystallizing dish, heated on a hot plate, containing 1–1.5 cm of washed sand (Figure 6.4b). A thermometer is inserted in the sand with the thermometer bulb completely submerged. The heating of a reaction vessel can be closely controlled by raising or lowering the vessel to a different depth in the sand, as well as by changing the heat supplied by the heating mantle or hot plate.

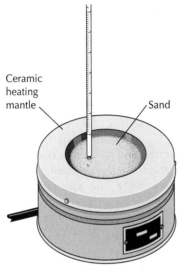

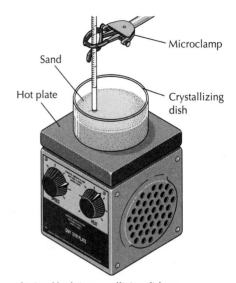

(a) Sand bath in ceramic heating mantle

(b) Sand bath in crystallizing dish on hot plate

**FIGURE 6.4**
Sand baths.

---

SAFETY PRECAUTION

Sand contained in a crystallizing dish should not be heated above 200°C nor should the hot plate be turned to high heat settings. Either situation could cause the crystallizing dish to break.

---

**Steam Baths**

Steam bath

Steam cone

**FIGURE 6.5**
Steam baths.

Steam baths or steam cones provide a safe and efficient way of heating low-boiling-point flammable organic liquids (Figure 6.5). Steam baths are extensively used in the organic laboratory for heating liquids below 100°C and in situations where precise temperature control is not required. The concentric rings on the top of the steam bath can be removed to accommodate containers of various sizes. A round-bottomed flask should be positioned so that the rings cover the flask to the level of the liquid it contains. For an Erlenmeyer flask, remove only enough rings to create an opening that is slightly larger than one-half of the bottom diameter of the flask.

---

SAFETY PRECAUTION

Steam is nearly invisible and causes severe burns. Turn off the steam before placing a flask on a steam bath or removing it. Grasp the neck of a hot flask with flask tongs. Do not use a test tube holder or a towel.

---

Steam baths operate at only one temperature, approximately 100°C. Increasing the rate of steam flow does not raise the temperature, but it does produce clouds of moisture within the laboratory or hood and in your sample! Adjust the steam valve for a slow to moderate rate of steam flow when using a steam bath. (**Caution:** The metal screw on the steam valve may be hot enough to cause burns.)

A steam bath has two disadvantages. First, it cannot be used to boil any liquid with a boiling point above 100°C. Second, water vapor from the steam may contaminate moisture-sensitive compounds being heated on the steam bath unless special precautions are taken to exclude moisture.

*Water Baths*

When a temperature of less than 100°C is needed, a water bath allows for closer temperature control than can be achieved with the heating methods discussed previously. The water bath can be contained in a beaker or crystallizing dish. Once the desired temperature of the water bath is reached, the water temperature can be maintained by using a low heat setting on a hot plate.

The thermometer used to monitor the temperature of the water bath should **always** be clamped, as shown in Figure 6.6, so that it is not touching the wall or bottom of the beaker holding the water.

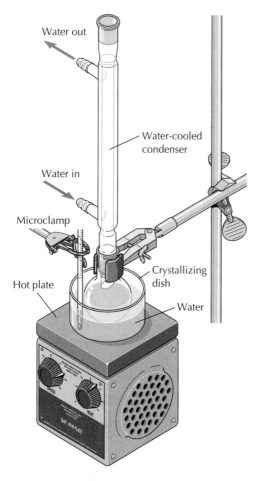

Heating miniscale reflux apparatus in a
crystallizing dish

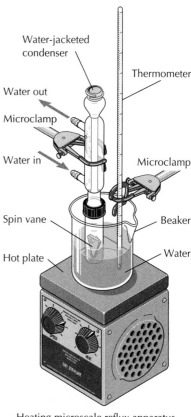

Heating microscale reflux apparatus
in a water bath

**FIGURE 6.6** Water baths.

It is very easy to bump a thermometer that is merely set in the beaker and propped against the lip, thereby breaking it or upsetting the water bath. The reaction vessel should be submerged in the water bath no farther than the depth of the reaction mixture it contains.

If magnetic stirring of a reaction mixture is needed, the reaction vessel should be clamped as near to the stirring motor as possible and centered on the hot plate/stirrer surface. A crystallizing dish may be a better choice than a beaker for the water bath, particularly if the reaction vessel is a round-bottomed flask. The wide, shallow crystallizing dish allows a round-bottomed flask to be clamped closer to the magnetic stirrer than is possible in a beaker.

| 6.3 | ## Cooling Methods |
|---|---|

Cooling baths are frequently needed in the organic laboratory to control exothermic reactions, to cool reaction mixtures before the next step in a procedure, or to promote recovery of the maximum amount of crystalline solid from a recrystallization. Most commonly, cold tap water or an ice-water mixture serves as the coolant. Effective cooling with ice requires the addition of just enough water to provide complete contact between the flask or vial being cooled and the ice. Even crushed ice does not pack well enough against a flask for efficient cooling because the air in the spaces between the ice particles is a poor conductor of heat.

Temperatures from $0°$ to $-10°C$ can be achieved by mixing solid sodium chloride into an ice-water mixture. The amount of water mixed with the ice should be only enough to make good contact with the vessel being cooled. A cooling bath of 2-propanol and chunks of solid carbon dioxide (dry ice) can be used for temperatures from $-10°$ to $-78°C$. (**Caution:** Foaming occurs as solid carbon dioxide chunks are added to 2-propanol.)

## References

1. James A. Kaufman & Associates. *The Laboratory Safety Newsletter* 1993, 1(4), 15.

2. Lodwig, S. N. *J. Chem. Educ.* **1989,** *66,* 77–84.

## Questions

1. A student continually increases the heat applied to a flask set in a heating mantle until it appears that the liquid in the flask is very hot, yet no boiling is observed. The student's neighbor mentions that either a boiling stick or boiling stones are needed. What is the *first* thing the student should do to correct the situation? Why?

2. Why must a heating mantle be plugged into a variable transformer instead of directly into a 120-V outlet?

3. Steam causes more severe burns than boiling water, even though both have a temperature of 100°C. Explain.

4. Explain why it takes longer to carry out a reaction on a steam bath in Denver, Colorado, than in New York City.

# 7

# ASSEMBLY OF REACTION APPARATUS

When carrying out organic reactions, it may be necessary to prevent loss of volatile reagents while maintaining the reaction mixture at the boiling point, keep atmospheric moisture from entering a reaction setup, make additions of reagents, or prevent noxious vapors from entering the laboratory. Typical apparatus for all these reaction conditions is described in Technique 7.

## 7.1 Refluxing a Reaction Mixture

Most organic reactions do not occur quickly at room temperature but require a period of heating. If the reaction were heated in an open container, the solvent and other liquids in the system would soon evaporate; if the system were closed, pressure could build up and it could explode when heated. Chemists have developed a simple method of heating a reaction mixture for extended time periods without loss of reagents. This process is called *refluxing,* which simply means boiling a solution while continually condensing the vapor by cooling and returning the liquid to the reaction flask.

The rate of heating a reflux apparatus is not critical as long as the liquid boils at a moderate rate. With more heat, faster boiling occurs, but the temperature of the liquid in the flask cannot rise above the boiling point of the solvent or solution. If the system is boiling at too rapid a rate, the capacity of the condenser to cool the vapors may be exceeded and reagents (or product!) lost from the top of the condenser.

A condenser mounted vertically above the reaction flask provides the means of cooling the vapor so that it condenses and flows back into the reaction flask. Condensers are available for either water cooling or air cooling. When the boiling point of the reaction mixture is less than 150°C, a water-jacketed condenser is used. For reaction mixtures with boiling points above 150°C, an air condenser is sufficient because the vapor loses heat rapidly enough to the surroundings to condense before it can escape from the top of the condenser.

*Miniscale Reflux Apparatus*

Begin the assembly of a reflux apparatus by firmly clamping the round-bottomed flask to a ring stand or vertical support rod. Position the clamp holder far enough above the bench top so that a ring or a lab jack can be placed underneath the flask to hold the heating mantle. Add the reagents with the aid of a conical funnel for liquids and a powder funnel for solids. Add a boiling stone or magnetic stirring bar to the flask. Grease the lower joint of the condenser and fit it into the top of the flask. Attach rubber tubing to the water-jacket outlets as shown in Figure 7.1a. **Water must flow into**

*A funnel keeps the reagents off the inside of the ground glass joint.*

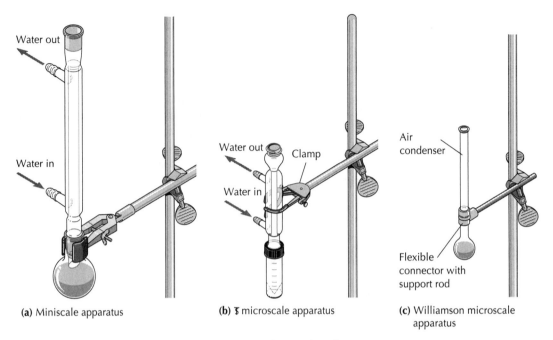

(a) Miniscale apparatus  (b) ℑ microscale apparatus  (c) Williamson microscale apparatus

**FIGURE 7.1** Apparatus for simple reflux.

*Water flows into the condenser at the bottom outlet and out at the top outlet.*

**the water jacket at the bottom outlet and out at the top outlet** to ensure that a column of water without any air bubbles surrounds the inside tube. Raise the heating mantle, supported on an iron ring or a lab jack, until it is touching the bottom of the round-bottomed flask. At the end of the reflux period the heating mantle is lowered away from the reaction flask.

*Standard Taper Microscale Glassware*

Place the reagents for the reaction in a conical vial or 10-mL round-bottomed flask sitting in a small beaker so that it will not tip over. Put a boiling stone or a magnetic spin vane into the reaction vessel. **Grease is not used on the joints of microscale glassware except when the reaction mixture contains a strong base such as sodium hydroxide.** Fit the condenser to the top of the conical vial or round-bottomed flask with a screw cap and an O-ring as shown in Technique 2, Figure 2.6. Fasten the apparatus to a vertical support rod or a ring stand with a microclamp attached to the condenser. Attach rubber tubing to the water jacket outlets (Figure 7.1b). **Water must flow into the water jacket at the bottom outlet and out at the top outlet** to ensure that a column of water without any air bubbles surrounds the inside tube.

*Water flows into the condenser at the bottom outlet and out at the top outlet.*

Lower the apparatus into an aluminum heating block, a sand bath, or a water bath heated on a hot plate or into a sand-filled Thermowell heater. At the end of the reflux period, raise the apparatus out of the heat source.

*Williamson
Microscale Glassware*

Place a 5- or 10-mL round-bottomed flask in a 30-mL beaker and use the plastic funnel to add the reagents to the flask. Add a boiling stone or magnetic stirrer. Attach the air condenser to the flask using the flexible connector with the support rod. Clamp the apparatus to a vertical support rod or a ring stand as shown in Figure 7.1c. We have found it necessary to wrap the air condenser with a wet paper towel or wet pipe cleaners to prevent loss of vapor when refluxing reaction mixtures containing solvents or reagents that boil under 120°C. Lower the apparatus into an aluminum heating block, a sand bath, or a water bath that is heated with a hot plate or into a sand-filled Thermowell heater. At the end of the reflux period, raise the apparatus out of the heat source.

| 7.2 | **Anhydrous Reaction Conditions** |

Sometimes it is necessary to prevent atmospheric moisture from entering the reaction chamber during the reflux period. In this case, a drying tube filled with a suitable drying agent, often anhydrous calcium chloride, is placed at the top of the condenser.

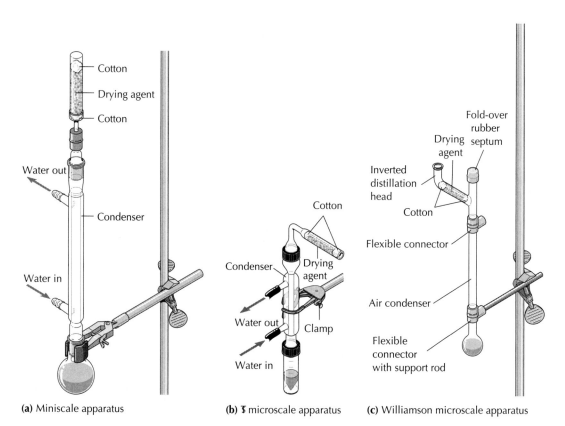

(a) Miniscale apparatus          (b) 🜨 microscale apparatus          (c) Williamson microscale apparatus

**FIGURE 7.2** Refluxing under anhydrous conditions.

*Miniscale Glassware*　　　For miniscale glassware, a thermometer adapter with a rubber sleeve serves to hold the drying tube (Figure 7.2a). A small piece of cotton is placed at the bottom of the drying tube to prevent drying agent particles from plugging the outlet of the tube; a piece of cotton is also placed over the drying agent at the top of the drying tube to keep the particles from spilling.

*Standard Taper*　　　The L-shaped standard taper microscale drying tube has a ground
*Microscale Glassware*　glass inner joint that fits into the outer ground glass joint at the top of the condenser and is secured with an O-ring and screw cap (Figure 7.2b). A small piece of cotton is pushed into the drying tube to prevent the drying agent particles from falling into the reaction vessel; cotton is also placed near the open end of the drying tube to hold the drying agent in place.

*Williamson*　　　Figure 7.2c shows how the Williamson microscale Claisen adapter/
*Microscale Glassware*　distilling head can be used as a drying tube. A small piece of cotton is pushed to the bottom of the side arm, a suitable drying agent, such as anhydrous calcium chloride, is added, and a second piece of cotton is placed at the top to keep the drying agent from spilling. The other opening is closed with a fold-over rubber septum. The drying tube is fitted to the top of the air condenser with a flexible connector.

| 7.3 | **Addition of Reagents During a Reaction** |

*Miniscale Glassware*　　　When it is necessary to add reagents during the reflux period, a separatory funnel is used as a dropping funnel. If the round-bottomed flask has only one neck, a Claisen adapter provides a second opening into the flask, as shown in Figure 7.3a. For a three-necked flask, the third neck is closed with a ground glass stopper, as shown in Figure 7.3b. If it is also necessary to maintain anhydrous conditions during the reflux period, both the condenser and the separatory funnel can be fitted with drying tubes filled with a suitable drying agent.

*Standard Taper*　　　The addition of reagents to a microscale reaction being carried out
*Microscale Glassware*　under anhydrous conditions is done with a syringe. Figure 7.4a shows a standard taper microscale apparatus assembled for reagent addition using a syringe. The Claisen adapter provides two openings into the system. The opening used for the syringe can be capped either with a screw cap and Teflon septum or with a fold-over rubber septum.

*Williamson*　　　The addition of reagents to a microscale reaction being carried out
*Microscale Glassware*　under anhydrous conditions is done with a syringe. For Williamson

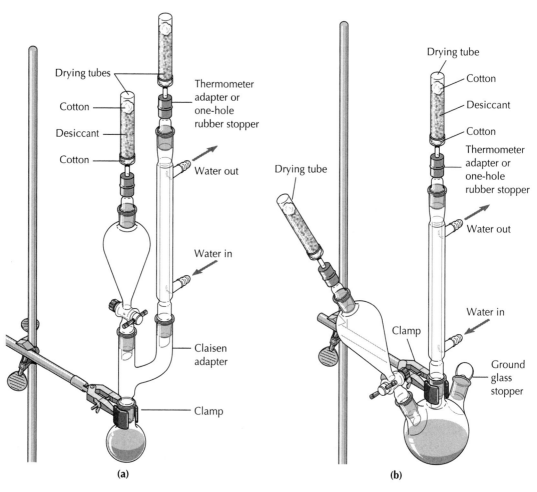

**FIGURE 7.3**
Maintaining anhydrous conditions while adding reagents to a reaction heated under reflux in (a) a one-necked reaction flask, (b) a three-necked flask.

microscale glassware, the Claisen adapter/distilling head provides two openings in the system and serves as the drying tube. The opening used for the syringe is capped with a fold-over rubber septum (sleeve stopper), as shown in Figure 7.4b.

## 7.4   Removal of Noxious Vapors

**Any reaction that emits noxious vapors should be performed in a hood.** When a noxious acidic gas such as nitrogen dioxide, sulfur dioxide, or hydrogen chloride forms during a reaction, it must be prevented from escaping into the laboratory. Acidic

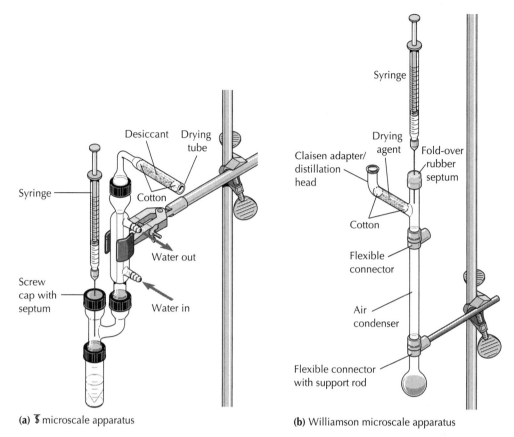

(a) ℥ microscale apparatus

(b) Williamson microscale apparatus

**FIGURE 7.4**
Using a syringe to add reagents to a microscale reaction being carried out under anhydrous conditions.

gases are readily soluble in water, so a gas trap containing either water or dilute aqueous sodium hydroxide solution effectively dissolves them.

**Miniscale Apparatus**

*Any reaction that emits noxious vapors should be performed in a hood.*

Attach a U-shaped piece of glass tubing to the top of a reflux condenser by means of a one-hole rubber stopper or a thermometer adapter. Carefully fit the other end of the U tube through a one-hole rubber stopper sized for a 125-mL filter flask. Place about 50 mL of ice water or dilute sodium hydroxide solution in the filter flask and position the open end of the U tube *just above* the surface of the liquid in the filter flask, as shown in Figure 7.5a.

In laboratories equipped with water aspirators, a gas trap can be made by placing a vacuum adapter at the top of a condenser. The side arm of the vacuum adapter is connected to the side arm of the water aspirator and the water turned on at a moderate flow rate. The noxious gases are pulled from the reaction apparatus and dissolved in the water passing through the aspirator (Figure 7.5b).

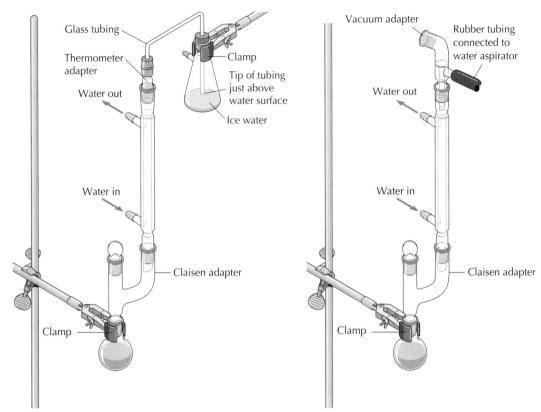

**(a)** Gas trap attached to reaction apparatus          **(b)** Noxious vapors exhausted through a water aspirator

**FIGURE 7.5** Miniscale apparatus to trap water-soluble noxious vapors.

*Standard Taper Microscale Apparatus*

*Any reaction that emits noxious vapors should be performed in a hood.*

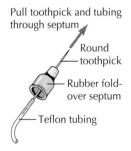

**FIGURE 7.6**
Threading Teflon tubing through a rubber septum.

A gas trap for microscale reactions can be prepared with fold-over rubber septa, Teflon tubing (1/16 inch in diameter), and a 25-mL filter flask. To insert the Teflon tubing through a rubber septum, carefully punch a hole in the septum with a syringe needle and push a round toothpick through the hole. Fit the tubing over the point of the toothpick and pull the toothpick (with tubing attached) back through the septum, as shown in Figure 7.6. Repeat this process to place a rubber septum on the other end of the tubing.

Half fill a 25-mL filter flask with ice water or dilute sodium hydroxide solution and close the top with one septum. Push the tubing down until the open end is *just above* the surface of the water or sodium hydroxide solution. Attach the other septum to the top of the condenser. The side arm of the filter flask serves as a vent (Figure 7.7a).

In laboratories equipped with water aspirators, a gas trap for standard taper microscale glassware can be made by placing a vacuum adapter at the top of a condenser. The side arm of the vacuum

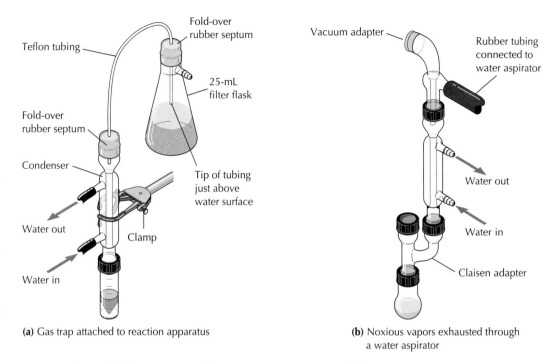

(a) Gas trap attached to reaction apparatus

(b) Noxious vapors exhausted through a water aspirator

**FIGURE 7.7** Standard taper microscale apparatus to trap water-soluble noxious vapors.

adapter is connected to the side arm of the water aspirator with heavy-walled rubber tubing and the water turned on at a moderate flow rate. The noxious gases are pulled from the reaction apparatus and dissolved in the water passing through the aspirator (Figure 7.7b).

**Williamson Microscale Glassware**

*Any reaction that emits noxious vapors should be performed in a hood.*

A gas trap for microscale reactions using Williamson glassware can be prepared with fold-over rubber septa, Teflon tubing (1/16 inch in diameter), and a 25-mL filter flask or a reaction tube. To insert the Teflon tubing through a rubber septum, carefully punch a hole in the septum with a syringe needle and push a round toothpick through the hole. Fit the tubing over the point of the toothpick and pull the toothpick (with tubing attached) back through the septum, as shown in Figure 7.6. Repeat this process to place a rubber septum on the other end of the tubing.

Half fill a 25-mL filter flask or a Williamson reaction tube with ice water or dilute sodium hydroxide solution and close the top with one septum. Push the tubing down until the open end is *just above* the surface of the water or sodium hydroxide solution. Attach the other septum to the top of the Claisen adapter. If a filter flask serves as the trap, the side arm provides a vent (Figure 7.8a); if the trap is a Williamson reaction tube, then a syringe needle must be inserted in the septum attached to the reaction tube to provide a vent (Figure 7.8b).

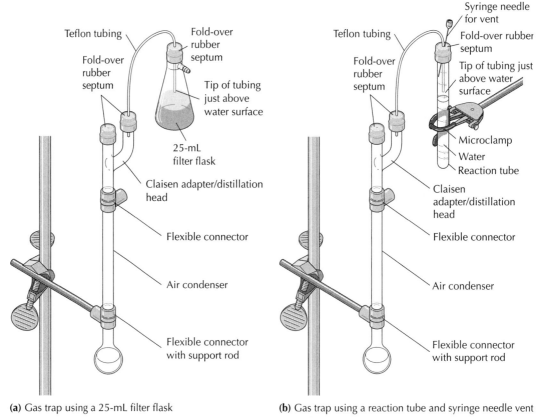

**(a)** Gas trap using a 25-mL filter flask        **(b)** Gas trap using a reaction tube and syringe needle vent

**FIGURE 7.8**
Williamson microscale apparatus to trap water-soluble noxious vapors.

# 8

# EXTRACTION AND DRYING AGENTS

Extraction is a technique used for separating a compound from a mixture. It is often used to separate a compound from the other materials present in a natural product, for example, to remove caffeine from coffee. *Liquid-liquid extraction* involves the distribution of a compound, called a *solute,* between two immiscible (nonsoluble) liquid *solvents,* or *phases.* Generally, although not always, one of the solvents in an extraction is water and the other is a much less polar organic solvent, such as dichloromethane, diethyl ether, ethyl acetate, or hexane. By taking advantage of the differing solubilities of a solute in a pair of solvents, compounds can be selectively transported from one phase to another. For example, inorganic salts, water-soluble acids and bases, and water-soluble organic compounds can be conveniently separated from a water-insoluble organic compound by extracting them from a nonpolar organic solvent into water.

In the extraction procedure, an *aqueous solution (water)* and an immiscible organic solvent are usually gently shaken in a separatory funnel (Figure 8.1). The solutes distribute themselves between the aqueous layer and the organic layer according to their relative solubilities. Inorganic salts generally prefer the aqueous layer, whereas most organic substances dissolve more readily in the organic solvent layer. Thus, two or three extractions of an aqueous mixture often suffice to quantitatively transfer a nonpolar organic compound, such as a hydrocarbon or a halocarbon, to an organic solvent. Separation of low-molecular-weight alcohols or other more polar organic compounds may require additional extractions or a different approach.

When an organic compound is distributed or partitioned between an organic solvent and water, the ratio of solute concentration in the organic solvent, $C_1$, to its concentration in water, $C_2$ is equal to the ratio of its solubilities in the two solvents. This distribution of an organic solute, either liquid or solid, can be expressed by

$$K = \frac{C_1}{C_2} = \frac{\text{g compound per mL organic solvent}}{\text{g compound per mL water}} \tag{1}$$

where $K$ is the *distribution coefficient*, or *partition coefficient*.

Any organic compound with a distribution coefficient greater than 1.5 can be separated from water by extractions with a water-insoluble organic solvent. As you will soon see, working through the mathematics of equation 1 shows that a series of extractions using small volumes of solvents is more efficient than a single large-volume extraction. A volume of solvent about one-third the volume of the aqueous phase is sufficient for most extractions. Commonly used extraction solvents are listed in Table 8.1.

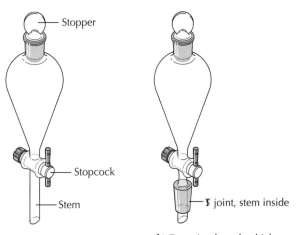

**FIGURE 8.1**
Funnels for extractions.

(a) Separatory funnel

(b) Dropping funnel, which can be used as a separatory funnel

| TABLE 8.1 | **Common extraction solvents** | | | | |
|---|---|---|---|---|---|
| Solvent | Boiling point, °C | Solubility in water, $g \cdot (100\ cc)^{-1}$ | Hazard | Density, $g \cdot mL^{-1}$ | Fire hazard[a] |
| Diethyl ether | 35 | 6 | Inhalation, fire | 0.71 | ++++ |
| Pentane | 36 | 0.04 | Inhalation, fire | 0.62 | ++++ |
| Petroleum ether[b] | 40–60 | low | Inhalation, fire | 0.64 | ++++ |
| Dichloromethane | 40 | 2 | $LD_{50},^{c}\ 1.6\ mL \cdot kg^{-1}$ | 1.32 | + |
| Chloroform | 61 | 0.5 | Banned for use in drugs and cosmetics | 1.48 | − |
| Hexane | 69 | 0.02 | Inhalation, fire | 0.66 | ++++ |
| Toluene | 111 | 0.06 | Inhalation, fire; less toxic than benzene | 0.87 | ++ |

a. Scale: extreme fire hazard = ++++.
b. Mixture of hydrocarbons.
c. $LD_{50}$, lethal dose orally in young rats.

## 8.1   Extraction: The Theory

If the distribution coefficient $K$ of a solute between water and an organic solvent is large (>100), a single extraction suffices to extract the compound from water into the organic solvent. Most often, however, the distribution coefficient is between 1 and 10, thereby making multiple extractions necessary.

Let us consider a simple case of extraction from water into ether, assuming a distribution coefficient of 5.0 for the organic compound being extracted. Just for illustration, let's use 1.0 g of compound dissolved in 50 mL of water being extracted into 15 mL of ether. If $C_1$ is the final concentration in ether (5 parts), $C_2$ is the final concentration in water (1 part), and $x$ is the amount of solute remaining in the water layer after extraction, then

$$5.0 = K = \frac{C_1}{C_2} = \frac{\dfrac{1.00\ g - x\ g}{15\ mL\ ether}}{\dfrac{x\ g}{50\ mL\ water}} = \frac{(1.00 - x)\,50}{15x}$$

$x = 0.40$ g remains in water layer after first extraction

$1.0 - x$ g = 0.60 g transferred to ether layer

So 0.40 g of solute remains in the water layer after the first extraction, and 0.60 g of solute was extracted into the ether layer.

A second extraction increases the overall efficiency of the extraction procedure. Thus, extracting the remaining water solution with a fresh 15-mL portion of ether removes more of the organic substance that still remains in the water.

Again, $C_1$ is the final concentration in ether (5 parts), $C_2$ is the final concentration in water (1 part), and $x$ is the amount of solute remaining in the water layer after the second extraction. As before,

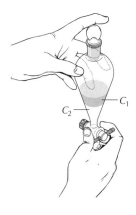

$$5.0 = K = \frac{C_1}{C_2} = \frac{\dfrac{0.40 \text{ g} - x \text{ g}}{15 \text{ mL ether}}}{\dfrac{x \text{ g}}{50 \text{ mL water}}} = \frac{(0.40 - x)\,50}{15x}$$

Thus, 0.16 g of solute remains in the water layer after the second extraction, and 0.24 g has been extracted into the second ether layer.

A third extraction of the residual water layer with a 15-mL portion of ether removes another 0.10 g of the organic compound from the water layer. In three ether extractions, a total of 0.94 g (0.60 g + 0.24 g + 0.10 g) of the organic substance is extracted from the water into the organic solvent. Only 6% of the organic substance remains in the water layer; most of it could be extracted with one more 15-mL portion of ether.

Suppose that, instead of extracting with three 15-mL portions of ether, we extract the original solution only once with a 45-mL portion of ether. Let $x$, $C_1$, and $C_2$ be defined as before and $K = 5.0$, also as before. Then

$$5.0 = K = \frac{C_1}{C_2} = \frac{\dfrac{1.00 \text{ g} - x \text{ g}}{45 \text{ mL ether}}}{\dfrac{x \text{ g}}{50 \text{ mL water}}} = \frac{(1.00 - x)\,50}{45x}$$

$$x = 0.18 \text{ g remains in water layer after one extraction}$$

$$1.0 - x \text{ g} = 0.82 \text{ g transferred to ether layer}$$

After one extraction with the same total volume of ether used for three extractions, 0.18 g of solute remains in the water layer, and 0.82 g of solute has been transferred to the ether layer. Compared with the total of 0.94 g separated by three 15-mL extractions, the single 45-mL extraction is less efficient because 18% of the organic solute remains behind in the water layer versus 6% with the same volume of solvent used in three portions.

In general, the fraction of the solute remaining in the original water solvent is given by

$$\frac{C_2, \text{ final}}{C_2, \text{ initial}} = \left( \frac{V_2}{V_2 + V_1 K} \right)^n$$

where $V_1$ = volume of ether in each extraction, $V_2$ = original volume of water, $n$ = number of extractions, and $K$ = distribution coefficient.

## 8.2    Miniscale Extractions

Place a separatory funnel big enough to hold two to four times the total solution volume in a metal ring firmly clamped to a ring stand or upright support rod (Figure 8.2, step 1). **Make sure that the**

**FIGURE 8.2**
Using a separatory
funnel.

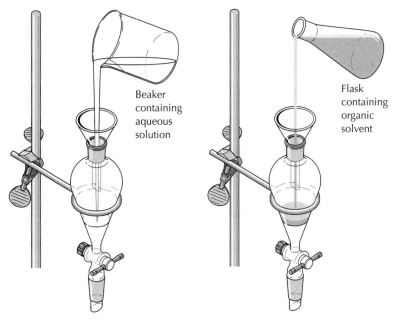

Beaker
containing
aqueous
solution

Flask
containing
organic
solvent

**1.** Add aqueous solution.

**2.** Add organic solvent.

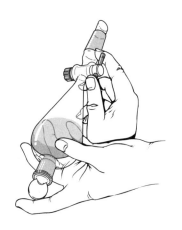

**3.** Insert stopper and hold
stopper with your finger.

**4.** Invert funnel and immediately
open stopcock to release pressure.

**stopcock fits tightly and is closed.** Pour the cooled aqueous solution to be extracted into the separatory funnel. Add a volume of organic solvent equal to approximately one-third the total volume of
the aqueous solution (Figure 8.2, step 2), and put the stopper in
place. Remove the funnel from the ring, hold the stopper firmly in
place with your index finger (Figure 8.2, step 3), invert the funnel,
and **open the stopcock immediately** to release the pressure from
solvent vapors (Figure 8.2, step 4). Close the stopcock, and shake the
mixture by inverting the separatory funnel four or five times before

**5.** Use a ring stand to hold separatory funnel until layers separate.

**6.** Draw off bottom layer.

**7.** Pour off top layer.

*Pour the top layer out of the top of the funnel so that it is not contaminated by the residual bottom layer adhering to the stopcock and tip.*

releasing the pressure by opening the stopcock; continue this shaking and venting process for five or six repetitions to ensure complete mixing of the two phases.

Place the separatory funnel in the ring once more and wait until the layers have completely separated (Figure 8.2, step 5). Remove the stopper and open the stopcock to draw off the bottom layer into an Erlenmeyer flask (Figure 8.2, step 6). Pour the remaining layer out of the funnel through the top (Figure 8.2, step 7).

Do this entire procedure each time an extraction is carried out. When extracting an aqueous solution several times with a solvent more dense than water, it is not necessary, of course, to pour the top layer out of the separatory funnel until all the extractions are done.

Finally, **label all flasks and do not discard any solution** until you have completed the entire extraction procedure and are certain which flask contains your desired product.

| 8.3 | Additional Information About Extractions |

*Densities*

Before you begin any extraction, look up the density of the organic solvent in a handbook to determine whether the solvent you are using for the extraction is more dense or less dense than water. The denser layer is always on the bottom. Organic solvents that are less dense than water form the upper layer in the separatory funnel

**FIGURE 8.3**
Solvent densities.

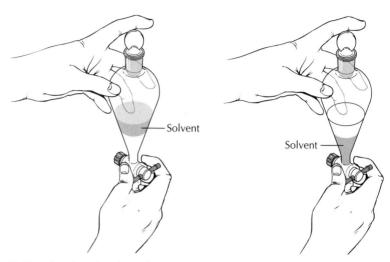

**(a)** Organic solvent less dense than water    **(b)** Organic solvent more dense than water

(Figure 8.3a), whereas those that are more dense form the lower layer (Figure 8.3b). Occasionally sufficient material is extracted from the aqueous phase to the organic phase or vice versa to change the relative densities of the two phases enough for them to exchange places.

It is always prudent to check which layer is the organic phase by adding a few drops of the layer in question to 1–2 mL of water in a test tube and observing whether it dissolves or not.

**Temperature of the Extraction Mixture**

Be sure that the extraction solution is at room temperature or slightly cooler before you add the organic extraction solvent. Most solvents used for extractions have low boiling points and boil if added to a warm solution. A few pieces of ice can be added to cool the aqueous solution.

**Venting the Funnel**

Be sure that you vent an extraction mixture by carefully inverting the funnel and immediately opening the stopcock (see Figure 8.2, step 4) before you begin the shaking process. If you do not do this, the stopper may pop from the funnel and liquids as well as gases will be released, as shown in Figure 8.4. Pressure buildup within the separatory funnel is always a problem when using low-boiling-point extraction solvents such as diethyl ether, pentane, and dichloromethane.

Venting extraction mixtures is especially important when you use a dilute sodium carbonate or bicarbonate solution to extract an organic phase containing traces of an acid. Carbon dioxide gas is given off in the neutralization process. The $CO_2$ pressure buildup can easily force the stopper out of the funnel, cause losses of your solutions (and product!), and cause injury to yourself or your neighbor. When using sodium carbonate or bicarbonate to extract or wash acidic contaminants from an organic solution, **vent the extraction mixture immediately after the first inversion and subsequently after every three or four inversions.**

**FIGURE 8.4**
Failure to vent the separatory funnel when washing with $Na_2CO_3$ or $NaHCO_3$ solution can cause the stopper to blow out.

*Washing the Organic Phase*

After the extraction is completed and the two immiscible liquids are separated, the organic layer is often *extracted* or *washed* with water or perhaps a dilute aqueous solution of an acid or a base. For example, a chemical reaction involving alkaline reagents often yields an organic extract that still contains some alkaline material. The alkaline material can be removed by washing the extract (organic phase) with a 5% solution of hydrochloric acid. Similarly, an organic extract obtained from an acidic solution should be washed with a 5% solution of sodium carbonate or sodium bicarbonate (see previous subsection on venting). The salts formed in these extractions are very soluble in water but not in typical organic extraction solvents, so they are easily transported into the aqueous phase. If acid or base washes are required, they are done in the same manner as any other extraction and are usually followed by a final water wash.

*Salting Out*

If the distribution coefficient for a substance to be extracted from water into an organic solvent is close to or lower than 1.0, a simple extraction procedure is not effective. In this case, a salting out procedure can help. Salting out is done by adding a saturated solution of NaCl (sometimes called brine) or $Na_2SO_4$, or the salt crystals themselves, to the aqueous layer. The presence of a salt in the water layer decreases the solubility of the organic compound in the aqueous phase. Therefore, the distribution coefficient increases because more of the organic compound is transferred from the aqueous phase to the organic layer. Salting out can also help to separate a homogeneous solution of water and a water-soluble organic compound into two phases.

A diethyl ether solution that has been extracted with aqueous solutions is saturated with water, which needs to be removed to recover a dry product. Doing a final extraction of an ether solution with a saturated solution of sodium chloride helps to "dry" or remove water from the ether solution. The affinity of the ions in a saturated sodium chloride solution for water molecules draws the water out of the organic phase. The ether solution still needs to be treated with an anhydrous drying agent [see Technique 8.7], but less drying agent is needed because of the sodium chloride extraction.

*Emulsions*

The formation of an *emulsion,* a suspension of insoluble droplets of one liquid in another liquid, between the organic and aqueous layers is sometimes encountered while doing extractions. When an emulsion forms, the entire mixture has a milky appearance, with no clear separation between the immiscible layers; or there may be a third milky layer between the aqueous and the organic phases. Emulsions are not usually formed during diethyl ether extractions, but they frequently occur when aromatic or chlorinated organic solvents are used. An emulsion often disperses if the separatory funnel and its contents are allowed to sit in a ring stand for a few minutes. However, be gentle when you use aromatic or chlorinated solvents to extract organic compounds from aqueous solutions; preventing emulsions is simpler than dealing with them. Instead of shaking the mixture vigorously, invert the separatory funnel and gently swirl the two layers together for 2–3 min. However, use of this swirling technique may mean that you need to extract an aqueous solution with an extra portion of organic solvent.

Should an emulsion occur, it can be dispersed by vacuum filtration through a pad of the filter aid Celite. Prepare the Celite pad by pouring a slurry of Celite and water onto a filter paper in a Buchner funnel. Remove the water from the filter flask before pouring the emulsion through the Buchner funnel. Return the filtrate to the separatory funnel and separate the two phases. Another method, useful when the organic phase is the lower layer, involves filtering the organic phase through a silicone-impregnated filter paper. For microscale extractions (discussed in Technique 8.6), centrifugation of the emulsified mixture effectively separates the phases.

**Caring for the Separatory Funnel**

When the entire operation is complete, clean up. Cleaning the funnel immediately and regreasing a glass stopcock prevents "frozen" stopcocks later. Grease is not necessary with Teflon stopcocks, but they also may freeze if not loosened during storage.

## 8.4        Summary of the Miniscale Extraction Procedure

1. With the stopcock closed, pour the aqueous mixture into a separatory funnel with a capacity at least twice as great as the amount of mixture.
2. Add a volume of immiscible organic solvent approximately one-third that of the aqueous phase. *You must know the density of the organic solvent.*
3. Invert the funnel with one hand on the stopper and neck of the funnel, and open the stopcock to release any pressure buildup.
4. Shake the funnel vigorously two or three times (however, see precautions about emulsions on page 63); vent the funnel.
5. Allow the two phases to separate.
6. *For an organic solvent less dense than water,* draw off the lower aqueous phase into a labeled Erlenmeyer flask; pour the organic phase out of the top of the funnel into a labeled Erlenmeyer flask; return the aqueous phase to the separatory funnel. *For an organic solvent more dense than water,* draw off the lower organic phase into a labeled Erlenmeyer flask; the upper aqueous phase remains in the separatory funnel.
7. Extract the original aqueous mixture twice more with fresh solvent.

8. Combine the organic solvent extracts in the same Erlenmeyer flask and pour this solution into the separatory funnel. Extract this organic solution with dilute acid or base, if necessary, to neutralize any bases or acids remaining from the reaction.

9. Wash the organic phase with water or saturated NaCl.

10. Dry the organic phase with an anhydrous drying agent [see Technique 8.7].

---

| **8.5** | **Using Pasteur Pipets to Transfer Liquids in Microscale Operations** |

Before we discuss microscale extractions, methods for transferring the small amounts of liquids involved in microscale operations without significant loss need to be considered. A liquid should never be poured in any microscale operation; too much is lost. Losses of even a few drops are significant in microscale work and need to be avoided. A Pasteur pipet serves as the tool of choice for transferring a liquid from one container to another. Volatile organic liquids, however, frequently drip out of a Pasteur pipet during such transfers because solvent vapor pressure increases as the rubber bulb warms from your fingers.

Two methods can be used to prevent volatile liquids from dripping out of a Pasteur pipet. In the first method, a small plug of cotton is pushed into the tip of the Pasteur pipet. In the second method, the Pasteur pipet is fitted with a syringe.

Airborne particles, such as dust and lint, also present problems when working with microscale volumes of liquids. A small cotton plug in the tip of a Pasteur pipet serves to filter the solution each time it is transferred.

---

**SAFETY PRECAUTION**

Glass Pasteur pipets are puncture hazards. They should be handled and stored carefully. Dispose of Pasteur pipets in a "sharps" box or in a manner that does not present a hazard to lab personnel or housekeeping staff. Check with your instructor about the proper disposal method in your laboratory.

---

*Pasteur Filter Pipets*

Pasteur filter pipets are made by using a piece of wire that has a diameter slightly less than the inside of the capillary portion of the pipet to push a tiny piece of cotton into the tip of a Pasteur pipet (Figure 8.5). A piece of cotton of the appropriate size should offer only slight resistance to being pushed by the wire. If there is so much resistance that the cotton cannot be pushed into the tip of the pipet, then the piece is too large. Remove the wire and insert it through the tip to push the cotton back out of the upper part of the pipet, and take a bit off the piece of cotton before putting it back into the pipet. The finished cotton plug should be 2–3 mm long and should fit snugly but not too tightly. If the cotton is packed too tightly in the tip, liquid will not

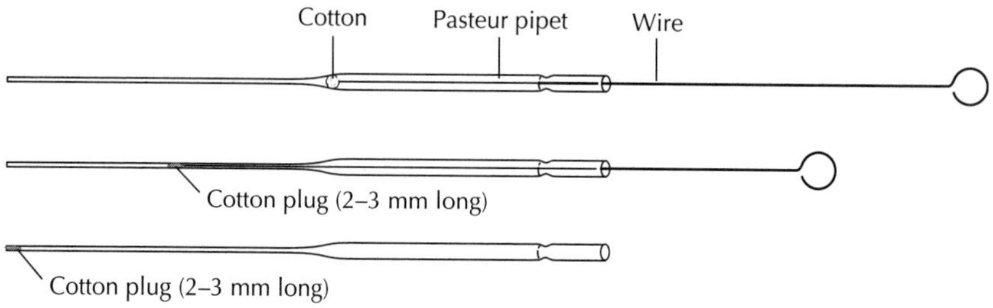

**FIGURE 8.5**
Preparing a
Pasteur filter pipet.

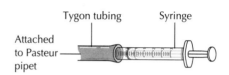

**FIGURE 8.6**
Syringe fitted to
a Pasteur pipet.

flow through it; if it fits too loosely, it may be expelled with the liquid. With a little practice, you should be able to prepare a filter pipet easily.

**Pasteur Pipet Fitted with a Syringe**

When transferring volatile liquids, dripping from the tip of a Pasteur pipet can be minimized by substituting a 3-mL syringe with a Luer-Lok fitting for the rubber bulb.* The syringe is attached to the Pasteur pipet with a short piece of 4-mm-interior-diameter latex or Tygon tubing (Figure 8.6).

**Temporary Storage of Pasteur Pipets**

When using a Pasteur pipet (or any other pipet with a bulb or syringe attached), you should set the pipet in a test tube or Erlenmeyer flask to keep it upright. Laying the pipet on the bench top or other horizontal surface allows the rubber bulb or syringe to be contaminated with the liquid being transferred (Figure 8.7).

---

\* Becton-Dickinson 3-mL syringes with a Luer-Lok fitting (B-D product # 309585) are available from VWR and Thomas Scientific.

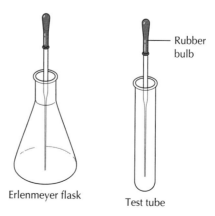

**FIGURE 8.7**
Correct ways to
temporarily store
Pasteur pipets.

## 8.6      Microscale Extractions

*Read Techniques 8.1 and 8.3 before undertaking a microscale extraction for the first time.*

The small volumes of liquids used in microscale reactions cannot be handled in a separatory funnel because most of the material would be lost on the glassware. Instead, a conical vial or a centrifuge tube holds the two-phase system, and a Pasteur pipet serves as the tool for separating one phase from the other (Figure 8.8). The V-shaped bottom of a conical vial or a centrifuge tube enhances the visibility of the interface between the two phases in the same way that the conical shape of a separatory funnel enhances the visibility of the interface just above the stopcock. Centrifuge tubes are particularly useful for extractions with combinations of organic and aqueous phases that form emulsions. The tubes can be spun in a centrifuge to produce a clean separation of the two phases.

### 8.6a   Equipment and Techniques Common to All Microscale Extractions

Before discussing specific types of extractions, we should consider the equipment and techniques common to all microscale extractions.

*Conical Vials*

Conical vials, with a capacity of 5 mL, work well for extractions in which the total volume of both phases does not exceed 4 mL. Conical vials tip over very easily. **Always place the vial in a small**

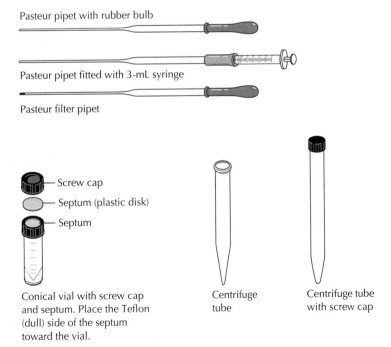

FIGURE 8.8
Equipment for microscale extractions.

**beaker.** The plastic septum used with the screw cap on a conical vial has a chemically inert coating of Teflon on one side. The Teflon looks dull and should be placed toward the vial. (The shiny side of the septum is not inert to all organic solvents.)

*Centrifuge Tubes*

Centrifuge tubes with a 15-mL capacity and tight-fitting caps serve for extractions involving a total volume of up to 12 mL. Set centrifuge tubes in a test tube rack.

*Mixing the Two Phases*

As with extractions performed in a separatory funnel, thorough mixing of the two phases is essential for complete transfer of the solute from one phase to the other. Mix the two phases by capping the conical vial or centrifuge tube and shaking it vigorously. Slowly loosen the cap to vent the vial or centrifuge tube. Repeat the shaking and venting process four to six times. Alternatively, or for a centrifuge tube without a screw cap, draw the two phases into a Pasteur pipet (no cotton plug in the tip) and squirt the mixture back into the centrifuge tube five or six times to mix the two phases thoroughly. The use of a vortex mixer, if available, is another way of mixing the two phases.

*Separation of the Phases with a Pasteur Pipet and Syringe*

The liquid in the lower layer is more easily removed from a conical vial or centrifuge tube than the upper layer. A Pasteur pipet, fitted with a syringe [see Technique 8.5] can be used to remove the lower layer. Place the tip of the pipet on the bottom of the V in the conical vial or centrifuge tube. Draw the lower layer into the pipet with a steady pull on the syringe plunger until the interface between the layers reaches the bottom of the vial or tube. **Do not exceed the capacity of the Pasteur pipet (approximately 2 mL); no liquid should be drawn into the syringe.** Remove the Pasteur pipet from the extraction vessel and transfer the contents of the pipet to the receiving container—another conical vial, a centrifuge tube, or a test tube. Hold the receiving container close to the extraction vessel so that the transfer can be accomplished quickly without any loss of liquid (Figure 8.9). Depress the syringe plunger to empty the pipet.

*Separation of the Phases with a Pasteur Filter Pipet*

The liquid in the lower layer is more easily removed from a conical vial or centrifuge tube than the upper layer. A Pasteur filter pipet provides better control for transferring volatile solvents such as dichloromethane or ether during a microscale extraction than does a Pasteur pipet without the cotton plug. **Expel the air from the rubber bulb before inserting the pipet to the bottom of the conical vial or centrifuge tube.** Slowly release the pressure on the bulb and draw the lower layer into the pipet. Maintain a steady pressure on the rubber bulb while transferring the liquid to another container. Hold the receiving container close to the extraction vial or centrifuge tube so that the transfer can be accomplished quickly without any loss of liquid (see Figure 8.9).

*If the Upper Phase Is Drawn into the Pasteur Pipet*

The interface between the two phases can be difficult to see in some instances, and a small amount of the upper layer may be drawn into the Pasteur pipet. Maintain a steady pressure on the Pasteur pipet with the rubber bulb or syringe and allow the two phases in the pipet to separate. Slowly expel the lower layer into the receiving container until the interface between the phases is at the bottom of the pipet. Then move the pipet to the original container and add the upper layer in the pipet to the remaining upper phase.

*Label All Receiving Containers*

Extractions always involve multiple solutions. Careful labeling of all containers holding aqueous and organic solutions is essential. **Do not discard any solution until you have completed the entire extraction procedure and know with certainty which vessel contains your product.**

**FIGURE 8.9**
Holding vials while transferring solutions.

## 8.6b Extractions with an Organic Phase More Dense Than Water

Extraction of an aqueous solution with a solvent that is more dense than water, such as dichloromethane ($CH_2Cl_2$), and washing a dichloromethane/organic product solution with water are examples of this type of extraction. The lower-phase dichloromethane needs to be removed from the conical vial or centrifuge tube in order to separate the layers.

Place the aqueous solution and the specified amount of organic solvent in a labeled conical vial or centrifuge tube (Figure 8.10, step 1). Tightly cap the vial or tube and shake the mixture thoroughly. Loosen the cap slightly to release the pressure. Alternatively, use the squirt method (five or six squirts) or a vortex mixer to mix the phases. Allow the layers to separate completely.

Put a Pasteur filter pipet or a Pasteur pipet fitted with a syringe [see Technique 8.5] into the conical vial or centrifuge tube with the

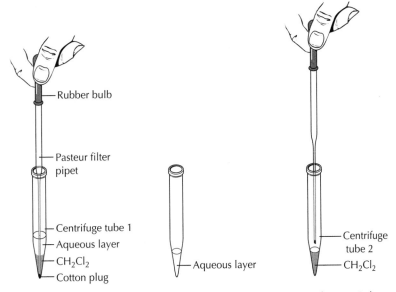

**FIGURE 8.10**
Extracting an aqueous solution with an organic solvent (dichloromethane) more dense than water.

1. Expel air from rubber bulb and insert Pasteur pipet to bottom of centrifuge tube. Draw lower layer into pipet.

Aqueous layer remains in centrifuge tube.

2. Transfer organic layer to another centrifuge tube or test tube.

Labels in figure: Rubber bulb; Pasteur filter pipet; Centrifuge tube 1; Aqueous layer; $CH_2Cl_2$; Cotton plug; Aqueous layer; Centrifuge tube 2; $CH_2Cl_2$

*In any extraction, no material should be discarded until you are certain which container holds your desired product.*

tip touching the bottom of the cone. Slowly draw the lower layer into the pipet until the interface between the two layers is exactly at the bottom of the V. Transfer the pipet to another centrifuge tube, conical vial, or test tube and expel the dichloromethane solution into the second container (Figure 8.10, step 2). The aqueous layer remains in the extraction tube and can be extracted a second time with another portion of $CH_2Cl_2$. The second dichloromethane solution is added to centrifuge tube 2 after the separation.

**Washing an Organic Liquid**

If the organic phase transferred to tube or vial 2 is being washed with an aqueous solution, the aqueous reagent is added to tube 2. Cap the tube or vial, shake it to mix the phases, then loosen the cap to release any pressure buildup. The lower organic phase is separated and transferred to another centrifuge tube (or conical vial) if more washings are necessary. Otherwise, the organic phase is transferred to a dry test tube for treatment with a drying agent [see Technique 8.7].

## 8.6c  Extractions with an Organic Phase Less Dense Than Water

The microscale extraction of an aqueous solution with an organic solvent that is less dense than water, such as diethyl ether or hexane, and washing an ether solution with aqueous reagents are examples of this type of extraction.

Two centrifuge tubes or conical vials are needed for the extraction of an aqueous solution with a solvent less dense than water. Place the aqueous solution in the first centrifuge tube or conical vial, and add the specified amount of organic solvent—diethyl ether in this example. Cap the tube or vial and shake it to mix the layers. Vent the tube by slowly releasing the cap and allow the phases to separate. Alternatively, use the squirt method (five or six squirts) or a vortex mixer to mix the phases. Allow the layers to separate completely.

Put a Pasteur filter pipet or a Pasteur pipet fitted with a syringe [see Technique 8.5] into the tube or vial with the tip touching the bottom of the cone (Figure 8.11, step 1). Slowly draw the aqueous layer into the pipet until the interface between the ether and aqueous solution is at the bottom of the V. Transfer the aqueous solution to the second centrifuge tube or conical vial (Figure 8.11, step 2). The ether solution remains in the first tube. Add a second portion of ether to the aqueous phase in the second tube, cap the tube, and shake it to mix the phases. After the phases separate, again remove the lower aqueous layer and place it in a test tube (Figure 8.11, step 3). Transfer the ether solution in the first tube to the ether solution in the second tube with the Pasteur pipet (Figure 8.11, step 4). Repeat the procedure if a third extraction is necessary.

*In any extraction, no material should be discarded until you are certain which container holds your desired product.*

**Washing an Organic Liquid**

If the experiment specifies washing an ether or other organic solution that is less dense than water with an aqueous solution, place the organic solution in a centrifuge tube or conical vial. Add the specified amount of water or aqueous reagent solution, cap the tube (or vial), and shake it to mix the phases. Open the cap to release any vapor pressure built up and allow the layers to separate. Transfer the lower aqueous layer to a test tube with a Pasteur filter pipet or a Pasteur pipet fitted with a syringe [see Technique 8.5]. The upper organic phase remains in the extraction tube (or conical vial) ready for the next step, which may be washing with another aqueous reagent solution or drying with an anhydrous salt (Figure 8.12).

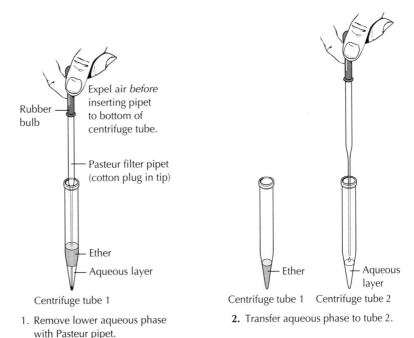

Rubber bulb

Expel air *before* inserting pipet to bottom of centrifuge tube.

Pasteur filter pipet (cotton plug in tip)

Ether

Aqueous layer

Centrifuge tube 1

1. Remove lower aqueous phase with Pasteur pipet.

Ether

Aqueous layer

Centrifuge tube 1    Centrifuge tube 2

**2.** Transfer aqueous phase to tube 2.

Aqueous layer

Aqueous layer

Second portion of ether

Test tube

Centrifuge tube 2

3. Remove aqueous phase and transfer to a test tube.

Centrifuge tube 1    Centrifuge tube 2

4. Combine ether solution from tube 1 with ether solution in tube 2.

**FIGURE 8.11**
Extracting an aqueous solution with an organic solvent (ether) less dense than water.

**FIGURE 8.12**
Washing an organic
phase less dense than
water.

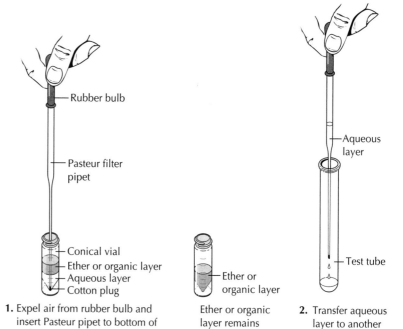

Rubber bulb

Pasteur filter
pipet

Aqueous
layer

Conical vial
Ether or organic layer
Aqueous layer
Cotton plug

Ether or
organic layer

Test tube

**1.** Expel air from rubber bulb and
insert Pasteur pipet to bottom of
vial. Draw aqueous layer into pipet.

Ether or organic
layer remains
in vial.

**2.** Transfer aqueous
layer to another
vial or test tube.

## 8.7  Drying Organic Liquids

Most organic separations involve extractions from an aqueous
solution; therefore, no matter how careful you are, some water is
usually present in the organic liquid. A small amount of water dissolves in most extraction solvents (see Table 8.1), and possibly the
physical separation of the layers in the extraction process was
incomplete. As a result, the organic layer usually needs to be dried
before performing additional operations on it.

   The most common way to **dry (remove the water from)** an
organic agent is to add an anhydrous chemical drying agent that

**TABLE 8.2  Common anhydrous chemical drying agents**

| Drying agent | Acid-base properties | Capacity value[a] | Capacity[b] | Efficiency,[c] mg · L$^{-1}$ | Speed |
|---|---|---|---|---|---|
| $MgSO_4$ | neutral | 7 | high | 2.8 | fairly rapid |
| $CaCl_2$ | neutral | 6 | high | 1.5 | fairly slow |
| $Na_2SO_4$ | neutral | 10 | high | 25 | slow |
| $K_2CO_3$ | basic | 2 | low | moderate | fairly rapid |
| $CaSO_4$ | neutral | 2 | low | 0.004 | fast |
| KOH | basic | high | high | 0.1 | fast |
| $H_2SO_4$ | acidic | very high | high | 0.003 | fast |

a. Capacity value = (maximum number moles of $H_2O$)/(moles of drying agent).
b. Capacity = relative amount of water taken up per unit weight of drying agent.
c. Efficiency = measure of equilibrium residual water [mg · (L air)$^{-1}$] at 25°C.

binds with the water. Chemical drying agents react with water to form hydrates insoluble in the organic phase:

$$nH_2O + \text{drying agent} \longrightarrow \text{drying agent} \cdot nH_2O$$

Drying agents for organic liquids are usually anhydrous inorganic salts. Table 8.2 lists common drying agents used for organic liquids.

Several factors need to be considered in selecting a drying agent, namely, its capacity, its efficiency, the speed with which it works, and its chemical inertness. Capacity value refers to the number of moles of water bound in the hydrated form of the salt. Capacity corresponds to the amount of water taken up per unit weight of drying agent. Efficiency expresses how much water the drying agent leaves behind in the organic liquid. (**Note:** The higher the efficiency value the *less* efficient the drying agent.) The speed with which the hydrate forms determines how long the drying agent needs to be in contact with the organic solution. Table 8.3 lists suitable drying agents to use with various classes of organic compounds. Drying agents must be unreactive toward both the organic solvent and any organic compound dissolved in the solvent. For example, a base such as $K_2CO_3$ is not suitable for drying an acidic organic compound because it would undergo a chemical reaction with the acid.

As you can see from Table 8.2, some drying agents have a high capacity but leave water in the organic solution; $Na_2SO_4$ is a good example. It is particularly useful as a preliminary drying agent, but it is also widely used as a general-purpose drying agent because it is inexpensive and can be used with many types of compounds. $CaSO_4$, on the other hand, is very efficient but has a low capacity, which means that it works better after a preliminary drying of the liquid with $Na_2SO_4$ or $MgSO_4$. Also some drying agents, such as $Na_2SO_4$ and $CaCl_2$, do not react quickly with water, so they must be given time to form the hydrate. Of the drying agents listed in Tables 8.2 and 8.3, magnesium sulfate is used most often because it very rarely reacts with solutes or solvents, it has a good capacity for water, it is quite efficient, and it has a reasonable speed of removing water. However, its exothermic reaction with the water in the solution being dried sometimes causes the solvent to boil if the drying agent is added too rapidly. Slow addition of the drying agent prevents this problem.

*Using a Drying Agent*    To remove water from an organic liquid, add a small quantity of the powdered drying agent directly to the solution to be dried (Figure 8.13a). If an organic solution has water in it, the first bit of drying

---

T A B L E  8 . 3   **Preferred drying agents**

| Class of compounds | Recommended drying agents |
| --- | --- |
| Alkanes and alkyl halides | $MgSO_4$, $CaCl_2$, $CaSO_4$, $H_2SO_4$ |
| Hydrocarbons and ethers | $CaCl_2$, $MgSO_4$, $CaSO_4$ |
| Aldehydes, ketones, and esters | $Na_2SO_4$, $MgSO_4$, $CaSO_4$, $K_2CO_3$ |
| Alcohols | $MgSO_4$, $K_2CO_3$, $CaSO_4$ |
| Amines | KOH, $K_2CO_3$ |
| Acidic compounds | $Na_2SO_4$, $MgSO_4$, $CaSO_4$ |

**FIGURE 8.13**
Adding drying agent
to solution.

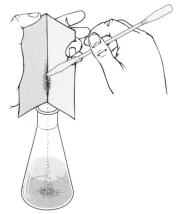

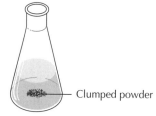

Clumped powder

(a) Adding powdered drying agent
to solution

(b) Drying agent clumped at bottom
of flask

*Always place the
organic liquid being
treated with drying
agent in an Erlenmeyer
flask closed with a cork
to prevent evaporation
losses.*

agent you add will clump together (Figure 8.13b). As soon as some
of the drying agent moves freely in the mixture while the flask is
gently swirled, you have added enough. The solution may be
stirred with a magnetic stirring bar or simply swirled occasionally
by hand to ensure as much contact with the surface of the drying
agent as possible. The best drying efficiency is obtained if the liquid
stands over the drying agent for at least 10 min. Often a preliminary drying period of 30–60 s, followed by removal of the drying
agent, then by the addition of a second portion of drying agent that
stands in the liquid for 10 min, removes the water more completely
than the use of a single portion of drying agent.

## 8.8 Methods for Separating Drying Agents and Organic Liquids

After the drying agent has absorbed the water present in the
organic liquid, it must be separated from the organic liquid. The
container receiving the liquid should be clean and dry and have a
volume about twice the volume of the organic liquid.

*Miniscale Separation
of Drying Agents*

If the solvent is going to be evaporated, place filter paper in a small
funnel and set the funnel in an Erlenmeyer flask (Figure 8.14). If the
solvent is being distilled, use a round-bottomed flask, set on a cork
ring, as the receiver. Decant the solution slowly into the filter paper,
leaving most of the drying agent in the flask. Rinse the drying agent
in the flask with 2–3 mL of dry solvent and also pour this rinse into
the filter paper. The extraction procedure is now complete, and the
organic liquid is ready for the removal of the solvent.

In some extraction procedures, the organic liquid is not dissolved in a solvent. In this situation, you must minimize the loss of
liquid product during the removal of the drying agent. Instead of
filter paper, a tightly packed cotton plug about 5–6 mm in diameter

**FIGURE 8.14**
Filtration of the drying agent from a solution when the solvent will be evaporated. (Replace the Erlenmeyer flask with a round-bottomed flask if the solvent is to be removed by distillation or with a rotary evaporator.)

***Microscale Separation of Drying Agents***

**FIGURE 8.15**
Filtration of drying agent from an organic liquid when no solvent is present.

put in the outlet of a funnel traps the drying agent and absorbs only a small amount of the organic liquid (Figure 8.15). Slowly pour the liquid from the drying agent.

If the drying agent is granular or chunky, the cotton plug can be omitted and the liquid carefully decanted into the funnel, leaving all the drying agent in the original flask. The drying agent may or may not be rinsed with a few milliliters of solvent in this procedure. The organic liquid is ready for the final distillation or analysis.

In microscale extractions, the organic liquid is usually dried in a centrifuge tube or a test tube. A Pasteur filter pipet [see Technique 8.5] is used to remove the liquid from the drying agent and to transfer it to a clean, dry container (Figures 8.16a and 8.17a). An alternative procedure, useful for a powdered drying agent such as magnesium sulfate, uses one Pasteur pipet as a funnel and a second Pasteur pipet to transfer the liquid. Here the cotton is packed tightly at the top of the tip (Figures 8.16b and 8.17b).

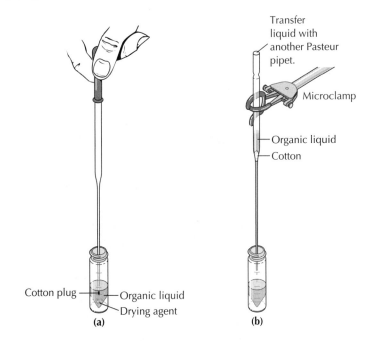

**FIGURE 8.16**
Using standard taper microscale equipment and a Pasteur pipet (a) with a cotton plug in the tip to separate an organic liquid from the drying agent and (b) as a filtering funnel.

**FIGURE 8.17**
Using Williamson
microscale equipment
and a Pasteur pipet
(a) with a cotton plug in
the tip to separate an
organic liquid from the
drying agent and (b) as
a filtering funnel.

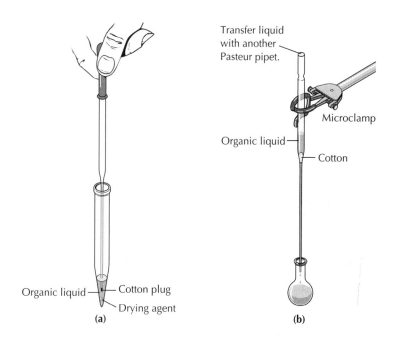

Transfer liquid
with another
Pasteur pipet.

Microclamp

Organic liquid

Cotton

Organic liquid — Cotton plug
Drying agent

(a)                                          (b)

---

## 8.9    Recovery of the Organic Product from the Dried Extraction Solution

Once the extraction solution has been dried, it is necessary to remove the solvent to recover the desired organic product. In experiments in which the amount of solvent is small (less than 25 mL), the solvent could be removed by evaporating it on a steam bath in a hood or by blowing it off with a stream of nitrogen or air in a hood. Your instructor will advise you whether these methods for removing solvents are followed in your laboratory. Concern for the environment and environmental laws now limit and sometimes prohibit using these methods. Removing solvents by distillation or with a rotary evaporator are alternatives to evaporation; both methods allow the solvents to be recovered.

*Evaporation*

Place a boiling stick or boiling stone in the Erlenmeyer flask containing the solution to be evaporated and heat the flask on a steam bath in a hood. Your product will be the liquid or solid residue left in the flask when the boiling ceases.

*Distillation*

Assemble the simple distillation apparatus shown in Technique 11, Figure 11.6. If the solvent is ether, pentane, or hexane, use a steam bath as a heat source to eliminate the fire hazard an electric heating mantle poses with the very flammable vapors from these solvents. Continue the distillation until the solvent has completely distilled, an end point indicated by a drop in the temperature reading on the thermometer. The drop in temperature occurs because there is no longer enough hot vapor surrounding the thermometer bulb. Your product and a small amount of solvent remains in the boiling flask.

**Using a Rotary Evaporator**

*Read Technique 11.7 on vacuum distillation before using a rotary evaporator.*

A rotary evaporator is an apparatus for removing solvents rapidly in a vacuum (Figure 8.18). No boiling stones or sticks are necessary because the rotation of the flask minimizes bumping. Rotary evaporation is done in a round-bottomed flask that is no more than half filled with the solution being evaporated. A trap is placed between the round-bottomed flask and the vacuum source so that the evaporating solvent can be recovered.

The following protocol is a generalized outline of the steps in using a rotary evaporator, but you should consult your instructor about the exact operation of the rotary evaporator in your laboratory. Select a round-bottomed flask that will be half-full or less with the solution undergoing evaporation. Connect the flask to the rotary evaporator with a joint clip. Use an empty trap and be sure that it is also clipped tightly to the rotary evaporator housing. Position a room-temperature water bath under the flask containing the solution so that the flask is approximately one-third submerged in the water bath. Turn on the water to the condenser, then turn on the vacuum source. As the vacuum develops, turn on the motor that rotates the evaporating flask. When the vacuum stabilizes at 20–30 Torr or lower, begin to heat the water bath. A water bath temperature of 50°–60°C will quickly evaporate solvents with boiling points under 100°C.

When solvent no longer condenses in the trap (receiving flask), the evaporation is complete. Stop the rotation of the flask and lower the water bath away from the evaporation flask. Open the stopcock and allow air to bleed slowly into the system. Hold the flask with one hand and remove the clip holding it to the evaporator. Turn off the vacuum source and the condenser water. Disconnect the trap from the rotary evaporator housing and empty the solvent in the trap into the appropriate waste container or recovered solvent container.

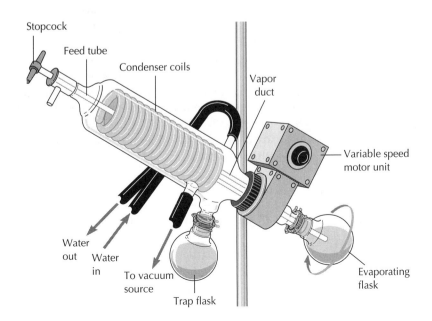

**FIGURE 8.18**
Schematic diagram of a rotary evaporator.

Stopcock

Feed tube

Condenser coils

Vapor duct

Variable speed motor unit

Water out    Water in

To vacuum source

Trap flask

Evaporating flask

## Questions

1. An extraction procedure specifies that an aqueous solution containing dissolved organic material be extracted twice with 10-mL portions of diethyl ether. A student removes the lower layer after the first extraction and adds the second 10-mL portion of ether to the upper layer remaining in the separatory funnel. After shaking the funnel, the student observes only one liquid phase with no interface. Explain.

2. A crude nonacidic product mixture dissolved in diethyl ether contains acetic acid. Describe an extraction procedure that could be used to remove the acetic acid.

3. What precautions need to be observed when an aqueous sodium carbonate solution is used to extract an organic solution containing traces of acid?

4. A poorly trained student forgets to record the densities of water and diethyl ether in his/her lab notebook. When the two layers form during an ether/water extraction, what would be a convenient, easy way to tell which layer is which?

5. You have 25 mL of a solution of benzoic acid in water, estimated to contain about 0.80 g of the acid. The distribution coefficient of benzoic acid in diethyl ether and water is approximately 10. Calculate the amount of acid that would be left in the water solution after four 15-mL extractions with ether. Do the same calculation using one 60-mL extraction with ether to determine which method is more efficient.

6. $K_2CO_3$ is an excellent drying agent for some classes of organic compounds. Would it be a better choice for an acid (R—COOH) or an amine (R—NH$_2$)? Why?

TECHNIQUE

# 9

# RECRYSTALLIZATION

A pure organic compound is one in which there are no detectable impurities. Because experimental work requires an immense number of molecules (Avogadro's number per mole), it is not necessarily true that 100% of the molecules in a given compound with consistent physical properties are identical to one another. Seldom is a pure compound purer than 99.99%. Even if it were that pure, one mole would still contain more than $10^{19}$ molecules of other compounds. Nevertheless, we want to work with compounds that are as pure as possible; therefore, we must have ways to purify impure materials.

## 9.1    Theory of Recrystallization

Recrystallization is a common method of purifying organic solids. The technique of recrystallization depends on the increased solubility of a compound in a hot solvent. A saturated solution at a higher temperature normally contains more solute than the same solute-solvent pair at a lower temperature. Therefore, the solute precipitates when a warm saturated solution cools. *Recrystallization* is a process whereby a crystalline material dissolves in a hot solvent, then returns to a solid again by crystallizing in the cooled solvent. Because the total concentration of impurities in the solid of interest is usually significantly lower than the concentration of the substance being purified, as the mixture cools the impurities remain in solution while the highly concentrated product crystallizes.

*Crystal Formation*

Crystal formation of a solute from a solution is a selective process. When a solid crystallizes at the right speed under the appropriate conditions of concentration and solvent, an almost perfectly pure crystalline material can result, because only molecules of the right shape fit into the crystal lattice. In recrystallization, dissolution of the impure solid in a suitable hot solvent destroys the impure crystal lattice, and crystallization from the cold solvent selectively produces a new, more perfect (purer) crystal lattice. Slow cooling of the saturated solution promotes the formation of pure crystals because the molecules of the impurities do not fit properly into the newly forming lattice. Crystals that form slowly are larger and purer than ones that form quickly. Indeed, rapid crystal formation traps the impurities again, because the lattice grows so rapidly that the impurities are simply surrounded by the crystallizing solute as the crystal forms.

*Choice of Solvent*

The most crucial aspect of a recrystallization procedure is the choice of solvent, because **the solute should have a maximum solubility in the hot solvent and a minimum solubility in the cold solvent.** Table 9.1 lists common recrystallization solvents.

In general, a solvent with a structure similar to that of the solute being dissolved is a better recrystallization solvent than solvents with dissimilar structures. Although the appropriate choice of solvent is a trial-and-error process, a relationship exists between the solvent's molecular structure and the solubility of the solute. This relationship is simply described as **like dissolves like.** Nonionic compounds dissolve in water only when they can associate with the water molecules through hydrogen bonding. Thus, all the hydrocarbons and the alkyl halides are virtually insoluble in water, whereas carboxylic acids, which readily form hydrogen bonds, are often recrystallized from water solution. Molecules that associate with water through hydrogen bonds include carboxylic acids, alcohols, and amines (Figure 9.1). Carboxylic acids hydrogen bond to a lone pair of electrons of water through the acidic proton; alcohols do likewise. Amines hydrogen bond through the lone pair of nitrogen to a hydrogen atom of water.

Polarity of the solvent is a crucial factor in solubility. One measure of polarity is the *dielectric constant,* $\varepsilon$ (see Table 9.1). A rough expectation is that solvents with higher dielectric constants should more readily dissolve polar (or ionic) compounds. The first four or five members (carbon content of $C_4$ to $C_5$) of a homologous series of carboxylic acids, alcohols, or amines are water soluble. But as the

**TABLE 9.1** **Common recrystallization solvents**

| Solvent | Formula | Boiling point, °C | Freezing point, °C |
|---|---|---|---|
| Diethyl ether | $(C_2H_5)_2O$ | 34.6 | −116 |
| Acetone | $(CH_3)_2CO$ | 56 | −95 |
| Petroleum ether[b] | — | 60–80 | — |
| Chloroform | $CHCl_3$ | 61 | −63 |
| Methanol | $CH_3OH$ | 65 | −98 |
| Hexane | $C_6H_{14}$ | 69 | −94 |
| Ethyl acetate | $CH_3COOC_2H_5$ | 77 | −84 |
| Ethanol | $C_2H_5OH$ | 78.5 | −117 |
| Water | $H_2O$ | 100 | 0 |
| Toluene | $C_6H_5CH_3$ | 110.6 | −95 |
| Acetic acid | $CH_3COOH$ | 118 | 16 |

a. Scale: Extreme fire hazard = ++++.
b. Petroleum ether (or ligroin) is a mixture of isomeric alkanes. The term "ether" refers to volatility, not the presence of oxygen.

molecular weight increases, the water solubility of the species decreases, because the hydrocarbon, or nonpolar, portion of the molecule begins to dominate its physical behavior.

The salts of low-molecular-weight carboxylic acids are quite water soluble. For example, potassium acetate, $CH_3COOK$, and sodium propionate, $CH_3CH_2COONa$, are largely ionic compounds with solubility characteristics similar to sodium chloride or potassium bromide.

Solvents of low polarity dissolve most nonionic organic compounds with ease. Because the solution-forming process is one of molecular mixing, even small polar organic compounds dissolve in nonpolar solvents if the ratio of polar functional groups per carbon atom is not too high. Thus, monofunctional lower members of all homologous series are soluble in nonpolar solvents. For example, methanol ($\varepsilon = 32.6$) is extremely soluble in diethyl ether, a relatively nonpolar solvent.

Among the low-polarity solvents, diethyl ether appears to provide the best solvent properties, although its extreme flammability and low boiling point (35°C) require careful attention to safety when using it. Ether in combination with hexane, methanol, or dichloromethane has excellent solvent properties for recrystallizations. Hexane is even more nonpolar than diethyl ether and has the advantage of a higher boiling point (69°C), yet it is still easy to remove from the recrystallized solid.

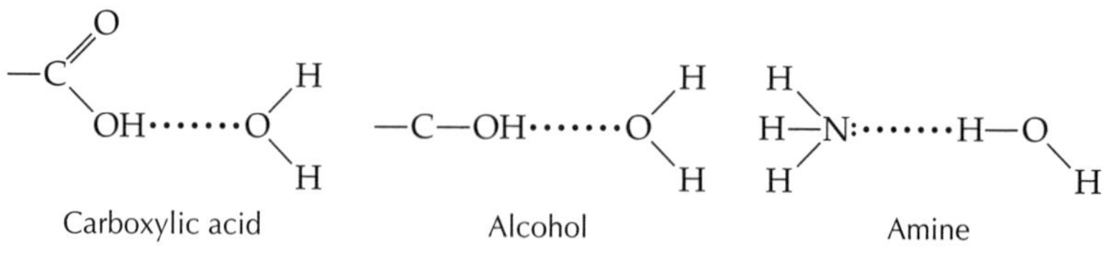

**FIGURE 9.1**
Compounds that associate with water by hydrogen bonding (dotted lines).

Carboxylic acid          Alcohol          Amine

| Miscibility in water | Solvent polarity | Dielectric constant ($\varepsilon$) | Fire hazard[a] | Inhalation toxicity |
|---|---|---|---|---|
| − | low | 4.3 | + + + + | − |
| + | intermediate | 20.7 | + + + | − |
| − | nonpolar | ca. 2 | + + + + | − |
| − | low | 4.8 | 0 | high |
| + | polar | 32.6 | + + | high |
| − | nonpolar | 1.9 | + + + + | − |
| − | intermediate | 6.0 | + + | − |
| + | polar | 24.3 | + + | − |
| | very polar | 80 | 0 | − |
| − | nonpolar | 2.4 | + + | high |
| + | intermediate | 6.15 | + | high |

**SAFETY PRECAUTION**

Both ether and hexane are very flammable and should be heated with a steam or hot-water bath. They should **never** be heated with a flame or on a hot plate.

Among the more polar solvents, both methanol and ethanol are commonly used for recrystallization because they evaporate easily, possess water solubility, and dissolve a wide range of both polar and nonpolar compounds.

## 9.2    Selecting a Proper Recrystallization Solvent

*Recrystallization from a Single Solvent*

To select a solvent for recrystallization, take a small sample (20–30 mg) of the compound to be recrystallized, place it in a test tube, and add 5–10 drops of a trial solvent. Shake the tube to mix the materials. If the compound dissolves immediately, it is probably too soluble in the solvent for a recrystallization to be effective. If no solubility is observed, heat the solvent to its boiling point. If complete solubility is observed, cool the solution to induce crystallization. The formation of crystals in 10–20 min suggests that you have a good recrystallization solvent.

*Recrystallization from Mixed Solvent Pairs*

When no single solvent seems to work, a pair of miscible solvents can sometimes be used. Usually mixed solvent pairs include one solvent in which a particular solute is very soluble and another in which solubility is marginal. Often such pairs consist of a polar solvent (with, for example, a high dielectric constant) mixed with a nonpolar solvent.

| TABLE 9.2 | Solvent pairs for mixed solvent recrystallizations[a] | | |
|---|---|---|---|
| **Solvent 1** | **Solvent 2** | **Solvent 1** | **Solvent 2** |
| Ethanol | Acetone | Methanol | Dichloromethane |
| Ethanol | Water | Methanol | Diethyl ether |
| Acetone | Water | Methanol | Water |
| Chloroform | Petroleum ether | Diethyl ether | Hexane (or petroleum ether) |
| Ethyl acetate | Hexane | | |

a. Properties of these solvents are given in Table 9.1.

To recrystallize a solid from a mixed solvent pair, first dissolve the solute in the solvent in which it is more soluble; warm the solvent before adding it to the solute. Then warm the solution nearly to its boiling point and add the other solvent dropwise until a slight cloudiness appears (indicating that the hot solution is saturated with the solute). Add some of the first solvent again until the cloudiness just disappears and then add a few more drops to ensure an excess (more if you are working with a large volume of solution). Let the solution cool slowly to initiate crystallization. Typical mixed solvent pairs are listed in Table 9.2.

Another technique uses a mixture of miscible solvents, such as ethanol and water, to dissolve the impure solid. This method sometimes requires testing several different proportions to find the optimum ratio of the two solvents. If the solid is more soluble in the solvent with the lower boiling point, the excess solvent can simply be boiled away until cloudiness is reached. Then the solution is allowed to cool.

## 9.3    What to Do If No Crystals Appear in the Cooled Solution

In many instances, recrystallization fails because too much solvent is used in the process. In these cases, you need to boil off a bit of the solvent and try the recrystallization again.

If recrystallization still does not occur from the supersaturated solution, or if oils form, there are several tricks that can be used to induce crystallization. **Always save a few crude crystals to use as seeds in the event that recrystallization does not occur.** Adding a few seed crystals to the cooled solution may induce crystallization by providing nuclei around which crystals can grow. Another way to promote crystal formation is to scratch the inside of the bottom of the flask vigorously with a stirring rod. The tiny particles of glass scratched from the flask serve as centers for crystallization to begin. It is said that Louis Pasteur accidentally initiated a significant recrystallization with dandruff from his beard, but we do not recommend this technique.

*Always save a few crude crystals to use as seeds in the event that recrystallization does not occur.*

The formation of oils is probably the most frustrating outcome of an attempted recrystallization. The presence of impurities lowers the melting point of substances, making "oiling out" especially prevalent during recrystallization of a solute with a melting point below the boiling point of the solvent being used. Impurities distribute themselves in the two liquids, the solvent and the oil, before crystallization occurs, so the impurities are trapped in the oil when it cools and hardens into a viscous glasslike substance. If you have an oil rather than crystals, you can add more solvent so that the compound does not come out of solution at so high a temperature, because the point of supersaturation may have been reached when the solvent temperature was still above the freezing (melting) point of the compound. It may also help to switch to a solvent with a lower boiling point (consult Table 9.1).

---

**9.4** | **Pointers for Successful Recrystallizations**

The suggestions given here apply to both miniscale and microscale recrystallizations. When you are recrystallizing a product, attention to these details will increase the purity and amount of your product.

Many students recover a smaller amount of product from a recrystallization than they should because of mechanical losses. Losses occur because (1) too much solvent is added, (2) too much charcoal is added to decolorize colored solutions, (3) premature crystallization occurs during a gravity filtration, or (4) the crystals are filtered or centrifuged before recrystallization is complete.

When a higher-boiling-point solvent, such as ethyl alcohol, water, or toluene, is used, the recrystallized product may not dry completely for a rather long time and should be allowed to dry at least overnight before determining its mass and melting point. If time permits, collecting a second crop of crystals by evaporating half the filtrate from the first crop will increase the yield. However, the purity of this second crop is likely to be lower than that of the first crop; determine the melting points of both crops before combining them.

---

**9.5** | **Miniscale Procedure for Recrystallizing a Solid**

S A F E T Y   P R E C A U T I O N

1. Most organic solvents used for recrystallizations are volatile and flammable. Therefore, they should be heated on a steam bath or in a hot-water bath, not on a hot plate or with an open flame.
2. Lift a hot Erlenmeyer flask with flask tongs. *Note:* Test tube holders are **not** designed to hold Erlenmeyer flasks securely and the flask may fall onto the bench top.

FIGURE 9.2
Two ways to add a solid
to an Erlenmeyer flask
for recrystallization.

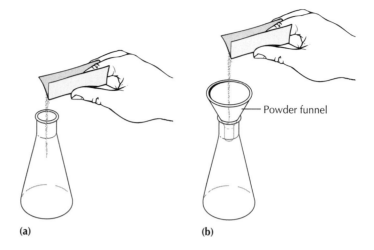

(a)                                                                    (b)

*Dissolving the Solid*

*Always save a few crude
crystals to use as seeds in
the event that recrystal-
lization does not occur.*

Place the solid to be recrystallized on a creased weighing paper
and carefully pour it into an Erlenmeyer flask (Figure 9.2a). Alternatively, a plastic powder funnel may be set in the neck of the
Erlenmeyer flask to prevent spillage (Figure 9.2b). Add one or two
boiling chips or a boiling stick. Heat an appropriate volume of the
solvent in another Erlenmeyer flask. Then add small portions of
hot (just below boiling) solvent to the solid being recrystallized.
Begin heating the solid-solvent mixture, allowing it to boil briefly
between additions, until the solid dissolves; then add a little
excess solvent. Remember that some impurities may be completely insoluble, so do not add too much solvent in trying to dissolve the last bit of solid. Bring the solution to a boil on a steam
bath (Figure 9.3).

With particularly volatile organic solvents, such as ether or
hexane, it is often easier to add a small amount of cold solvent, then
heat the mixture nearly to boiling. Slowly add more cold solvent to
the heated mixture until the solid just dissolves when the solution
is boiling; then add a slight excess of solvent.

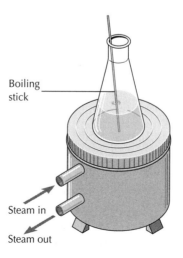

Boiling
stick

Steam in

FIGURE 9.3
Heating a solution on
a steam bath.

Steam out

*Cooling the Solution*

Filtration of the hot solution is sometimes necessary to remove insoluble impurities [see Technique 9.5a]. However, if there are no insoluble substances in your hot recrystallization mixture, you can proceed directly to the cooling step. The size of the crystals obtained will depend on the rate at which the solution cools: The slower the cooling, the larger the crystals. Cork the Erlenmeyer flask while the solution cools. Allowing the hot solution to stand on the bench top until crystal formation begins and the flask reaches room temperature, followed by final cooling in an ice-water bath, usually produces crystals of a reasonable purity and intermediate size. The whole cooling process should take at least 15 min; occasionally, it may take 30 min or more before crystals appear.

Careful attention to detail and slow cooling of the hot solution often results in the formation of beautiful, pure crystals. Beautiful crystals are to the organic chemist what a home run is to a baseball player!

*Collecting the Recrystallized Solid*

To complete the recrystallization procedure after all the crystals appear to have formed, collect the crystals by vacuum filtration, using a Buchner funnel, neoprene adapter, filter flask, and trap bottle or flask. The trap flask keeps backflow from a water aspirator out of the filter flask; with a house vacuum system, the trap flask keeps any overflow of the filter flask out of the vacuum line (Figure 9.4).

Choose the correct size of filter paper, one that will fit flat on the bottom of the Buchner funnel and just cover all the holes. Turn on the vacuum source and wet the paper with the recrystallizing solvent to pull it tight over the holes in the funnel. Pour the slurry of crystals and solvent into the funnel.

*Chemists sometimes refer to the supernatant liquid from a recrystallization as the "mother liquor."*

Wash the crystals on the Buchner funnel with a small amount of cold recrystallizing solvent (1 to 5 mL, depending on the amount of crystals) to remove any supernatant liquid adhering to them. To wash the crystals, allow air to enter the filtration system by removing the rubber tubing from the water aspirator nipple before turning off the water (to prevent backup of water into the

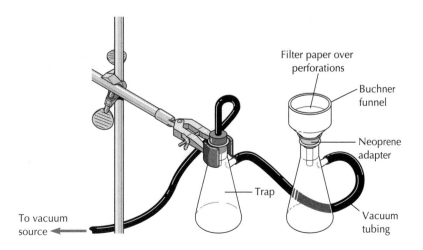

**FIGURE 9.4**
Apparatus for vacuum filtration. The second filter flask serves as a backflow trap.

Filter paper over perforations

Buchner funnel

Neoprene adapter

Trap

To vacuum source ←

Vacuum tubing

system), or turn off the vacuum line and loosen the neoprene adapter connecting the Buchner funnel to the filter flask. Then cover the crystals with the cold solvent, reconnect the vacuum, and draw the liquid off the crystals. Initiate the crystal drying process by pulling air through them for a few minutes. Again disconnect the vacuum as described earlier. You will probably need to leave the crystals open to the air in your desk for a time to dry them completely; place the crystals on a tared (weighed) watch glass. Remove any boiling chips or sticks before you weigh the crystals.

***A Second "Crop" of Crystals***

A second "crop" of crystals can sometimes be obtained by evaporating about half the solvent from the filtrate and again cooling the solution. This crop of crystals should be kept separate from the first crop of crystals until the melting points of both crops [see Technique 10] have been determined. If the two melting points are the same, indicating that the purity is the same, the crops may be combined. Usually the second crop has a slightly lower melting point and a larger melting range, indicating that some impurities crystallized with the desired product.

## Removing Impurities from the Recrystallization Solution

***Filtering Insoluble Impurities***

If you have some insoluble material that needs to be removed from a hot recrystallization solution—dust, bits of filter paper, or other insoluble impurities—a gravity filtration is necessary before cooling the recrystallization solution. If the solution needs filtration, prepare a fluted filter paper or obtain prefolded filter paper from the supply in the laboratory. Fluted filter paper provides a larger surface area than the usual filter paper cone, making for a faster filtration. Suction filtration under reduced pressure does not work well because the solution cools rapidly during this process and premature crystallization can occur. Also, small particles may pass through the filter paper when suction is used.

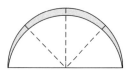

**1.** Crease filter paper.

To make a fluted filter, crease a regular filter paper in half four times (Figure 9.5, step 1). Then fold each of the eight sections of the filter paper inward so that the paper looks like an accordion (Figure 9.5, step 2). Finally, open the paper to form a fluted cone (Figure 9.5, step 3). Alternatively, filter paper already folded in this manner is commercially available.

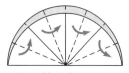

**2.** Fold each quarter inward.

Place the fluted filter paper in a clean, short-stemmed funnel (a plastic powder funnel works well) and put the funnel in a second clean Erlenmeyer flask. Add a small amount (1–3 mL) of the recrystallization solvent to the second (receiving) flask and heat the flask, funnel, and solvent on your steam bath (Figure 9.6, step 1). The boiling solvent warms the funnel and helps prevent premature crystallization of the solute during filtration. If your steam bath is large enough, keep both flasks hot during the filtration process; if it

**3.** Fluted filter paper.

**FIGURE 9.5**
Fluting filter paper.

**FIGURE 9.6**
Filtering a
recrystallization
solution.

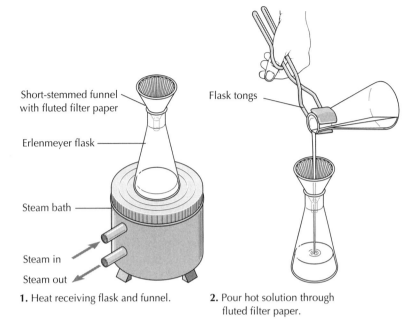

Short-stemmed funnel
with fluted filter paper

Flask tongs

Erlenmeyer flask

Steam bath

Steam in

Steam out

**1.** Heat receiving flask and funnel.

**2.** Pour hot solution through
fluted filter paper.

is too small for both, keep the unfiltered solution hot and set the receiving flask on the bench top. Next, pour the hot recrystallization solution through the fluted filter paper (Figure 9.6, step 2).

**SAFETY PRECAUTION**

Lift a hot Erlenmeyer flask with flask tongs.

Be sure that the hot solution is added in small quantities to the fluted filter paper, because cooling at this stage may cause crystallization in the filter paper. Keep the unfiltered solution hot during this step so that all the solid remains in solution. If you have difficulty in keeping the solution from crystallizing on the filter paper, add extra hot solvent to the flask containing the unfiltered solution and reheat it to the boiling point before continuing the filtration. Then, after the filtration is complete, boil away the extra solvent you added.

When all the hot solution has filtered through the paper, check to see whether any crystallization occurred in the Erlenmeyer receiving flask during the rapid cooling of the filtration step. If it has, reheat the mixture to dissolve the solid. Then allow the solution to cool slowly. While the solution is cooling, the Erlenmeyer flask should be loosely stopped or covered.

***Using Activated Charcoal to Remove Colored Impurities***

If the solution is deeply colored after the solid dissolves in the recrystallization solvent, treatment with activated charcoal (Norit or Darco, for example) may remove the colored impurities. Particles of activated charcoal have a large surface area and a strong affinity for highly conjugated colored compounds, but too much charcoal will also adsorb the compound you are purifying and reduce your yield. Add 20–30 mg of the activated carbon to the hot **but not boiling** solution.

---

**SAFETY PRECAUTION**

Adding charcoal to a boiling solution invariably causes the solution to foam out of the flask. **Cool the hot solution briefly before adding the charcoal.**

---

Now heat the solution to boiling for a few minutes. While the solution is still very hot, gravity filter it according to the procedure described in the previous section.

## 9.6 Summary of the Miniscale Recrystallization Procedure

*Always save a few crude crystals to use as seeds in the event that recrystallization does not occur.*

1. Dissolve the solid sample in a minimum volume of hot solvent with a boiling chip or boiling stick present.
2. If the color of the solution reveals impurities, add a small amount of activated charcoal to the hot but not boiling solution (optional).
3. If insoluble impurities are present or charcoal treatment is used, gravity filter the hot solution through a fluted filter paper.
4. Cool the solution slowly to room temperature and then in an ice-water bath to induce crystallization.
5. Remove the crystals from the solvent by vacuum filtration.
6. Wash the crystals with a small amount of cold solvent.
7. Allow the crystals to air dry completely on a watch glass before weighing them and determining their melting point.

## 9.7 Microscale Recrystallization Methods

*Read Techniques 9.1–9.4 before you undertake your first microscale recrystallization.*

Microscale methods are used for recrystallizations of less than 300 mg of solid. Microscale recrystallizations generally follow the steps outlined in Technique 9.5, using smaller equipment. For the recrystallization of less than 150 mg of material, a specialized apparatus called a Craig tube may be employed.

## 9.7a  Recrystallizing 300 Milligrams or Less of a Solid

A test tube or 10-mL Erlenmeyer flask holds the recrystallization solution and a Hirsch funnel replaces the Buchner funnel for collecting the crystals. The following steps outline the procedure for a microscale recrystallization.

*Always save a few crude crystals to use as seeds in the event that recrystallization does not occur.*

1. Place the solid in a 13 × 100 mm test tube, a reaction tube, or a 10-mL Erlenmeyer flask; add a boiling stick. With a Pasteur pipet, add enough solvent to just cover the crystals.

2. Boil the contents of the test tube or flask, then add additional solvent dropwise, allowing the mixture to boil briefly after each addition. Continue this process until just enough solvent has been added to dissolve the solid. Be aware that some impurities may not dissolve.

3. If colored impurities are present, cool the mixture slightly and add 10 mg of Norit carbon-decolorizing pellets (about 10 pellets). Boil the mixture briefly. If the color is not removed after 1–2 min, add a few more Norit pellets and boil briefly. Prepare a Pasteur filter pipet [see Technique 8.5]. Warm the Pasteur pipet by immersing it in a test tube of hot solvent and drawing the hot solvent into it several times; use the heated pipet to separate the hot recrystallization solution from the Norit pellets and transfer it to another test tube or flask. If crystallization begins in the solution with the carbon pellets during this process, add a few more drops of solvent and warm the mixture to boiling to redissolve the crystals before completing the transfer.

4. If the recrystallization mixture contains insoluble impurities, use a Pasteur filter pipet as outlined in step 3 to separate the solution from the insoluble impurities.

5. Allow the solution to cool slowly to room temperature, then chill it in an ice-water bath to complete crystallization.

6. Collect the crystals by vacuum filtration, using a Hirsch funnel (Figure 9.7).

7. Allow the crystals to air dry completely on a watch glass before weighing them and determining their melting point.

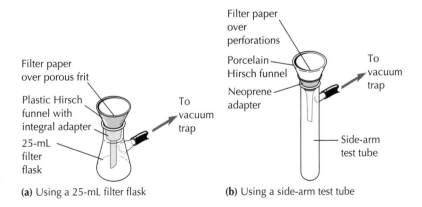

**FIGURE 9.7**
Vacuum filtration using a Hirsch funnel.

(a) Using a 25-mL filter flask

(b) Using a side-arm test tube

## 9.7b  Recrystallizing up to 150 Milligrams of a Solid in a Craig Tube

Recrystallizations of up to 150 mg may be done in a Craig tube, a small device that serves as both recrystallization vessel and filtration apparatus. The Craig tube eliminates the losses associated with separation of the crystals from the solution by vacuum filtration. It also allows multiple recrystallizations to be performed without removing the crystals from the tube, thus preventing significant losses of material on the glassware.

The Craig tube consists of a glass tube that has a band with a rough surface at the point where the tube widens and a separate plug made of glass or Teflon (Figure 9.8). If the plug is glass, the large end has a rough surface. When the plug is inserted into the tube and the apparatus inverted as described later, the rough glass surface of the tube forms an incomplete seal with the plug. This imperfect seal allows the crystallization solution (mother liquor) to flow out during centrifugation, leaving the crystals in the tube. This process separates the crystals from the recrystallization solution.

*Chemists sometimes refer to the supernatant liquid from a recrystallization as the "mother liquor."*

Heat for a Craig tube recrystallization is supplied by an aluminum block heated to 10°–20°C above the boiling point of the solvent; one of the holes in the block is sized for the Craig tube. Bumping and boiling over occur very quickly with a Craig tube; prevent this problem by using a boiling stick or by stirring rapidly by rolling a microspatula inserted in the Craig tube between your fingers while the mixture is heating. Boiling chips should not be used with a Craig tube because their subsequent removal is difficult.

***Craig Tube Recrystallization Procedure***

*Always save a few crude crystals to use as seeds in the event that recrystallization does not occur.*

Place the material to be recrystallized in the outer tube of the Craig assembly. If the amount of solvent is not specified, cover the crystals with a few drops of solvent and begin heating the mixture on the aluminum block; continue adding solvent dropwise and boiling the mixture gently between additions until the solid dissolves completely. When dissolution is complete, insert the plug into the tube and set the assembly inside a 25-mL Erlenmeyer flask to cool slowly and undisturbed. When crystallization appears to be complete, cool the tube in an ice-water bath for 5 min.

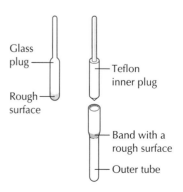

Glass plug
Rough surface
Teflon inner plug
Band with a rough surface
Outer tube

**FIGURE 9.8**
Craig tube.

*Slow cooling promotes the formation of the larger and purer crystals essential for good recovery of pure product.*

Recover the crystals in the following manner. Obtain a centrifuge tube. If the centrifuge tube and the Craig tube plug are glass, place a small piece of cotton in the bottom of the centrifuge tube; cotton is unnecessary with either a plastic centrifuge tube or a Teflon plug. Slip a loop of thin copper wire over the stem of the plug (Figure 9.9, step 1) and invert the centrifuge tube over the Craig apparatus (Figure 9.9, step 2). Then turn the centrifuge tube upright and bend the end of the wire over the lip of the centrifuge tube (Figure 9.9, step 3).

*During centrifugation, the recrystallization solvent flows through the small pores in the rough etched ring of the Craig tube, while the crystals remain on the end of the inner plug.*

Set the tube in a centrifuge and balance its weight in the opposite hole with another centrifuge tube that is two-thirds to three-fourths full of water; centrifuge for 1–2 min. (If you are having difficulty balancing the centrifuge, weigh the centrifuge tube containing the Craig apparatus in a small beaker, then adjust the amount of water in the other centrifuge tube until the mass approximately equals that of the tube with the Craig apparatus.) Remove the tube from the centrifuge and carefully remove the Craig tube from the centrifuge tube by pulling on the copper wire. Turn the Craig apparatus upright. Remove the plug from the outer tube and set the tube upright in your desk to allow the crystals to dry. Carefully scrape any crystals adhering to the plug onto a small watch glass and store them until the crystals in the Craig tube are dry. The loss of crystals adhering to the inside of the tube is less if the crystals have dried before they are removed from the tube. The solvent in the bottom of the centrifuge tube should be treated as directed in the experimental cleanup procedure.

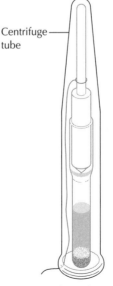

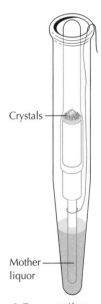

**FIGURE 9.9**
Removing solvent from a Craig tube by centrifugation.

**1.** Place loop of copper wire over plug.

**2.** Place centrifuge tube over Craig apparatus, invert, and put in centrifuge.

**3.** Turn centrifuge tube upright after centrifugation.

## 9.8 Summary of Microscale Recrystallization Procedures

**Recrystallizing 300 Milligrams or Less of a Solid**

*Always save a few crude crystals to use as seeds in the event that recrystallization does not occur.*

1. Dissolve the solid in a minimum volume of hot solvent in a small test tube or 10-mL Erlenmeyer flask; use a boiling stick or boiling chip to prevent bumping.
2. If colored impurities are present, boil the mixture briefly with 8–10 Norit pellets.
3. If insoluble impurities are present or Norit pellets were used, transfer the hot recrystallization solution to another test tube or flask, using a heated Pasteur filter pipet.
4. Cool the solution slowly to room temperature to induce crystallization, then complete cooling in an ice-water bath.
5. Collect the crystals by vacuum filtration on a Hirsch funnel.
6. Allow the crystals to air dry completely on a watch glass before weighing them.

**Recrystallizing up to 150 Milligrams of Solid in a Craig Tube**

*Always save a few crude crystals to use as seeds in the event that recrystallization does not occur.*

1. Dissolve the solid in a minimum volume of hot solvent in the outer tube of the Craig apparatus; use a boiling stick or continuously roll a stirring rod between your fingers to prevent bumping.
2. Insert the plug into the tube and set the assembly inside a 25-mL Erlenmeyer flask to cool undisturbed to room temperature. When crystallization is complete, cool the tube in an ice-water bath.
3. Place an inverted centrifuge tube over the Craig apparatus, then turn the centrifuge upright. Separate the solvent from the crystals by spinning the centrifuge tube in a centrifuge.
4. Allow the crystals to dry in the outer tube before removing them for weighing.

## Questions

1. Describe the characteristics of a good recrystallization solvent.
2. The solubility of a compound is 59 g per 100 mL in boiling methanol and 30 g per 100 mL in cold methanol, whereas its solubility in water is 7.2 g per 100 mL at 95°C and 0.22 g per 100 mL at 2°C. Which solvent would be better for recrystallization of the compound? Explain.
3. Explain how the rate of crystal growth can affect the purity of a recrystallized compound.
4. In what circumstances is it necessary to filter a hot recrystallization solution?
5. Why should a hot recrystallization solution be filtered by gravity rather than by vacuum filtration?
6. Low-melting solids often "oil out" of a recrystallization solution rather than crystallizing. If this were to happen, how would you change the recrystallization procedure to ensure good crystals?
7. An organic compound is quite polar and is thus much more soluble in methanol than in pentane (bp 36°C). Why would methanol and pentane be an awkward solvent pair for recrystallization? Consult Table 9.1 to assist you in deciding how to change the solvent pair so that recrystallization will proceed smoothly.
8. For a variety of reasons, *N,N*-dimethylformamide (DMF) is not usually utilized as a recrystallization solvent even

though it has a moderate dielectric constant ($\varepsilon = 38$). Included among its properties are an unpleasant odor, a high boiling point (153°C), and the fact that it is highly hygroscopic. Refer to Table 9.1 and discuss what other solvent(s) might serve as an alternative.

# 10 MELTING POINTS AND MELTING RANGES

Molecules in a crystal are arranged in a regular pattern. Melting occurs when the fixed array of molecules in the crystalline solid rearranges to the more random, freely moving liquid state. The transition from solid to liquid requires energy in the form of heat to break down the crystal lattice. The temperature at which this transition occurs is the solid's *melting point,* an important physical property of any solid compound. The melting point of a compound is useful in establishing its identity and as a criterion of its purity.

## 10.1 Melting Point Theory

A solid at any temperature has a finite vapor pressure. As the temperature of a solid increases, its vapor pressure increases as well. Both the solid and the liquid phases are always in equilibrium with the vapor and, at the melting point, are also in equilibrium with each other (Figure 10.1). Figure 10.2 shows a vapor pressure-temperature diagram for both the solid and liquid phases of a chemical compound.

The temperature at which a compound melts is a physical characteristic of the substance, and for pure compounds it is generally reproducible. However, the presence of even a small quantity of impurity usually depresses the melting point a few degrees and causes melting to occur over a relatively wide temperature range. Because the melting point is the temperature at which the vapor

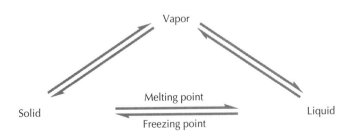

**FIGURE 10.1**
Solid and liquid in equilibrium.

**FIGURE 10.2**

Vapor pressure-
temperature diagram for
solid and liquid phases
of a pure compound.

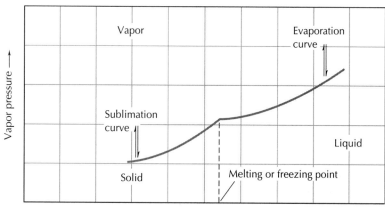

**FIGURE 10.2**

Vapor pressure-temperature diagram for solid and liquid phases of a pure compound.

pressures of the pure liquid and the pure solid are equal, the presence of an impurity that is soluble in the liquid can change the temperature at which this equilibrium state occurs. Energetically favorable solute (impurity)-solvent (compound) interactions lower the melting point.

The phase diagram depicted in Figure 10.3 shows the observed melting curve for various mol % mixtures of compounds A and B ranging from 100 mol % of A with 0 mol % B to 0 mol % A with 100 mol % B. A pure sample of compound A melts at temperature $T_a$ whereas pure compound B melts at temperature $T_b$. At $T_a$ and $T_b$, pure samples of A and B melt sharply over a very narrow temperature range of 1° or less.

Mixtures of A and B exhibit different melting behavior. Consider the behavior of a solid consisting of 80% of compound A and 20% of

**FIGURE 10.3**

Melting-point composition diagram for the binary mixture A + B. The presence of an impurity in a solid not only decreases the melting point but also produces melting over a relatively wide temperature range. In this diagram, $T_a$ is the melting point of pure solid A, $T_b$ of pure solid B, and $T_e$ of eutectic mixture E. The temperature range $T_e - T_m$ is the melting range of a solid containing 80 mol % A and 20 mol % B.

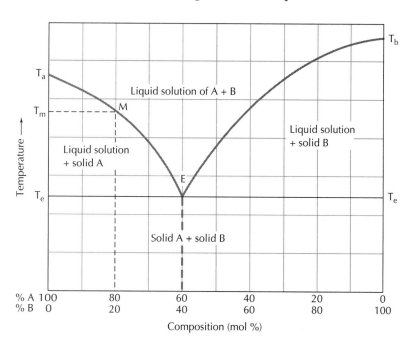

compound B. As compound A, with the lower melting point, begins to melt, compound B, still a solid, starts to dissolve in the liquid of compound A. The vapor pressure of the liquid solution of A and B is lower than that of pure liquid A at the melting point, whereas the vapor pressure of solid A at a particular temperature is virtually unchanged by the impurity B because the solids do not mix together intimately. Thus, the vapor pressures of the liquid and the solid are equal at a lower temperature and the temperature at which solid A melts is lower when B is present. An example of this behavior is the salting of roads to melt the ice at a temperature lower than 0°C.

There is a limit, however, to how far the melting point can be lowered. This limit is reached when the liquid solution becomes saturated in B, a condition causing some of solid B to remain after all of A has melted. In Figure 10.3, up to point E all of B dissolves in the melting A. After point E, when all of A is melted, a portion of solid B remains. Point E defines the composition of a saturated solution of B in liquid A and is called the *eutectic composition.* A solid mixture with the eutectic composition (60% A and 40% B) will melt sharply at the eutectic temperature, $T_e$. The eutectic point E can be thought of as the point at which increasing the percent of A means that B now acts as the impurity (right to left on the graph) and increasing the percent of B means that A acts as the impurity (left to right on the graph).

Again consider the melting of a solid mixture composed of 80% A and 20% B. As heat is applied, the temperature of the solid mixture rises. When it reaches temperature $T_e$, A and B will melt together at a constant ratio (the eutectic composition) and the temperature will remain constant. When the minor component, B, is completely liquefied and more heat is applied, solid A, in equilibrium with the eutectic composition of liquid A and B, continues to melt as the temperature increases. Because the vapor pressure of liquid A increases as the mole fraction of A in the liquid increases, the temperature required to melt A also rises. Melting occurs along curve EM in Figure 10.3, giving an observed temperature range of $T_e$–$T_m$. A heterogeneous microscopic composition distribution causes a melting range because the more impure regions melt first.

It is possible for a binary mixture like A + B to have a more complex melting-point composition diagram than the one shown in Figure 10.3. There can be two eutectic points when the two components interact to form a molecular compound of definite composition. Discussions of more complex melting phenomena are found in the reference given at the end of the chapter.

Relatively pure compounds normally melt over a temperature range of 0.5°–2.0°, whereas impure substances often melt over a much larger range and also appear to "sweat" before melting. Until the advent of thin-layer chromatography and modern spectroscopy, the melting point was the primary index of purity for an organic solid. A melting point is still used as an effective, quickly determined, preliminary indication of purity.

## 10.2  Apparatus for Determining Melting Ranges

Several types of electrically heated melting-point devices are commercially available. Many undergraduate laboratories use the Mel-Temp apparatus shown in Figure 10.4. A thin-walled glass capillary tube holds the sample. The capillary tube fits into one of three sample chambers in the heating block; multiple chambers allow simultaneous determinations of three melting points. A cylindrical cavity in the top of the heating block holds the thermometer, a light illuminates the sample chamber, and an eyepiece containing a small magnifying lens facilitates observation of the sample. A rheostat controls the rate of heating by allowing continuous adjustment of the voltage. The higher the rheostat setting, the faster the rate of heating. Figure 10.5 shows graphically how the rate of heating changes at different rheostat settings. Heating at any particular setting occurs more rapidly at the start, then slows as the temperature increases. The decreasing rate of heating at the higher temperatures allows for the slower rate of heating needed as one approaches the melting point.

The Fisher-Johns hot-stage apparatus represents a second type of apparatus for the determination of melting points (Figure 10.6). The sample is crushed between thin, circular, microscope coverslips

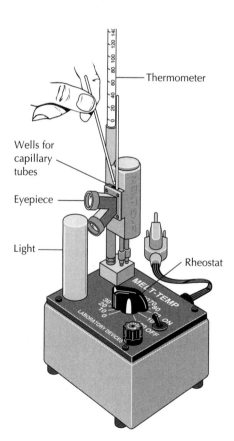

Thermometer

Wells for capillary tubes

Eyepiece

Light

Rheostat

**FIGURE 10.4**
Mel-Temp apparatus.
(Courtesy of Laboratory Devices, Inc., Holliston, MA.)

**FIGURE 10.5**
Heating rate curves for a Mel-Temp apparatus at various rheostat settings (indicated in volts). (Courtesy of Laboratory Devices, Inc., Holliston, MA.)

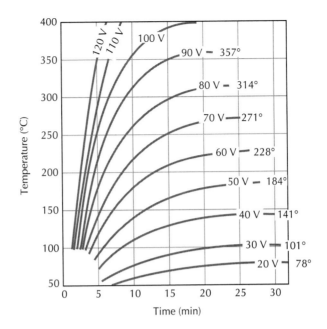

instead of being placed in a capillary tube. The coverslips fit in a depression in the metal block surface. A rheostat controls the rate of heating, and the lighted sample area is viewed through a small magnifying glass.

---

**SAFETY PRECAUTION**

If the heater on a Fisher-Johns apparatus is not turned off after the sample melts, the high heat may ruin the thermometer calibration or even break the thermometer. The latter event will lead to a spill of toxic mercury in the laboratory.

The Thomas-Hoover melting point apparatus is a third type of melting-point apparatus. In this instrument, capillary tubes holding the samples are submerged in an electrically heated oil bath. An

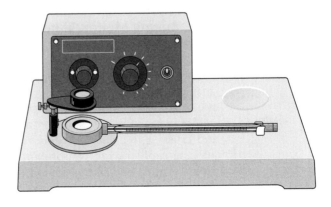

**FIGURE 10.6**
Fisher-Johns hot-stage melting-point apparatus. (Courtesy of Fisher Scientific, Pittsburgh, PA.)

**FIGURE 10.7**
Thomas-Hoover
melting-point
apparatus. (Courtesy
of Thomas Scientific,
Swedesboro, NJ.)

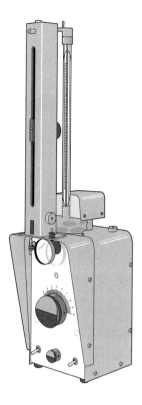

illuminated chamber holds the oil bath, and a magnifying glass
allows you to observe the samples (Figure 10.7).

**SAFETY PRECAUTION**

A Hoover-Thomas chamber with fouled oil (usually caused by
breakage of melting-point tubes) should not be used because toxic
fumes may be driven out by the heat.

## 10.3    Determining Melting Ranges

*Sample Preparation*

The melting range of an organic solid can be determined by intro-
ducing a small amount of the substance into a capillary tube with
one sealed end. Such tubes are commercially available. Place a few
milligrams of the dry solid on a piece of smooth-surfaced paper
and crush it to a fine powder with a spatula. Introduce the solid
into the capillary tube (approximately 1 mm in diameter) by tapping the open end of the tube in the powdered substance. A small
amount of material will stick in the open end. Invert the capillary

tube (the sealed end is now "down"), hold it very near the sealed end, and with quick motions tap the sealed end of the tube against the bench top. The solid will fall to the bottom of the tube. If the solid is still wet from a recrystallization, it will not fall to the bottom of the tube but will stick to the capillary wall. This failure to behave properly is probably a good thing, because melting points of wet solids are always low and thus nearly worthless. If your sample is still wet, allow it to dry completely before continuing with the melting range determination. The amount of solid in the tube should be about 1–2 mm in height. Melting-point determinations made with too much material produce a large melting range because more time is required to melt the complete sample as the temperature continues to rise.

*The ideal sample for a melting point is only 1–2 mm in height in the capillary tube.*

---

**SAFETY PRECAUTION**

Care must be taken while tapping the capillary tube against the bench top; it could break and cause a cut.

---

An alternative method for getting the solid to the bottom of the capillary tube is to drop it down a piece of glass tubing about 1 m in length or down the inside tube of your condenser, the bottom end of which is resting on the lab bench. After a few trips down the tube, the solid will have fallen to the bottom of the capillary tube.

Samples for the Fisher-Johns apparatus also need to be finely powdered. Put a few grains of the powdered sample on one coverslip and set the coverslip in the metal heating block. Place a second coverslip over the sample and gently flatten the powder until the two glass surfaces just touch each other; contact between the two coverslips ensures good heat transfer to the sample.

**Heating the Sample to the Melting Point**

The melting-point apparatus can be heated rapidly until the temperature is about 20°C below the expected melting point. Then decrease the rate of heating so the temperature rises only 1°–2° per minute and the sample has time to melt before the temperature rises above the true melting point. When you are taking successive melting points, remember that the apparatus needs to cool at least 20° below the expected melting point before it can be used for the next determination.

If you do not know the approximate melting point of a solid sample, you can make a quick preliminary determination by heating the sample rapidly and watching for the temperature at which melting begins. In a more accurate second determination, you can carefully control the temperature rise to 1°–2° per minute when you get to within 15°–20° of the expected melting point.

*Always prepare a fresh sample for each melting-point determination.*

Always prepare a fresh sample for each melting-point determination; many organic compounds decompose at the melting point, thus making reuse of the solidified sample invalid. Moreover, many low-melting-point samples (mp 25°–80°C) do not resolidify with cooling.

**Reporting
the Melting Range**

Unless you have an extraordinarily pure compound in hand, you will always observe and report a melting range—from the temperature where the first drop of liquid appears to the temperature where the solid is completely melted and only a clear liquid is present. This melting range is usually 1°–2° or slightly more. For example, salicylic acid that you have synthesized may give a melting range of 156°–159°C. An extremely pure sample of salicylic acid melting over less than a 1° range (for example, 160.0°–160.5°C) may have 160°C listed as its melting point. Published melting points are the highest values obtained after multiple recrystallizations; your observed values will probably be slightly lower.

**Sources of Error
in Melting-Point
Determinations**

When you heat a sample for a melting-point determination, you may see some strange and wonderful things happen before the first drop of liquid actually appears. The compound may soften and shrivel up as a result of changes in crystal structure. It may "sweat out" some solvent of crystallization. It may decompose, changing color as it does so. None of these changes should be called melting. **Only the appearance of liquid indicates the onset of true melting.** It can be difficult to distinguish exactly when melting does start. In fact, even with careful heating, two people may disagree by as much as 1°–2°.

*Only the appearance of
liquid indicates the
onset of true melting.*

There are other sources of error in a melting-point determination. Heating faster than 1°–2° per minute may lead to an observed melting range that is higher than the correct value. And if the rate of heating is extremely rapid (>10° per minute), you may observe thermometer lag, a condition caused by the failure of the mercury in the thermometer to rise as quickly as the temperature of the metal heating block. This error causes the observed melting range to be lower than it actually is. Determining accurate melting points requires patience.

*In electrical melting-
point devices, there is a
significant time lag
between a change in the
rheostat setting and
onset of the new rate
of heating.*

**Sublimation**

Another possible complication in melting-point determinations occurs if the sample sublimes. Sublimation is the change that occurs when a solid is transformed directly to a gas, without passing through the liquid phase [see Technique 12]: The sample in the capillary tube sublimes and disappears as heat is applied. Many common substances, such as camphor and caffeine, sublime, but you can determine their melting points by sealing the open end of the capillary tube in a Bunsen burner flame before it is placed in the melting-point apparatus (Figure 10.8a).

**Decomposition**

Some compounds decompose as they melt, a behavior usually indicated by a change in color of the sample to dark red or brown. The melting point of such a compound is reported in the literature with the symbol "d" after the temperature, for example, 186°C d, meaning that the compound melts at 186°C with decomposition. Sometimes decomposition occurs as a result of a reaction between the compound and oxygen in the air. If air is evacuated from the capillary tube and the tube is sealed, the melting point can be determined without decomposition (Figure 10.8b). Place the sample in

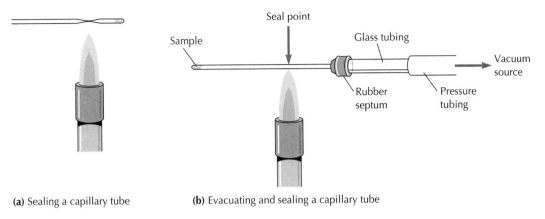

(a) Sealing a capillary tube    (b) Evacuating and sealing a capillary tube

**FIGURE 10.8** Methods for sealing a capillary tube with a Bunsen burner.

the capillary tube as directed earlier. Punch a hole in a rubber septum with a small nail, and from the inside insert the sealed end of the capillary tube through the septum. Fit the septum over a piece of glass tubing that is connected to a vacuum line. Turn on the vacuum source and hold and pull on the sample end of the capillary while heating the upper portion in a Bunsen burner flame until it seals.

---

**SAFETY PRECAUTION**

Be sure no flammable solvents are in the vicinity when you are using a Bunsen burner.

## 10.4    Mixture Melting Point

We have already discussed how impurities can lower the melting point of a compound. This behavior can be useful not only in evaluating a compound's purity but also in helping to identify it. Let us assume that two compounds have virtually identical melting ranges. Are the compounds identical? Possibly, but not necessarily, because the identical melting ranges may be simply a coincidence.

If roughly equal amounts of the two compounds are finely ground together with a spatula, the melting point of the mixture can provide useful information. If there is a melting-point depression or if the melting range is expanded by a number of degrees, it is reasonably safe to conclude that the two compounds are not identical. One of them has acted as an impurity toward the other by lowering the melting range. If there is no lowering of the mixture's melting range relative to the melting range of each compound, the two are very likely to be the same compound.

Sometimes only a modest melting-point depression is observed. To know whether this change is significant, the mixture melting point and the melting point of one of the two compounds should be determined simultaneously in separate capillary tubes. This experiment allows simultaneous identity and purity checks. Infrequently, a eutectic point (E; see Figure 10.3) can be equal to the melting point of the pure compound of interest. In a case where you have accidentally used the eutectic mixture, purity would be incorrectly suggested by a mixture melting point. Errors of this type can be discerned by testing various mixtures other than a 1:1 composition. Preparing 1:2 and 2:1 mixtures should allow you to avoid eutectic-point-induced misinterpretation.

Other ways of determining the identity of an organic compound involve spectroscopic methods [see Techniques 18–20] and thin-layer chromatography [see Technique 15].

## 10.5     Thermometer Calibration

The accuracy of your melting-point determinations can be no better than the accuracy of your thermometer. Often one simply assumes that the thermometer has been accurately calibrated. Although frequently this is the case, it is not always true. Thermometers can give high or low temperature readings of 1°–2° or more.

A thermometer can be calibrated with a series of compounds that are readily available in the pure state and whose melting points are easy to reproduce. A useful series of such compounds is given in the following table.

| Compound | Melting point, °C |
|---|---|
| Water | 0 |
| Benzoic acid | 122 |
| Salicylic acid | 160 |
| 3,5-Dinitrobenzoic acid | 205 |

The melting point of ice can be determined by simply measuring the temperature of a beaker of ice water; the others are done in the usual way in capillary tubes. The boiling points of acetone (56°C) and water (100°C) are also useful calibration reference points [see Technique 11.1], provided atmospheric pressure is taken into consideration. You may want to record the temperature deviation of your thermometer using the melting point or boiling point for a number of the suggested compounds and make a graph of thermometer corrections. Plot observed temperature against temperature correction and interpolate to correct future determinations with your thermometer. Usually these plots are linear.

| **10.6** | **Summary of Melting-Point Determination** |

1. Introduce the powdered, dry solid sample into a capillary tube that is sealed at one end.
2. Place the capillary tube in the melting-point apparatus.
3. Adjust the rate of heating so that the temperature rises at a moderate rate. The rate can be faster if, for example, the melting point is 170°C rather than 70°C.
4. When a temperature 15°–20° below the expected melting point is reached, decrease the rate of heating so that the temperature rises only 1°–2° per minute. **Note:** In electrically heated devices, there is a time lag before the rate of heating changes.
5. If the temperature is rising more than 1°–2° per minute at the time of melting, retake the melting point using a new sample.
6. Record the melting range as the range of temperatures that begins with the onset of melting and ends with the temperature at which only liquid remains in the tube. Do not confuse "sweating" with the onset of melting.

## Reference

1. Skau, E. L.; Arthur, J. C. Jr. In *Physical Methods of Chemistry*, A. Weissberger and B. W. Rossiter, eds.; Wiley-Interscience: New York, 1971, vol. 1, Part V.

## Questions

1. A student performs two melting-point determinations on a crystalline product. In one determination, the capillary tube contains a sample about 1–2 mm in height and the melting range is found to be 141°–142°C. In the other determination, the sample height is 4–5 mm and the melting range is found to be 141°–145°C. Explain the broader melting-point range observed for the second sample. The reported melting point for the compound is 143°C.

2. Another student reports a melting point of 136°–138°C for the melting range of the unknown in Question 1 and mentions in his/her notebook that the rate of heating is about 12° per minute. NMR analysis of this student's product does not reveal any impurities. Explain the low melting point.

3. A white crystalline compound melts at 111°–112°C and the melting-point capillary is set aside to cool. Repeating the melting-point analysis, using the same capillary, reveals a much higher melting point of 140°C. Yet repeated recrystallization of the original sample yields sharp melting points no higher than 114°C. Explain the behavior of the sample that was cooled then remelted.

4. A compound melts at 120°–122°C on one apparatus and at 128°–129°C on another. Unfortunately, neither apparatus is calibrated. How might you check the identity of your sample without calibrating either apparatus?

5. Why does sealing the open end of a melting-point capillary tube allow you to measure the melting point of a compound that sublimes?

# 11

# BOILING POINTS AND DISTILLATION

Distillation is a method for separating two or more liquid compounds on the basis of boiling-point differences. Unlike the liquid-liquid and solid-liquid separation techniques of extraction and crystallization, distillation is a liquid-gas separation in which vapor pressure differences are used to separate materials.

A liquid at any temperature exerts a pressure on its environment. This pressure, called the *vapor pressure*, results from molecules leaving the surface of the liquid to become vapor.

$$\text{molecules}_{\text{liquid}} \rightleftharpoons \text{molecules}_{\text{vapor}}$$

As a liquid is heated, its kinetic energy increases; the equilibrium shifts to the right and more molecules move into the gaseous state, thereby increasing the vapor pressure. Figure 11.1 shows the relationship between vapor pressure and temperature for benzene, water, and *tert*-butylbenzene.

## 11.1 Determination of Boiling Points

The *boiling point* of a pure liquid is defined as the temperature at which the vapor pressure of the liquid exactly equals the pressure exerted on it by the atmosphere. At an external pressure of 1.0 atm (760 Torr), the boiling point is reached when the vapor pressure equals 1.0 atm.

The boiling point of a liquid depends on the atmospheric pressure. Table 11.1 gives boiling points of several common solvents in Laramie, Wyoming (elevation 7520 ft), at the experimental station of the U.S. Department of Agriculture in Death Valley, California (elevation −285 ft), and in New York City (elevation 0 ft). When the boiling point of a substance is determined, both the atmospheric pressure and the boiling point need to be recorded.

Every pure and stable organic compound has a characteristic boiling point at 1 atm. The boiling point reflects its molecular structure, specifically the types of intermolecular interactions that bind the molecules together in the liquid state. Intermolecular interactions must be overcome for molecules to leave the liquid state and vaporize. Polar compounds have higher boiling points than nonpolar compounds of similar molecular weight. Increased molecular weight usually produces a larger molecular surface area and leads to a higher boiling point, provided the polarities remain constant.

*Miniscale Determination of Boiling Points*

The boiling point of 5 mL or more of a pure liquid compound can be determined by a simple distillation using miniscale standard taper glassware. The procedure for setting up a simple distillation

**FIGURE 11.1**
Examples of the
dependence of vapor
pressure on temperature.

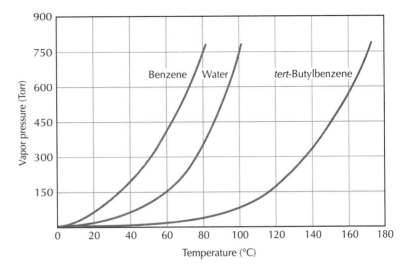

is described in Technique 11.3. When distillate is condensing
steadily and the temperature stabilizes, the boiling point of the substance has been reached.

Rather than use all the sample for a distillation, the microscale
procedure described next can be used to determine the boiling
point of any pure liquid and requires only 0.3 mL of the liquid.

*Microscale
Determination
of Boiling Points*

Place 0.3 mL of the liquid and a boiling stone in a Craig tube or a
reaction tube. Set the tube in the appropriate-size hole of an aluminum heating block [see Technique 6.2]. Alternatively, heat may
be supplied by a sand bath [see Technique 6.2], in which case the
tube and the thermometer need to be held by separate clamps.
Clamp the thermometer so that the bottom of the bulb is about
0.5 cm above the surface of the liquid, being sure that the thermometer does not touch the wall of the tube (Figure 11.2).

Gradually heat the sample to boiling and continue to increase
the rate of heating *slowly* until the ring of condensate is 1–2 cm
above the top of the thermometer bulb. When the temperature
reaches a maximum and stabilizes for at least 1 min, you have
reached the boiling point of the liquid. Rapid or excessive heating

| T A B L E   1 1 . 1 | **Boiling points of common compounds at different elevations (pressures)** | | |
|---|---|---|---|
| Compound | Death Valley, CA<br>Elevation = −285 ft<br>P = 1.01 atm | New York City<br>Elevation = 0 ft<br>P = 1.00 atm | Laramie, WY<br>Elevation = 7520 ft<br>P = 0.75 atm |
| Water | 100.3 | 100.0 | 93 |
| Diethyl ether | 35.0 | 34.6 | 27 |
| Benzene | 80.8 | 80.2 | 73 |
| Acetic acid | 119.0 | 118.8 | 110 |

**FIGURE 11.2**
Apparatus for
microscale boiling-point
determinations.

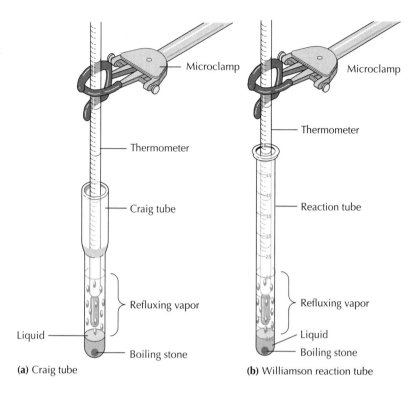

(a) Craig tube          (b) Williamson reaction tube

of the tube can lead to superheating of the vapor and can also radiate heat from the tube to the thermometer bulb, causing the observed boiling point to be too high.

---

## 11.2     Boiling Points and Separation of Mixtures

The boiling point of a mixture depends on the vapor pressures of its components. Impurities can either raise or lower the observed boiling point of a sample, but in either case a substance that boils over a range of several degrees in temperature is usually not pure.

***Separation of a Binary Solution by Distillation***

Consider, for example, the boiling characteristics of a solution of pentane and hexane. Pentane and hexane are mutually soluble, and their molecules interact with one another only by van der Waals forces, which are the weakest intermolecular interactions. A solution composed of both pentane and hexane boils at temperatures intermediate between the boiling points of pentane (36°C) and hexane (69°C). If pentane alone were present, the vapor pressure above the liquid would be due only to pentane. However, with pentane as only a fraction of the solution, the partial pressure exerted by pentane ($P$) is equal to only a fraction of the vapor pressure of pure pentane ($P°$). The fraction is determined by $X_{pentane}$, the *mole fraction* of pentane, which is the ratio of moles of pentane to the total number of moles of pentane and hexane in the solution.

Mole fraction of pentane: $X_{pentane} = \dfrac{moles_{pentane}}{moles_{pentane} + moles_{hexane}}$

Partial pressure of pentane: $P_{pentane} = P°_{pentane}\, X_{pentane}$ (1)

The hexane present in the solution exerts its own independent partial vapor pressure.

Mole fraction of hexane: $X_{hexane} = \dfrac{moles_{hexane}}{moles_{pentane} + moles_{hexane}}$

Partial pressure of hexane: $P_{hexane} = P°_{hexane}\, X_{hexane}$ (2)

*Raoult's law applies only to liquids miscible in one another.*

The vapor pressure-mole fraction relationships expressed in equations 1 and 2 are valid only for ideal liquids in the same way that the ideal gas law strictly applies only to ideal gases. Equations 1 and 2 are applications of Raoult's law, named after the French chemist François Raoult, who studied the vapor pressures of solutions in the late nineteenth century.

Using Dalton's law of partial pressures, we can now calculate the total vapor pressure of the solution, which is the sum of the partial vapor pressures of the individual components:

$$P_{total} = P_{pentane} + P_{hexane} \quad (3)$$

Figure 11.3 shows the partial-pressure curves for pentane and hexane using Raoult's law and the total vapor pressure of the solution using Dalton's law. The boiling point of a pentane/hexane mixture is the temperature at which the individual vapor pressures of both pentane and hexane add up to the total pressure exerted on the liquid by its surroundings.

Being able to calculate the total vapor pressure of a solution can be extremely useful to a chemist; knowing the composition of the vapor above a solution can be just as important. Qualitatively, it is not hard to see that the vapor above a 50:50 pentane/hexane solution will be richer in pentane as a result of its greater volatility and

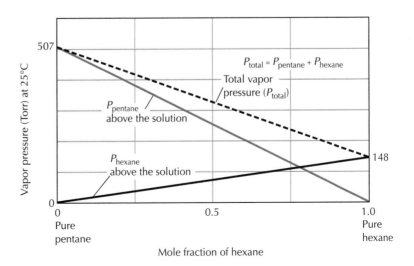

**FIGURE 11.3**
Vapor pressure-mole fraction diagram for pentane/hexane solutions at 25°C.

vapor pressure. Quantitatively, we can predict the composition of the vapor above a solution, for which Raoult's law is valid, simply by knowing the vapor pressures of its volatile components and the composition of the liquid solution.

Here is an illustration of how it is done. Applying the ideal gas law to the mixture of gases above a solution of pentane and hexane, we have equation 4. The quantity $Y_{pentane}$ is the *fraction of pentane molecules in the vapor* above the solution.

$$Y_{pentane} = \frac{P_{pentane}}{P_{total}} \qquad (4)$$

A single expression for the total vapor pressure (equation 5) can be derived easily from equations 1, 2, and 3, because $X_{hexane} = 1.0 - X_{pentane}$.

$$P_{total} = X_{pentane} (P°_{pentane} - P°_{hexane}) + P°_{hexane} \qquad (5)$$

Finally, the combination of equations 1, 4, and 5 allows the calculation of the mole fraction of pentane in the vapor state.

$$Y_{pentane} = \frac{P°_{pentane} X_{pentane}}{X_{pentane} (P°_{pentane} - P°_{hexane}) + P°_{hexane}} \qquad (6)$$

Thus, if the vapor pressures of pure pentane and pure hexane at various temperatures and the composition of the liquid are known, the fraction of pentane in the vapor above the solution can be calculated. This kind of calculation can be used to construct a temperature-composition diagram (sometimes called a phase diagram) like the one shown in Figure 11.4. A similar diagram can also be constructed directly from experimental data.

It is useful to follow the dashed line in Figure 11.4, moving from initial liquid composition $L_1$ to initial vapor composition $V_1$ to $L_2$, and so on. Point $L_1$ indicates a boiling point of 44°C at atmospheric pressure for a solution containing a 1:1 molar ratio of pentane to hexane. Analysis of vapor composition at $V_1$ reveals a molar composition of 87% pentane and 13% hexane, as indicated by point

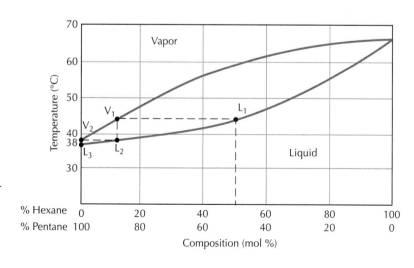

**FIGURE 11.4**
Calculated temperature-composition diagram for pentane/hexane solutions at 1.0 atm pressure.

$V_1$. The mole fraction of the component with the lower boiling point is greater in the vapor than in the liquid. Now, if the vapor at $V_1$ condenses, the liquid that collects ($L_2$) will have the same composition as the vapor ($V_1$). If the condensed liquid ($L_2$) is vaporized, the new vapor will be even richer in pentane, point $V_2$. Repeating the boiling and condensing process several more times allows us to obtain essentially pure pentane.

As pentane-enriched vapor is removed, the remaining liquid contains a decreasing proportion of pentane. The liquid, originally at $L_1$, now is richer in hexane (the component with the higher boiling point). As the mole fraction of hexane in the liquid increases, the boiling point of the liquid also increases until the boiling point of pure hexane, 69°C, is reached. In this way pure hexane can also be collected. The process of repeated vaporizations and condensations, called *fractional distillation,* allows us to separate components of a mixture by exploiting the vapor pressure differences of the components.

## 11.3    Simple Distillation

In a *simple distillation,* only one or two vaporizations and condensations occur, corresponding to points $L_1$ and $V_1$ in Figure 11.4. The condensed liquid is called the *distillate* or *condensate.* Simple distillation would not effectively separate a mixture such as 1:1 molar pentane and hexane. As the distillation proceeds, the remaining pentane/hexane mixture becomes increasingly more concentrated in hexane and less concentrated in pentane. Consequently, the boiling point of the mixture continues to increase. Figure 11.5 shows a distillation curve of vapor temperature versus volume of distillate for the simple distillation of a 1:1 pentane/hexane solution. The initial distillate is collected at a temperature above the boiling point of pure pentane and the final distillate never reaches the boiling point of pure hexane, a result indicating a poor separation of the two compounds.

Even though simple distillation does not effectively separate a mixture of liquids whose boiling points differ by less than

**FIGURE 11.5**
Distillation curve for simple distillation of a 1:1 molar solution of pentane and hexane.

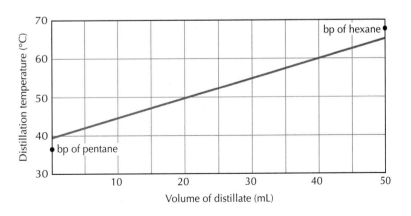

60°–70°C, organic chemists use simple distillations in two commonly encountered situations: (1) when the last step in the purification of a liquid compound involves a simple distillation to obtain the pure product and determine its boiling point; (2) when simple distillation is used to remove a low-boiling solvent from a dissolved organic compound with a high boiling point.

In a simple distillation, **the distilling flask should be only between one-third and one-half full of the liquid being distilled.** With a flask that is too full, liquid can easily bump over into the condenser. If the flask is nearly empty, a substantial fraction of the material, called the *holdup volume,* will be needed just to fill the flask and distilling head with vapor along with a thin liquid film on the glass surfaces. When the desired liquid is dissolved in a large quantity of a solvent with a lower boiling point, the distillation should be interrupted after the solvent has distilled, and the liquids with higher boiling points should be poured into a smaller flask before continuing the distillation. Figure 11.6 shows the miniscale apparatus for a simple distillation.

*Steps in Assembling
a Miniscale Apparatus
for Simple Distillation*

*A funnel keeps the
ground glass joint from
becoming coated with
the liquid and prevents
loss of product.*

1. Select a round-bottomed flask of a size that will be one-third to one-half filled with the liquid being distilled. Place a clamp firmly on the neck of the flask and attach the clamp to a ring stand or support rod. Using a conical funnel, pour the liquid into the flask. Add one or two boiling stones.

---

SAFETY  PRECAUTION

> Boiling stones should **never** be added to a hot liquid because this may cause a superheated liquid to boil violently out of the flask.

2. *Lightly* grease the bottom joint and the side-arm joint [see Technique 2.2] on the distilling head. Fit the distilling head to the round-bottomed flask and twist the joint to achieve a tight seal. Finish assembling the rest of the apparatus *before* inserting the thermometer adapter and thermometer. **Note: The distilling flask and distilling head need to be in a completely vertical position and the condenser should be positioned with a downward slant.**

3. Attach rubber tubing to the outlets on the condenser jacket. Wire hose clamps are often used to prevent water hoses from being blown off the outlets by a surge in water pressure. Grease the inner joint on the bottom of the condenser, attach the vacuum adapter, and while the pieces are lying on the desk top, place a joint clip over the joint. Clamp the condenser to another ring stand or upright support rod, as shown in Figure 11.6. Fit the upper joint of the condenser to the distilling head, twist to spread the grease, and place a joint clip over the joint.

*The use of joint
clips ensures that
ground glass joints do
not come apart.*

4. Figure 11.6 shows a round-bottomed flask serving as the receiving vessel. Depending on the particular procedure being

**FIGURE 11.6**
Simple distillation apparatus. The enlargement shows the correct placement of the thermometer bulb for accurate measurement of the boiling point.

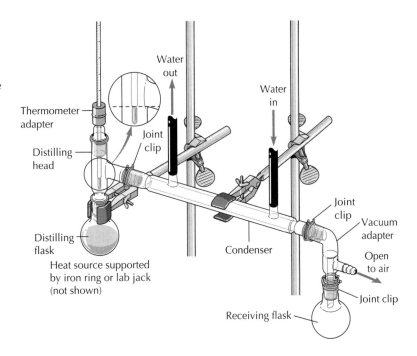

carried out, an Erlenmeyer flask or a graduated cylinder may be substituted for the round-bottomed flask. Position the Erlenmeyer flask or graduated cylinder so that the outlet of the vacuum adapter is slightly inside the mouth of the receiving vessel. A beaker should not be used as the receiving vessel because its wide opening readily allows vapor (product) to escape. It is usually necessary to have at least two receiving vessels at hand; the first container is for collecting the initial distillate that consists of impurities with low boiling points before the expected boiling point of the desired fraction is attained.

5.  Gently push the thermometer through the rubber sleeve on the thermometer adapter.

---

**SAFETY PRECAUTION**

Holding the thermometer by the upper part of the stem while inserting it through the rubber sleeve of thermometer adapter could break the thermometer and force a piece of broken glass into your hand. Grasp the thermometer close to the bulb and push it gently 1–2 cm into the adapter. Move your hand several centimeters up the thermometer stem and repeat the pushing motion; continue this process until the thermometer is properly positioned.

*The position of the thermometer bulb is crucial in obtaining an accurate boiling point. If it is positioned too high, it is not completely surrounded by vapor and the observed boiling point will be lower than the true boiling point.*

6.  Grease the joint on the thermometer adapter and fit it into the top joint of the distilling head. Adjust the position of the thermometer to **align the top of the thermometer bulb with the**

bottom of the side arm on the distilling head (see detail in Figure 11.6). Alternatively, a thermometer with a standard taper fitting may be used instead of the thermometer and rubber-sleeved adapter.

*A slow to moderate flow rate for the condenser water is usually sufficient and lessens the chance of blowing the rubber tubing off the condenser.*

7.  Check to ensure that the rubber tubing is tightly attached to the condenser and that **water flows in at the bottom and out at the top.** Turn on the water.

8.  Place a heating mantle under the distillation flask, using an iron ring or lab jack to support the mantle, and begin heating the flask.

### Carrying Out the Distillation

The expected boiling point of the liquid being distilled determines the heat input, controlled by a variable transformer [see Technique 6.2]; a liquid with a high boiling point requires more heat to vaporize it than a liquid with a low boiling point. Heat the liquid slowly to a gentle boil. A ring of condensate begins to move up the inside of the flask and then up the distilling head. The temperature observed on the thermometer does not rise appreciably until the ring of vapor reaches the thermometer bulb because it is measuring the vapor temperature, not the temperature of the boiling liquid. If the ring of vapor stops moving before it reaches the thermometer, increase the setting on the variable transformer.

When the vapor reaches the thermometer, the temperature reading increases rapidly. Collect any liquid that condenses below the expected boiling point as the first fraction, or *forerun*—which is usually discarded—then change to a second receiving vessel to collect the desired fraction when the temperature stabilizes at or slightly below the expected boiling point of the liquid. Record the temperature at which you begin to collect the desired fraction. Adjust the heat input to maintain a distillate collection rate of 1 drop every 1–2 s. It may be necessary to increase the heat input during the distillation if the rate of distillate collection slows. **It is essential to stop the distillation by lowering the heat source before the distillation flask reaches dryness** or when the temperature either begins to climb above the expected boiling range or begins to drop. Record the temperature at which the last drop of distillate is collected; the initial and final temperatures for the main fraction are the boiling range.

---

**SAFETY PRECAUTION**

A distillation flask should **never** be allowed to reach dryness. By leaving a small residue of liquid in the boiling flask, you will not overheat the flask and break it, nor will you char the last drops of residue, which causes cleaning difficulty. Moreover, some compounds such as ethers, secondary alcohols, and alkenes, form peroxides by air oxidation. If a distillation involving one of these compounds is carried to dryness, the peroxides could explode.

### 11.3a Miniscale Short-Path Distillation

When only 4–6 mL of liquid are distilled, a simple distillation apparatus is modified to a short path by omitting the condenser, as shown in Figure 11.7. The short path reduces the *holdup volume,* the amount of space that must be filled with vapor, and also prevents distillate (product) from being lost on the walls of the condenser. A beaker or crystallizing dish of water surrounding the receiving flask replaces the condenser. If the liquid boils between 50°C and 100°C, the beaker should contain an ice-water mixture; in this case, it may be necessary to attach a drying tube to the side arm of the vacuum adapter to prevent moisture from condensing inside the receiving flask. If the liquid boils above 100°C, tap water provides sufficient cooling. For liquids that boil above 150°C, air cooling of the receiving flask will suffice.

Carry out the distillation as described in Technique 11.3 for a simple distillation, but do the short-path distillation at a rate of less than 1 drop per second. If the receiving flask is being cooled by a water bath, it may be necessary to stop the distillation by removing the heat source while changing receiving flasks.

### 11.3b Microscale Distillation Using Standard Taper Apparatus

Microscale apparatus is required when the volume of a liquid to be distilled is only a few milliliters. Standard taper microscale glassware can be assembled into a short-path distillation apparatus with

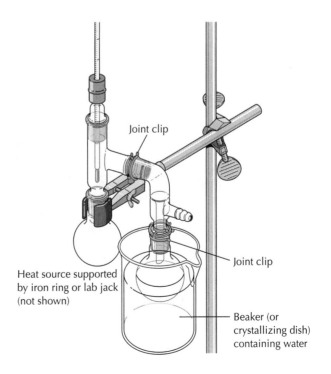

Joint clip

Joint clip

Heat source supported by iron ring or lab jack (not shown)

Beaker (or crystallizing dish) containing water

**FIGURE 11.7**
Short-path distillation apparatus.

**FIGURE 11.8**
Thermometer adapters for ⊤ 14/10 microscale glassware.

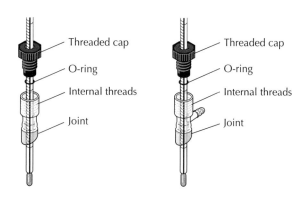

**(a)** Thermometer adapter      **(b)** Thermometer/vacuum

a 14/10 distillation head, thermometer adapter (Figure 11.8), and bent vacuum adapter, as shown in Figure 11.9.* For the distilling vessel, use a conical vial for 1–3 mL of liquid or a 10-mL round-bottomed flask for 4–5 mL of liquid. For distillation of very volatile liquids, a water-jacketed condenser can be inserted between the

* Standard taper 14/10 distillation heads, thermometer adapters, and vacuum take-off adapters are commercially available, but not part of basic microscale kits.

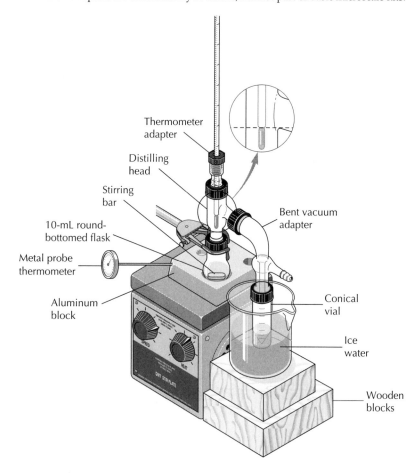

**FIGURE 11.9**
Short-path standard taper microscale distillation apparatus.

**FIGURE 11.10**
Hickman distilling heads. The condensate collects in the well at the bottom of the head in both versions.

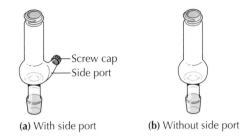

**(a)** With side port    **(b)** Without side port

distilling head and the vacuum adapter in the same manner as used in a simple distillation apparatus.

*Hickman Distilling Head*

Another type of standard taper microscale distillation apparatus consists of a Hickman distilling head (Figure 11.10) and a 3-mL or 5-mL conical vial or a 10-mL round-bottomed flask. The Hickman distilling head serves as both condenser and receiving vessel, an arrangement that considerably reduces the holdup volume. Vapors condense on the upper portion of the Hickman still and drain into the bulbous collection well. One version of the Hickman still has a port at the side for easy removal of the condensate (Figure 11.10a).

To carry out a microscale distillation, select a conical vial or 10-mL round-bottomed flask appropriate for the volume of liquid to be distilled; the vessel should be no more than two-thirds full.

*Grease is not used on the ground glass joints of microscale glassware because its presence could contaminate your product.*

Place the liquid in the vial and add a magnetic spin vane or a boiling stone. Attach the Hickman distilling head to the vial with a screw cap and O-ring. Usually an air condenser or a water-cooled condenser (for particularly volatile liquids) is placed above the Hickman distilling head to minimize the loss of vapor (Figure 11.11).

Clamp the assembled apparatus at the Hickman distilling head, with the vial placed in an aluminum heating block. If you are using a spin vane, turn on the magnetic stirrer. Begin heating the aluminum block slowly to a temperature 20°–30°C above the boiling point of the liquid being distilled. Position a thermometer inside the condenser and the Hickman distilling head, with the top of the thermometer bulb aligned with the bottom of the head's collection well, as shown in Figure 11.11. Clamp the thermometer firmly above the condenser. (**Note:** The inside diameter of some water-jacketed condensers is too small to accommodate a thermometer; in this case, you can ascertain the boiling point of the distillate by a microscale boiling-point determination [see Technique 11.1].)

After the liquid in the vial boils, you should notice a ring of condensate slowly moving up the vial and into the Hickman still.

*It may be necessary to wrap the distillation vial loosely with glass wool to prevent rapid heat loss, but do not cover the well of the Hickman still.*

The temperature observed on the thermometer rises as the vapor reaches the thermometer bulb. You may also see the upper neck of the Hickman still become wet and shiny as the vapor condenses and begins to fill the well. The distillation should be done at a rate slow enough to allow the vapor to condense and not evaporate out of the system.

The collection well has a capacity of about 1 mL, so the distillate may need to be removed once or twice during a distillation.

**FIGURE 11.11**
Standard taper apparatus for a microscale distillation using a Hickman distilling head with a side port.

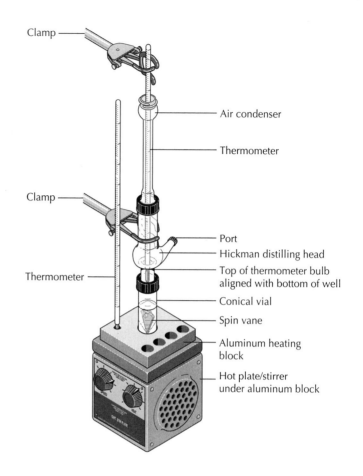

Clamp

Air condenser

Thermometer

Clamp

Port
Hickman distilling head
Top of thermometer bulb aligned with bottom of well
Conical vial
Spin vane
Aluminum heating block
Hot plate/stirrer under aluminum block

Thermometer

Open the port and remove the distillate with a clean Pasteur pipet or a syringe inserted through the plastic septum in the screw cap of the port.

## 11.3c Microscale Distillation Using Williamson Apparatus

Microscale apparatus is required when the volume of liquid to be distilled is only a few milliliters. The Williamson microscale distillation apparatus is essentially a miniature version of the standard taper short-path distillation apparatus. The apparatus consists of a 5-mL or 10-mL round-bottomed flask and a distillation head connected by a flexible connector with a support rod. The thermometer is held in place by the flexible thermometer adapter, as shown in Figure 11.12. The distillate is collected in a small vial that is at least three-fourths submerged in a 50-mL beaker of ice and water.

To carry out a microscale distillation, select a 5-mL or 10-mL round-bottomed flask appropriate for the volume of liquid to be distilled; the flask should be no more than two-thirds full. Using a Pasteur pipet, transfer the liquid to the flask and add a magnetic

**FIGURE 11.12**
Williamson microscale
distillation apparatus.

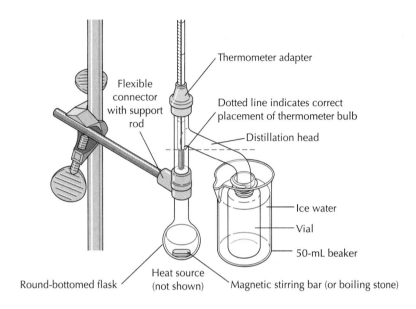

stirring bar or a boiling stone. Attach the flexible connector with support rod to the flask and clamp the rod to a vertical support rod or ring stand.

Fit the flexible thermometer adapter to the top of the distilling head and carefully push a thermometer through the adapter.

---
**SAFETY PRECAUTION**

Holding the thermometer by the upper part of the stem while inserting it through the rubber sleeve of the thermometer adapter could break the thermometer and force a piece of broken glass into your hand. Grasp the thermometer close to the bulb and push it gently 1–2 cm into the adapter. Move your hand several centimeters up the thermometer stem and repeat the pushing motion; continue this process until the thermometer is properly positioned.

---

The top of the thermometer bulb should be placed just below the side arm, as shown by the dashed line drawn across the distillation head in Figure 11.12. Fit the distillation head into the flexible connector holding the distillation flask. Place the receiving vial in a 50-mL beaker of ice and water, and position the vial under the outlet of the distillation head as far as it will go. Put a sand bath or an aluminum heating block with a flask depression under the round-bottomed flask. The temperature of the sand bath or aluminum block needs to be 20°–50°C above the boiling point of the liquid being distilled.

After the liquid in the flask boils, you should notice a ring of condensate slowly moving up the flask and into the distillation head. The temperature observed on the thermometer rises as the vapor reaches the thermometer bulb. The distillation should be done at a slow enough rate for the vapor to condense and not evaporate out of the system. It may be necessary to wrap a wet pipe cleaner or a wet paper towel around the side arm of the distillation head to increase its cooling efficiency, particularly for the distillation of compounds boiling below 100°C.

| **11.4** | **Fractional Distillation** |

*Fractionating Columns*

*Fractional distillation,* a process in which repeated vaporizations and condensations occur, allows us to separate components of a mixture by utilizing the vapor pressure differences of the components. In a fractional distillation a fractionating column is inserted between the distillation flask and the distilling head of a simple distillation apparatus. The use of a fractionating column in the distillation apparatus provides a large surface area over which a number of separate liquid-vapor equilibria can occur. As vapor travels up a column, it cools, condenses into a liquid, then vaporizes again after it comes into contact with hotter vapor rising from below. The process can be repeated many times. If the fractionating column is efficient, the vapor that finally reaches the distilling head at the top of the column will be composed entirely of the component with the lower boiling point. Thus we have carried out the process shown by the dashed line in Figure 11.4.

The efficiency of a fractionating column is expressed as its number of *theoretical plates*—a term best defined with reference to Figure 11.4. Assume that the original solution being distilled has a 1:1 molar ratio of pentane to hexane. A fractionating column would have one theoretical plate if the liquid that distills from the top of the column has the composition $L_2$. In other words, a column has one theoretical plate if one complete vaporization of the original solution followed by condensation of the vapor occurs in the column. The column would have two theoretical plates if the liquid that distills has the composition $L_3$; notice that $L_3$ is already 98% pentane and only 2% hexane. Starting with a 1:1 solution, Figure 11.4 indicates that a column with three theoretical plates would seem sufficient to obtain essentially pure pentane, $L_3$, from the 1:1 pentane-hexane mixture present at the start of the distillation. However, as the distillation progresses, the residue in the boiling flask becomes richer in hexane, so more theoretical plates are required for complete separation of the two compounds.

Fractionating columns that can be used to separate two liquids boiling at least 25°C apart are shown in Figure 11.13. The larger the column surface area on which liquid-vapor equilibria can occur, the more efficient the column will be. The fractionating columns shown in Figure 11.13 have from two to eight theoretical plates. A fractionating column with two theoretical plates can be used to separate liquids with boiling points differing by about 70°C; an

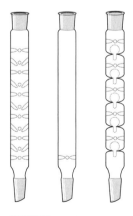

**FIGURE 11.13**
Examples of fractionating columns.

eight-theoretical-plate column can separate liquids boiling only 25°C apart.

More efficient columns can be made by packing a simple fractionating column with a wire spiral, glass helixes, metal sponge, or thin metal strips. These packings provide additional surface area on which liquid-vapor equilibria can occur. Care must be used in selecting packing materials to ensure that the packing does not undergo chemical reaction with the hot liquids in the fractionating column. Among the most efficient fractionating columns are those with helical bands of Teflon mesh that spin at thousands of rotations per minute. Spinning-band columns can have more than 150 theoretical plates and can be used to separate liquids having a boiling-point difference of only a degree or two.

In a fractional distillation, the composition of the vapor phase at the top of the column as it encounters the thermometer bulb and moves into the side arm determines the composition of the liquid that forms in the condenser and collects in the receiving flask. Let us illustrate this with a fractional distillation of a 1:1 molar solution of pentane and hexane. If the fractionating column has enough theoretical plates to completely separate the two compounds, the initial condensate will appear when the temperature is very close to 36°C, the boiling point of pure pentane. The observed boiling point will remain essentially constant while all the pentane distills into the receiving vessel. Then the boiling point will rise rapidly to 69°C, the boiling point of hexane. Figure 11.14 shows a distillation curve for the fractional distillation of pentane and hexane.

**Assembling a Fractional Distillation Apparatus**

The distilling flask capacity should be two to three times as large as the volume of liquid being distilled. When the desired material is contained in a large quantity of a solvent with a lower boiling point, the distillation should be interrupted after the solvent has distilled, and the liquids with higher boiling points (the solution that remains in the boiling flask) should be transferred to a smaller flask before continuing the distillation.

Figure 11.15 shows the apparatus for a fractional distillation. Follow the steps listed in Technique 11.3 for assembling a simple distillation apparatus with the addition of the fractionating column between

**FIGURE 11.14**

Distillation curve for the fractional distillation of a 1:1 molar solution of pentane and hexane. The dashed line represents the distillation curve for a simple distillation of the same solution. The abrupt temperature increase at approximately 26–28 mL of distillate demonstrates the greater efficiency of fractional distillation.

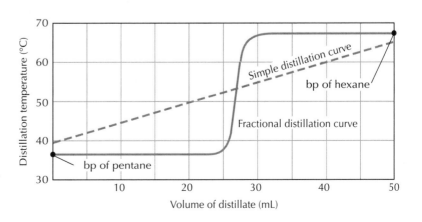

**FIGURE 11.15**
Fractional distillation
apparatus. The
fractionating column
is inserted between
the distilling flask and
the distilling head.

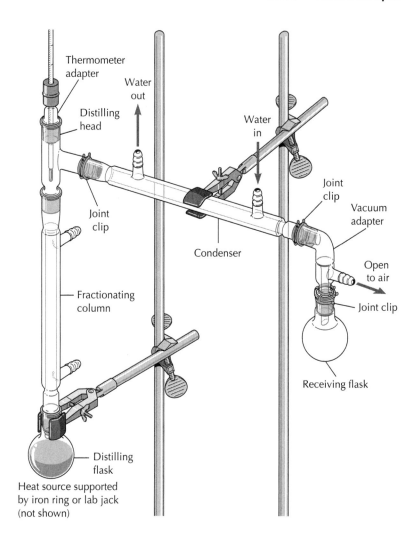

the distilling flask and the distilling head. Be sure that you have added one or two boiling stones to the distilling flask and that the thermometer is placed correctly, as shown in the detail in Figure 11.6.

*Carrying Out a*
*Fractional Distillation*

Heat the distilling flask slowly. Control of heating in a fractional distillation is extremely important; the rate of heating needs to be increased gradually as the distillation proceeds. However, applying too much heat causes the distillation to occur so quickly that the repeated liquid-vapor equilibria required to bring about maximum separation on the surfaces of the fractionating column cannot occur. On the other hand, if too little heat is applied, the column may lose heat faster than it can be warmed by the vapor, thus preventing the vapor from reaching the top of the column. Therefore, too little heat during the distillation causes the thermometer reading to drop below the boiling point of the liquid, simply because vapor is no longer reaching the top of the column. The thermometer temperature may also drop during a fractional distillation when a compound with a lower boiling point has completely distilled and not enough heat is being supplied to force the vapor of the compound with the next-highest boiling point up to the top of the column. The addition of more heat corrects this situation.

The rate of distillation is always a compromise between the speed and the efficiency of the fractionation. For an easy separation, 1–2 drops per second can be collected. Generally a slow, steady distillation where 1 drop is collected every 2–3 s is a more reasonable rate. Difficult separations (when the boiling points of the distilling compounds are close together) require a slower distillation rate as well as more efficient fractionating columns—those with more theoretical plates. The distillation rate can be increased during collection of the last fraction, when all the compounds with lower boiling points have already been distilled.

*Collecting the Fractions*

You will need a labeled receiving vessel (round-bottomed flask, vial, or Erlenmeyer flask) for each fraction you plan to collect. The cutoff points for the fractions are the boiling points (at atmospheric pressure) of the substances being separated. For example, in a fractional distillation of the 1:1 solution of pentane (bp 36°C) and hexane (bp 69°C) described in Figure 11.4, the first fraction would be collected when the temperature at the distilling head reached 35°–36°C. The temperature would stay at 36°C for a period of time while the pentane distilled.

Eventually the temperature either rises or drops several degrees, the latter change indicating that there is no longer enough pentane vapor to maintain the boiling-point temperature at the thermometer bulb. When this happens, increase the heat input and change to the second receiving flask. Liquid then begins to distill again. Leave the second receiver in place until the temperature reaches 69°C, the boiling point of hexane; then change to the third receiving flask. Thus, the second receiver contains only a small amount of distillate. Continue collecting fraction 3 (hexane) until only 1–2 mL of liquid remain in the boiling flask.

---

**SAFETY PRECAUTION**

A distillation flask should **never** be allowed to boil dry.

---

*Summary of Fractional Distillation Procedure*

1. Use a round-bottomed flask that has a capacity two or three times the volume of the liquid mixture you wish to distill. Clamp the flask to a ring stand or upright support rod. Pour the liquid into the flask and add one or two boiling chips.
2. Set up the rest of the apparatus as shown in Figure 11.15.
3. Heat the mixture to boiling and collect the distillate in fractions based on the boiling points of the individual components in the mixture. Use a separate receiving vessel for each fraction.

## 11.5 Azeotropic Distillation

The systems described up to this point are solutions whose components interact only slightly with one another and thus approximate the behavior of an ideal solution. As discussed in Technique 11.2, the behavior of such solutions follows Raoult's law:

$$P = P° X$$

where $P$ = partial pressure of component, $P°$ = its vapor pressure, and $X$ = mole fraction of component.

Most liquid solutions, however, deviate from ideality. Such deviations result from intermolecular interactions (for example, hydrogen bonding) in the liquid state. In the distillation of some solutions, mixtures that boil at a constant temperature are produced. Such constant-boiling mixtures cannot be further purified by distillation and are called *azeotropes.*

One of the best-known binary mixtures that forms an azeotropic mixture during distillation is the ethanol/water system, shown in Figure 11.16. The azeotrope boils at 78.2°C and consists of 95.6% ethanol and 4.4% water by weight. Liquid of azeotropic composition will vaporize to a gas having exactly the same composition, because the liquid and vapor curves intersect at this point (Figure 11.16). No matter how many more liquid-vapor equilibria take place as the materials travel up the column, no further separation occurs. Continued distillation never yields a liquid that is higher than 95.6% in ethanol. Pure ethanol must be obtained by other means.

More detailed discussion about the formation of azeotropic distillation mixtures from nonideal solutions can be found in the references at the end of the chapter. Extensive tables of azeotropic data are available in references such as the *CRC Handbook of Chemistry*

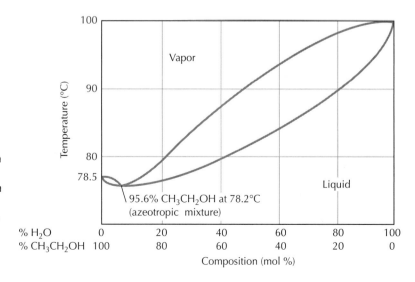

**FIGURE 11.16**

Temperature-composition diagram for ethanol/ water solutions at 1.0 atm pressure. The mixture of 95.6% ethanol and 4.4% water is an azeotrope with a boiling-point minimum.

TABLE 11.2 **Azeotropes formed by common solvents**

| Component X (bp) | % by wt | Component Y (bp) | % by wt | Azeotrope bp |
|---|---|---|---|---|
| Water (100) | 13.5 | Toluene (110.7) | 86.5 | 84.1 |
| Water (100) | 1.4 | Pentane (36.1) | 98.6 | 34.6 |
| Methanol (64.7) | 12.1 | Acetone (56.15) | 87.9 | 55.5 |
| Methanol (64.7) | 72.5 | Toluene (110.7) | 27.5 | 63.5 |
| Ethanol (78.3) | 68 | Toluene (110.7) | 32 | 76.7 |
| Water (100) | 1.3 | Diethyl ether (34.5) | 98.7 | 34.2 |

*and Physics*. Table 11.2 lists a few azeotropes formed by common solvents.

## 11.6 Vacuum Distillation

Many organic compounds with high boiling points decompose at temperatures below their atmospheric boiling points. Such compounds, and compounds with boiling points above 200°C, can be distilled at temperatures lower than their atmospheric boiling points when a partial vacuum is applied to the distillation apparatus. Distillation at reduced pressure, called *vacuum distillation*, takes advantage of the fact that the boiling point of a liquid is a function of the pressure under which the liquid is contained [see Technique 11.1]. It should be noted, however, that vacuum distillation is inherently less efficient than fractional distillation at atmospheric pressure.

A partial vacuum can be obtained in the laboratory with either a vacuum pump or a water aspirator. Vacuum pumps can easily produce pressures of less than 0.5 Torr. The pressure obtained with a water aspirator can be no lower than the vapor pressure of water, which is 13 Torr at 15°C and sea level. In practice, an efficient water aspirator produces a partial vacuum of 15–25 Torr.

The boiling point of a compound at any given pressure other than 760 Torr is difficult to calculate exactly. As a rough estimate, a 50% drop in pressure lowers the boiling point of an organic liquid 15°–20°C. Below 25 Torr, reducing the pressure by one-half lowers the boiling point approximately 10°C (Table 11.3).

A nomograph provides another way of estimating the boiling points of relatively nonpolar compounds at either reduced or

TABLE 11.3 **Boiling points (°C) at reduced pressures**

| Pressure (Torr) | Water | Benzaldehyde | Diphenylether |
|---|---|---|---|
| 760 | 100 | 179 | 258 |
| 100 | 51 | 112 | 179 |
| 40 | 34 | 90 | 150 |
| 20 | 22 | 75 | 131 |

**FIGURE 11.17**
Nomograph for estimating boiling points at different pressures.

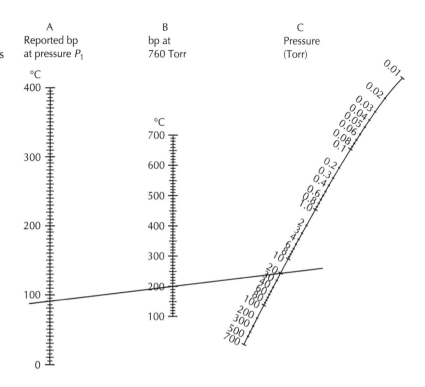

atmospheric pressure (Figure 11.17). For example, if the boiling point of a compound at 760 Torr is 200°C and the vacuum distillation is being done at 20 Torr, the approximate boiling point is found by aligning a straightedge on 200 in column B with 20 in column C; the straightedge intersects column A at 90°C, the approximate boiling point of the compound at 20 Torr, as shown by the line on Figure 11.17. Similarly, one can estimate the boiling point at atmospheric pressure if the boiling point at a reduced pressure is known. By aligning the boiling point in column A with the pressure in column C, a straightedge intersects column B at the approximate atmospheric boiling point. The graph gives a less than accurate estimate of boiling points for polar compounds that associate strongly in the liquid phase.

*Apparatus for Miniscale Vacuum Distillation*

The vacuum distillation apparatus shown in Figure 11.18 works adequately for most vacuum distillations, although a fractionating column may need to be included to provide satisfactory separation of some mixtures. Because liquids often boil violently at reduced pressures, a Claisen connecting adapter is always used in a vacuum distillation to lessen the possibility of liquid bumping up into the condenser. If undistilled material jumps through the Claisen adapter into the condenser, you must begin the distillation again.

If a satisfactory vacuum is to be maintained, each connecting surface must be completely greased with **high-vacuum silicone grease,** and the rubber tubing to the aspirator or vacuum pump should be thick-walled so that it does not collapse. If the partial vacuum is not as low as expected, carefully check all connections for possible leaks.

**FIGURE 11.18**
Vacuum distillation
apparatus.

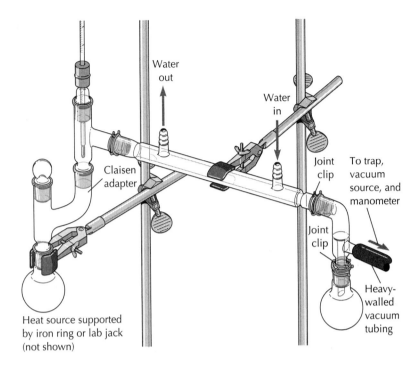

Uncontrolled bumping during a vacuum distillation can be lessened by using a large distillation flask and by adding small pieces of wood splints in place of boiling stones or by magnetic stirring. Instead of wood splints or boiling stones, a very finely drawn out Pasteur pipet capillary tip can provide a steady stream of very small bubbles (Figure 11.19). The bottom of the capillary

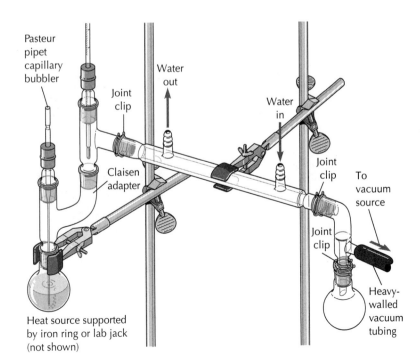

**FIGURE 11.19**
Vacuum distillation
apparatus fitted with
a capillary bubbler.

tube bubbler should be just above the bottom surface of the distilling flask and must always be below the liquid's surface. Do not use wood splints or boiling stones when you use a capillary bubbler; their violent motions may break the fragile tip of the bubbler, rendering it useless. A capillary bubbler should not be used with air-sensitive compounds unless an inert gas is fed into the top of the Pasteur pipet.

**Standard Taper Microscale Apparatus for Vacuum Distillation**

*The well in the Hickman distilling head has a capacity of only 1 mL.*

For a volume of 2–5 mL of liquid a 10-mL round-bottomed flask and the microscale ℑ 14/10 apparatus shown in Figure 11.20 can be used for a vacuum distillation.* If the volume of liquid to be distilled is less than 2 mL, the microscale apparatus shown in Figure 11.21 can be used for a vacuum distillation.

The ground glass joints of microscale glassware should not be greased. Usually clean standard taper joints are completely sealed by compression of the O-ring when the cap is screwed down tightly. Only if the requisite reduced pressure cannot be obtained should microscale joints be greased with **high-vacuum silicone**

---

* Standard taper 14/10 distillation heads, thermometer adapters, and vacuum take-off adapters are commercially available, but they are not part of basic microscale kits.

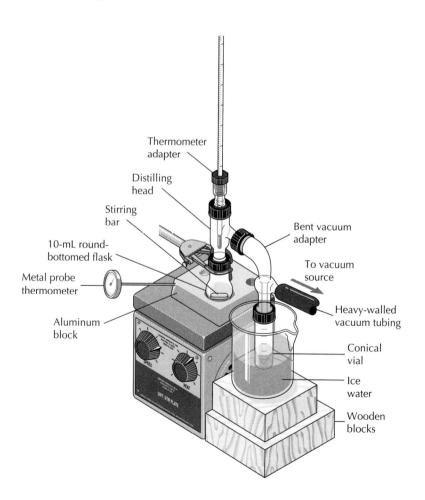

Thermometer adapter

Distilling head

Stirring bar

10-mL round-bottomed flask

Metal probe thermometer

Bent vacuum adapter

To vacuum source

Heavy-walled vacuum tubing

Conical vial

Ice water

Wooden blocks

Aluminum block

**FIGURE 11.20**
Short-path standard taper microscale apparatus for vacuum distillation.

**FIGURE 11.21**
Standard taper
microscale apparatus
for distillation with
a Hickman distilling
head.

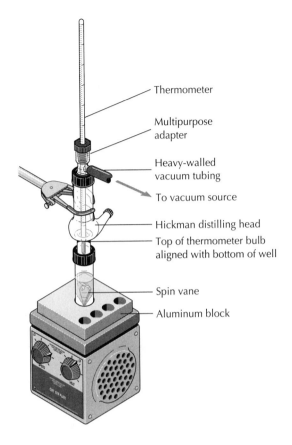

Thermometer

Multipurpose
adapter

Heavy-walled
vacuum tubing

To vacuum source

Hickman distilling head

Top of thermometer bulb
aligned with bottom of well

Spin vane

Aluminum block

**grease.** Care must be exercised to use a very thin film of grease applied only at the top of the inner joints. No grease should be allowed to seep from the bottom of any joint because the grease might severely contaminate the liquid being distilled.

*Monitoring the Pressure During a Vacuum Distillation*

The pressure can be continuously monitored with a manometer (Figure 11.22) or read periodically with a McLeod gauge (Figure 11.23). If a water aspirator is used as the source of the vacuum, the trap bottle or flask is used to prevent any backflow of water from entering the distillation apparatus. When a vacuum pump is used, a cold trap, kept at isopropyl alcohol/dry ice ($-77°C$) or liquid nitrogen ($-196°C$) temperature, must be placed between the distillation system and the pump. The trap collects any volatile materials that could get into the pump oil and cause a rise in the vapor pressure of the oil, which would decrease the efficiency of and possibly damage the pump. A pressure relief valve serves to close the system from the atmosphere and to release the vacuum after the system has cooled following the distillation.

A *McLeod gauge* is used to measure pressures below 5 Torr. It works by compressing the gas inside the system into a closed capillary tube with a pressure great enough to be measured with a mercury column. Initially the gauge must be in the horizontal, resting position with the mercury in the reservoir (Figure 11.23a). When the

**FIGURE 11.22**
Two types of closed-end manometers used in vacuum distillation.

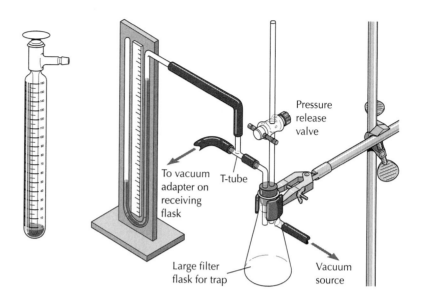

Pressure release valve

To vacuum adapter on receiving flask

T-tube

Large filter flask for trap

Vacuum source

pressure inside the distillation system has stabilized, the gauge is slowly rotated until the open-ended reference capillary tube is in the vertical position. The pressure is indicated by the scale on the closed-end capillary tube when the mercury level in the reference capillary tube reaches the calibration mark (Figure 11.23b). After the pressure has been read, the gauge must be returned to the horizontal, resting position.

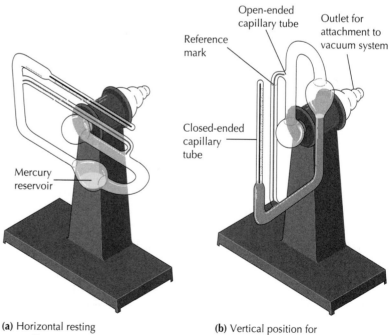

Open-ended capillary tube

Outlet for attachment to vacuum system

Reference mark

Closed-ended capillary tube

Mercury reservoir

**FIGURE 11.23**
McLeod gauge used in vacuum distillation.

**(a)** Horizontal resting position

**(b)** Vertical position for reading pressure

*Steps in a Vacuum Distillation*

---

S A F E T Y   P R E C A U T I O N

> Safety glasses must be worn at all times while carrying out a vacuum distillation because of the danger of an implosion shattering the glassware.

1. Add the liquid to be distilled to a round-bottomed flask sized so that it will be less than half filled. Then add some wood splints, boiling chips, or a magnetic stirrer (if you use the apparatus in Figure 11.18) and set up the apparatus as shown in Figure 11.18 or 11.19 for a miniscale distillation, Figure 11.20 or 11.21 for a microscale distillation.
2. Attach the apparatus to the vacuum system and monitor the pressure with a manometer (Figure 11.22) or a McLeod gauge (Figure 11.23).
3. Close the pressure release valve and turn on the vacuum.
4. When the vacuum has reached an appropriate level, heat the distilling flask cautiously to obtain a moderate distillation rate. Periodically monitor the pressure during the distillation.
5. When the distillation is complete, remove the heat source and allow the apparatus to cool nearly to room temperature before breaking the vacuum. **Turn off the aspirator or vacuum pump only after the vacuum is broken.**

---

## 11.7    Steam Distillation

Codistillation with water, called *steam distillation,* allows distillation of relatively nonvolatile organic compounds without complex vacuum systems. Steam distillation can be thought of as a special kind of azeotropic distillation; it is especially useful for separating volatile organic compounds from nonvolatile inorganic salts or from the leaves and seeds of plants. Indeed, the process has found wide application in the flavor and fragrance industries as a means of separating essences or flavor oils from plant material.

Steam distillation depends on the mutual insolubility or immiscibility of many organic compounds with water. In such two-phase systems, at any given temperature each component exerts its own full vapor pressure independently of the other compounds present. The total vapor pressure above the two-phase mixture is equal to the sum of the vapor pressures of the pure components independent of their relative amounts.

Consider the codistillation of iodobenzene (bp 188°C) and water (bp 100°C). The vapor pressures ($P°$) of both substances increase with temperature, but the vapor pressure of water will

always be higher than that of iodobenzene because water is more
volatile. At 98°C,

$$P^{\circ}_{\text{iodobenzene}} = 46 \text{ Torr}$$

$$P^{\circ}_{\text{water}} = 714 \text{ Torr}$$

$$P^{\circ}_{\text{iodobenzene}} + P^{\circ}_{\text{water}} = 760 \text{ Torr}$$

Therefore, a mixture of iodobenzene and water codistills at 98°C.

From an ideal gas law calculation, the mole fraction of iodobenzene in the vapor at the distilling head is 0.06 (46 Torr/760 Torr),
and therefore the mole fraction of water in the vapor is 0.94. However, because iodobenzene has a much higher molecular weight
than water (204 g · mol$^{-1}$ versus 18 g · mol$^{-1}$), its weight percentage in the vapor is much larger than 0.06, as the following calculation shows:

$$\frac{\text{moles}_{\text{iodobenzene}}}{\text{moles}_{\text{water}}} = \frac{P^{\circ}_{\text{iodobenzene}}}{P^{\circ}_{\text{water}}}$$

$$\frac{g_{\text{iodobenzene}}/\text{MW}_{\text{iodobenzene}}}{g_{\text{water}}/\text{MW}_{\text{water}}} = \frac{P^{\circ}_{\text{iodobenzene}}}{P^{\circ}_{\text{water}}}$$

Rearranging the previous expression and substituting the molecular weights and vapor pressures allows us to calculate the
weight ratio of iodobenzene to water in the distillate from the
steam distillation.

$$\frac{g_{\text{iodobenzene}}}{g_{\text{water}}} = \frac{0.73 \ g_{\text{iodobenzene}}}{1.0 \ g_{\text{water}}}$$

In other words, the distilling liquid contains 42% iodobenzene and
58% water by weight. In any steam distillation, a large excess of
water is used in the distilling flask so that virtually all the organic
compound (iodobenzene in this example) can be distilled from the
mixture at a temperature well below the boiling point of the pure
compound.

The temperature in any steam distillation of a reasonably
volatile organic compound will never rise above 100°C, the boiling
point of water, unless your laboratory is below sea level. The steam
distillation of most compounds occurs between 80°C and 100°C.
For example, at 1.0 atm, octane (bp 126°C) steam distills at 90°C,
and 1-octanol (bp 195°C) steam distills at 99°C. The lower distillation temperature has the added advantage of preventing decomposition of organic compounds with higher boiling points during
distillation.

***Apparatus for Steam***      Use more water than the amount of organic mixture being distilled
***Distillation***                 and select a distilling flask that will be no more than half filled with
this organic/water mixture. Add one or two boiling stones to the

**FIGURE 11.24**
Steam distillation apparatus for internal generation of steam. During the distillation, additional water can be added to the distilling flask from the dropping funnel.

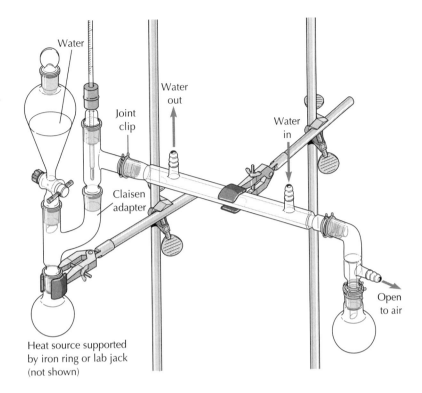

Water

Water out

Joint clip

Water in

Claisen adapter

Open to air

Heat source supported by iron ring or lab jack (not shown)

flask. For a steam distillation, modify a simple distillation apparatus by adding a Claisen connecting tube or adapter between the boiling flask and the distilling head. This adapter provides a second opening into the system to accommodate a source of steam or the addition of water.

Steam can be generated simply by boiling a large amount of water with the mixture in the distillation flask (Figure 11.24). If the codistillation of an organic compound with low volatility requires a large volume of steam, a separatory funnel placed in the second opening of the Claisen connecting tube provides a way of adding more water to the system without stopping the distillation. Figure 11.25 shows the apparatus for a steam distillation that uses externally generated steam, such as a steam line or a flask of boiling water.

*Steps in a Steam Distillation*

1.  Set up the apparatus as shown in Figure 11.24 or 11.25.
2.  Add the organic mixture and an excess of water to a firmly clamped distilling flask at least twice as large as the combined organic-water volume. Add one or two boiling stones.
3.  Heat the mixture until all the organic layer has distilled into the receiving flask. Sometimes it is worthwhile to collect an additional 10–15 mL of water after organic material is no longer detected to ensure complete recovery.
4.  Separate the organic phase of the distillate from the aqueous phase in a separatory funnel.

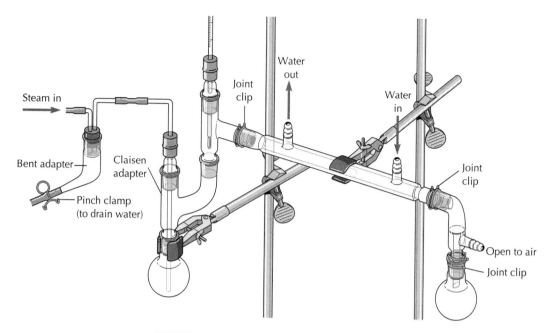

**FIGURE 11.25**
Steam distillation apparatus for use with an external steam source.

## References

1. Lide, D. R. (Ed.) *Handbook of Chemistry and Physics;* 82nd ed. CRC Press: Boca Raton, FL, 2001.

2. Perry, E. S.; Weissberger, A. (Eds.) *Techniques of Organic Chemistry;* 2nd ed.; Wiley-Interscience: New York, 1965, Vol. 4.

## Questions

1. Explain why the observed boiling point for the first drops of distillate collected in the simple distillation of a 1:1 molar solution of pentane and hexane illustrated in Figure 11.5 must be above the boiling point of pentane.

2. A mixture contains 80% hexane and 20% pentane. Use the phase diagram shown in Figure 11.4 to estimate the composition of the vapor over this liquid. This vapor is condensed and the resulting liquid is heated. What is the composition of the vapor above this liquid?

3. A student carried out a simple distillation on a compound known to boil at 124°C and reported an observed boiling point of 116°–117°C. Gas chromatographic analysis of the product showed that the compound was pure, and a cal-

ibration of the thermometer indicated that it was accurate. What procedural error might the student have made in setting up the distillation apparatus?

4. The directions in an experimental procedure specify that the solvent, diethyl ether, be removed from the product by using a simple distillation. Why should the heat source for this distillation be a steam bath, not an electrical heating mantle?

5. Azeotropes can be used to assist chemical reactions. Treatment of 1-butanol with acetic acid in the presence of a nonvolatile acid catalyst results in formation of the ester butyl acetate and water. The mixture of 1-butanol/butyl acetate/water forms a ternary azeotrope that boils at 90.7°C. This azeotrope separates into two layers,

the upper being largely ester and the lower, water. The ester forms by an equilibrium reaction that does not especially favor product formation. Describe an apparatus that you could use to take advantage of azeotrope formation to drive the equilibrium toward the products, thus maximizing the yield of ester.

6. A compound has a boiling point of 300°C at atmospheric pressure. Use the nomograph (Figure 11.17) to determine the pressure at which the compound would boil at about 200°C.

## TECHNIQUE

# 12    SUBLIMATION

The evaporation of most solid organic compounds first requires melting, a process that usually takes a reasonably high temperature. However, some substances, such as iodine, camphor, and 1,4-dichlorobenzene (mothballs) exhibit appreciable vapor pressure below their melting points. You may already have seen iodine crystals evaporate to a purple gas during gentle heating and smelled the characteristic odors of camphor or mothballs. These substances all change directly from the solid phase to the gas phase without forming an intermediate liquid phase by a process called *sublimation.*

The process of sublimation seems somewhat unusual in that, unlike normal phase changes from solid to liquid to gas, no liquid phase forms between the solid and gas phases. The interconversion of the gaseous and solid forms of carbon dioxide (also called "dry ice") may be the best-known example of sublimation. Carbon dioxide does not have a melting point at atmospheric pressure. More than 5 atm of pressure are necessary before dry ice melts at −57°C. The sublimation point for $CO_2$ at atmospheric pressure is −78°C, well below room temperature and below the melting point of dry ice.

In the laboratory we can use sublimation as a purification method for an organic compound (1) if it can vaporize without melting, (2) if it is stable enough to vaporize without decomposition, (3) if the vapor can be condensed back to the solid, and (4) if the impurities present do not also sublime. Many organic compounds that do not sublime at atmospheric pressure sublime appreciably at reduced pressure, thus enabling their purification by sublimation. Use of reduced pressure, supplied by a vacuum source, also makes decomposition and melting less likely to occur during the sublimation.

## 12.1    Assembling the Apparatus for a Sublimation

The apparatus for a sublimation consists of an outer vessel, connected to a vacuum source, that holds the sample being purified. An inner container, sometimes called a "cold finger," provides a

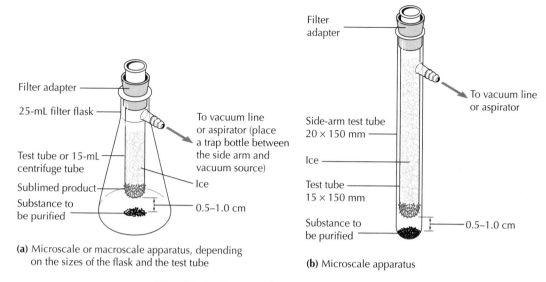

**FIGURE 12.1** Two simple apparatuses for sublimation.

cold surface on which the vaporized compound can condense as a solid.

Two simple arrangements for sublimation under reduced pressure are shown in Figure 12.1. The inner test tube, which contains cold water or ice and water, serves as a condensation site for the sublimed solid. The outer vessel, a side-arm test tube or a filter flask, holds the substance being purified, and the side arm provides a connection to the vacuum source. The inner and outer vessels are sealed together by a neoprene filter adapter. The distance between the bottom surfaces of the inner tube and the outer tube or filter flask should be 0.5–1.0 cm. If the vapor has to travel a long distance, a higher temperature is needed to keep it in the gas phase, and decomposition of the solid sample may very well occur. If the surfaces are too close, impurities can spatter up to contaminate the con-

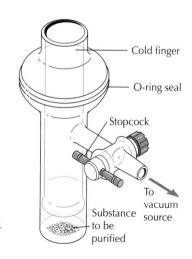

**FIGURE 12.2**
Sublimation apparatus for gram quantities
of material.

densed solid on the surface of the inner tube. Connect the side arm of the test tube or filter flask to a water aspirator or vacuum line, using a safety flask between the aspirator and the sublimation apparatus.

The side-arm test tube apparatus serves well for 10–150 mg of material. The filter flask apparatus can be sized to suit the amount of material being purified. For example, microscale quantities of 10–150 mg can be sublimed in a 25-mL filter flask, whereas 1 g of material would require a 125-mL filter flask with a correspondingly larger test tube for the cold finger. The apparatus shown in Figure 12.2 is a commercially available sublimation apparatus used for gram quantities of material.

| 12.2 | ## Carrying Out a Sublimation |

SAFETY PRECAUTION

The lip of the inner test tube must be large enough to prevent it from being pushed through the bottom of the filter adapter by the difference in pressure created by the vacuum. Slippage of the inner test tube could cause both vessels to shatter as the inner test tube hits the outer test tube or flask. Placing a microclamp on the inner test tube above the filter adapter helps to keep the test tube from moving once it is positioned in the filter adapter.

*The ice and water are added after the vacuum is applied so that moisture from the air does not condense on the inner tube before sublimation takes place.*

Place the sample (10–150 mg) to be sublimed in a 25-mL filter flask or a side-arm test tube. Fit the inner test tube through the filter adapter and adjust the position of the inner tube so that it is 0.5–1.0 cm above the bottom of the flask or side-arm test tube. Turn on the water aspirator or vacuum line. After a good vacuum has been achieved, fill the inner test tube with ice and water, then proceed to heat the sublimation tube gently using a sand bath [see Technique 6.2]. If a filter flask (25-mL for 10–150 mg, 125-mL for 0.25–1.0 g) is used as the outer container, heat it gently on a hot plate or with a sand bath.

During sublimation, you will notice material disappearing from the bottom of the outer vessel and reappearing on the cool outside surface of the inner test tube. If the sample begins to melt, briefly withdraw the apparatus from the heat source. If all the ice melts, remove half the water from the inner test tube with a Pasteur pipet, then add additional ice. After sublimation is complete, remove the heat source, slowly let air back into the system by gradually removing the rubber tubing from the water aspirator, and then turn off the water flow in the aspirator or turn off the vacuum source and slowly disconnect the rubber tubing from the side arm. Carefully remove the inner test tube and scrape the purified solid onto a tared weighing paper. After weighing the sublimed solid, store it in a tightly closed vial.

## Questions

1. Which of the following three compounds would most likely be amenable to purification by sublimation: polyethylene, menthol, or benzoic acid?

2. A solid compound has a vapor pressure of 65 Torr at its melting point of 112°C. Give a procedure for purifying this compound by sublimation.

3. Hexachloroethane has a vapor pressure of 780 Torr at its melting point of 186°C. Describe how the solid would behave while carrying out a melting-point determination at atmospheric pressure (760 Torr) in a capillary tube open at the top.

**TECHNIQUE**

# 13

# REFRACTOMETRY

A beam of light traveling from a gas into a liquid undergoes a decrease in its velocity. If the light strikes the horizontal interface between gas and liquid at an angle other than 90°, the beam bends downward as it passes from the gas into the liquid. Application of this phenomenon allows the determination of a physical property known as the *refractive index*, a measure of how much light is bent, or *refracted*, as it enters the liquid. The refractive index can be determined quite accurately to four decimal places, making this physical property useful for assessing the purity of liquid compounds. The closer the experimental value approaches the value reported in the literature, the purer the sample. Even trace amounts of impurities (including water) change the refractive index, so unless the compound has been extensively purified, the experimentally determined value may not agree with the literature value past the second decimal place.

## 13.1  Theory of Refractometry

The bending of a light beam as it passes from one medium to another is known as refraction. The refractive index, $n$, represents the ratio of the velocity of light in a vacuum (or in air) to the velocity of light in the liquid being studied. The velocities of light in both media are related to the angles that the incident (incoming) and the refracted (outgoing) beams make with a theoretical line drawn 90° to the liquid surface (Figure 13.1):

$$n = \frac{V_{vac}}{V_{liq}} = \frac{\sin \theta}{\sin \theta'}$$

where $V_{vac}$ = velocity of light in a vacuum (or air), $V_{liq}$ = velocity of light in the liquid, $\theta$ = angle of incident light in a vacuum (or air), and $\theta'$ = angle of refracted light in the liquid. The velocity of

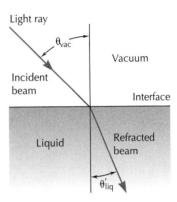

**FIGURE 13.1**
Refraction of light in a liquid.

light in a liquid sample is always less than that of light in air, so refractive index values are numerically greater than 1.

The variables of temperature and the wavelength of the light being refracted influence the refractive index for any substance. The temperature of the sample affects its density; and the density change, in turn, affects the velocity of the light beam as it passes through the sample. Therefore, the temperature (20°C in the following example) at which the refractive index was determined is always specified by a superscript in the notation of $n$:

$$n_D^{20} = 1.3910$$

The wavelength of the light used also affects the refractive index, because light of differing wavelengths refracts at different angles. The bright yellow line of the sodium spectrum at 589 nm, commonly called the sodium D line, usually serves as the standard wavelength for refractive index measurements and is indicated by the D in the subscript of the symbol $n$. If light of some other wavelength is used, the specific wavelength in nanometers appears in the subscript.

## 13.2     The Refractometer

The instrument used to measure the refractive index of a compound is called a *refractometer*. Figure 13.2 shows an Abbe refractometer, an example of the type commonly found in undergraduate organic laboratories. The instrument includes a built-in thermometer for measuring the temperature at the time of the refractive index reading and a system for circulating water at a constant temperature around the sample holder. This type of refractometer uses a white light source instead of a sodium lamp and contains a series of compensating prisms that give a refractive index equal to that obtained with the D line of sodium at 589 nm.

A few drops of sample are introduced between a pair of hinged prisms (Figure 13.3). The light passes through the sample and is reflected by an adjustable mirror. When the mirror is properly aligned, the light is reflected through the compensating prisms and, finally, through a lens with crosshairs to the eyepiece.

**FIGURE 13.2**
Abbe refractometer.
(Courtesy of Leica, Inc.,
Optical Products
Division, P.O. Box 123,
Buffalo, NY 14240.)

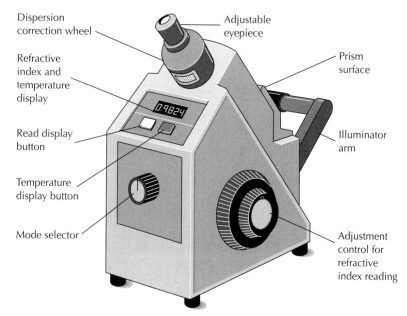

Dispersion correction wheel

Adjustable eyepiece

Refractive index and temperature display

Prism surface

Read display button

Temperature display button

Mode selector

Illuminator arm

Adjustment control for refractive index reading

**FIGURE 13.3**
Hinged prisms of
the Abbe refractometer,
shown in the open
position. (Courtesy
of Leica, Inc., Optical
Products Division,
P.O. Box 123, Buffalo,
NY 14240.)

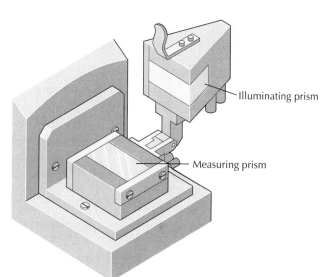

Illuminating prism

Measuring prism

## 13.3

### Steps in Determining a Refractive Index

*Do not use acetone to
clean prisms because it
dissolves the adhesive
holding the prism.*

1.  Check the surface of the prisms for residue from previous determinations. If the prisms need cleaning, place a few drops of methanol on the surfaces and blot (do not rub) the surfaces with lens paper. Allow the residual methanol to evaporate completely before placing the sample on the lower prism.
2.  With a Pasteur pipet held 1–2 cm above the prism, place 4–5 drops of the sample on the measuring (lower) prism. Do not touch the prism with the tip of the dropper because the highly polished surface scratches very easily, and scratches ruin the

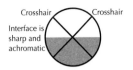

Crosshair     Crosshair

Interface is sharp and achromatic

**FIGURE 13.4**
The view through the eyepiece when the refractometer is adjusted correctly.

instrument. Lower the illuminating (upper) prism carefully so that the liquid spreads evenly between the prisms.

3. Rotate the adjustment control until the dark and light fields are exactly centered on the crosshairs in the eyepiece (Figure 13.4). If color (usually red or blue) appears as a horizontal band at the interface of the fields, rotate the chromatic adjustment, or dispersion correction wheel, until the interface is sharp and uncolored (achromatic). Occasionally the sample evaporates from the prisms, making it impossible to produce a sharp achromatic interface between the light and dark fields. If evaporation occurs, apply more sample to the prism and repeat the adjustment procedure.

4. Press the display button and record the refractive index in your notebook. Then press the temperature button and record the temperature.

5. Open the prisms, blot up the sample with lens paper, and follow the cleaning procedure with methanol outlined in step 1.

## 13.4     Temperature Correction

Values reported in the literature are often determined at a number of different temperatures, although 20°C has become the standard. To compare an experimental refractive index with a value reported at a different temperature, a correction factor must first be calculated. The refractive index for a typical organic compound increases by $4.5 \times 10^{-4}$ for each 1-degree decrease in temperature. Refractive index values vary inversely with temperature because the density of a liquid almost always decreases as the temperature increases. This decrease in density produces an increase in the velocity of light in the liquid, causing a corresponding decrease in the refractive index.

To compare an experimental refractive index measured at 25°C to a reported value at 20°C, a temperature correction needs to be calculated:

$$\Delta n = 4.5 \times 10^{-4} \times (T_1 - T_2)$$

where $T_1$ is the observed temperature in degrees Celsius and $T_2$ is the temperature reported in the literature in degrees Celsius. Add the correction factor, including its sign, to your experimentally determined refractive index. For example, if your experimental refractive index is 1.3888 at 25°C, then you obtain a corrected value at 20°C of 1.3911 by adding the correction factor of 0.0023 to the experimental refractive index.

$$\Delta n = [4.5 \times 10^{-4} \times (25 - 20)] = 0.00225 \text{ (rounds to 0.0023)}$$

$$n^{20} = n^{25} + 0.0023 = 1.3888 + 0.0023 = 1.3911$$

This correction needs to be applied before comparing the experimental value to a literature value reported at 20°C.

If an experimental refractive index is determined at a temperature lower than that of the literature value to which it is being compared, the correction has a negative sign and the corrected refractive index would be lower than the experimental value.

## Questions

1. A compound has a refractive index of 1.3191 at 20.1°C. Calculate its refractive index at 25.0°C.

2. It is usually recommended that the glass surfaces of a refractometer be cleaned with ethanol or methanol, not with acetone. Why?

TECHNIQUE

# 14

# OPTICAL ACTIVITY AND POLARIMETRY

Optical activity, the ability of substances to rotate plane-polarized light, played a crucial role in the development of chemistry as the link between the molecular structures that chemists write and the real physical world. A major development in the structural theory of chemistry was the concept of the three-dimensional shape of molecules. When Jacobus van't Hoff and Joseph le Bel noted the asymmetry possible in tetrasubstituted carbon compounds, they claimed that their "chemical structures" were identical to the "physical structures" of the molecules. Not only was the fanciful structural theory of the organic chemist useful in explaining the facts of chemistry; it also happened to be "true." Van't Hoff and le Bel could make this claim because their theories of the tetrahedral carbon atom accounted not only for chemical properties but also for the optical activity of substances, a physical property.

## 14.1    Mixtures of Optical Isomers: Separation/Resolution

A molecule that possesses no internal mirror plane of symmetry and whose mirror image is not superimposable is said to be *chiral,* or "handed." A preliminary suggestion of chirality, a molecular property, can be gained from a *stereocenter* (sometimes called a chiral or asymmetric center), that is, a carbon center bearing four different substituents.

Chiral compounds possess the property of *enantiomerism.* Enantiomeric pairs are stereoisomers that are nonsuperimposable mirror images. Chiral compounds such as 2-butanol and the amino acid alanine, which contain only one stereocenter, are simple examples of enantiomers.

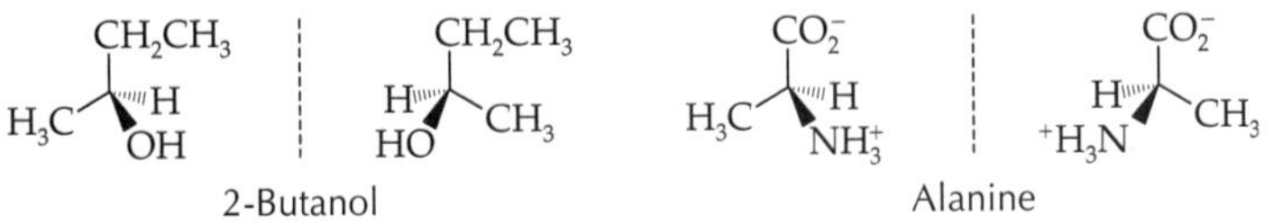

2-Butanol                                    Alanine

The enantiomers of 2-butanol have identical physical properties, including boiling points, IR spectra, NMR spectra, refractive indices, and TLC $R_f$ values, except for the direction in which they rotate plane-polarized light. Both enantiomers are optically active—one rotates polarized light in a clockwise direction and is called the **(+)-isomer**. The other enantiomer rotates polarized light counterclockwise and is called the **(−)-isomer.** The rotational power of (+)-2-butanol is exactly the same in the clockwise direction as that of (−)-2-butanol in the counterclockwise direction. Unfortunately, there is no simple theoretical way to predict the direction of the rotation of plane-polarized light on the basis of the absolute configuration at a carbon stereocenter. Thus, it is not apparent whether the structure on the left in each pair is a (+) or a (−) enantiomer.

Usually simple compounds obtained from the stockroom are optically inactive, even when the molecule is chiral. Thus you would normally find that a bottle of 2-butanol is optically inactive. This paradox is explained by the fact that the reduction of 2-butanone with, for example, sodium borohydride has two ways to proceed. It undergoes reaction both ways at equal rates, thus giving rise to a 50:50 mixture of the product enantiomers, a product that is optically inactive:

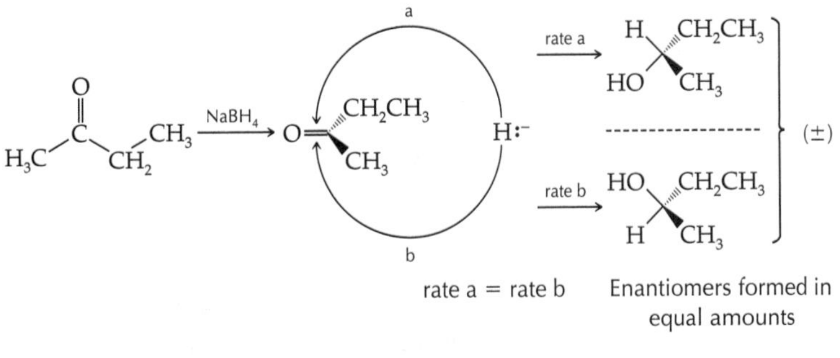

rate a = rate b        Enantiomers formed in
                       equal amounts

An equal mixture of (+)- and (−)-enantiomers is called a *racemic mixture.* Before such a mixture can be separated, the enantiomers must be transformed into a pair of *diastereomers* having different physical and chemical properties. A mixture of diastereomers is prepared from a racemic mixture by reaction with an optically active substance. The diastereomers can be separated by recrystallization because of the differential solubility of the two diastereomers.

**Resolution with Acids or Bases**

The simplest reaction for preparing diastereomers from racemic mixtures is that of an acid with a base to form a salt. For resolution to occur, the added reagent in the acid-base reaction must be optically active. Neutralization of a racemic amine, for example, by an optically active carboxylic acid is a method for resolving the amine; neutralization of a carboxylic acid by an optically active amine is a way of resolving the acid. In such neutralizations, two salts are produced. These salts are diastereomers and, as such, differ in their solubilities in various solvents. Therefore, they can be separated by fractional crystallization and the less soluble diastereomeric salt is the more easily obtained. The process for resolution of an amine is represented in the following steps:

**(a)** Formation of diastereomeric salts from racemic amine

(+)-mandelic acid

(±)-α-Phenylethylamine

and → separate →

Mixture of diastereomeric salts with different solubility properties

**(b)** Separation of diastereomeric salts by fractional crystallization

and

**(c)** Isolation of resolved enantiomers

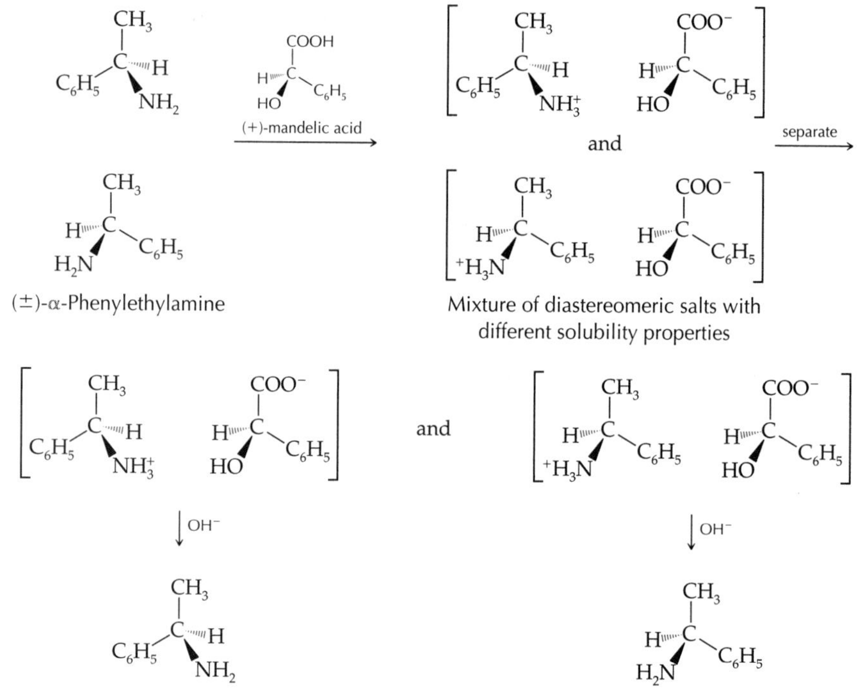

If you examine the diastereomeric salts in (a), you will see that each compound has two stereocenters. When you compare their structures, you will find that the carbon stereocenters bearing the —OH group have identical configurations in each compound, whereas the stereocenters bearing the —NH$_3^+$ groups have opposite configurations. Thus the two salts fit the definition of diastereomers as stereoisomers that are not mirror images.

Optically active acids and bases frequently used for the resolution of racemic mixtures are given in Table 14.1. The diastereomers

| **T A B L E   1 4 . 1** | **Optically active acids and bases used for resolutions** |
|---|---|
| **Bases** | **Acids** |
| Brucine | Tartaric acid |
| Strychnine | Mandelic acid |
| Quinine | Malic acid |
| Cinchonine | Camphor-10-sulfonic acid |
| α-Phenylethylamine | |

necessary for resolution do not need to be salts. For example, diastereomeric esters may be obtained by reaction of the enantiomers of an alcohol with an optically active acid.

***Enzymatic Resolution***  An increasingly useful method for the resolution of racemic mixtures utilizes an enzyme that selectively reacts with one of the enantiomers. Because all enzymes are chiral molecules, the transition states for the reaction of an enzyme with two enantiomers are diastereomeric and the energies for forming these two transition states differ. Thus one of the chiral enantiomers reacts faster than the other. In many cases an enzyme reacts specifically with only one enantiomer; this specificity provides an excellent method for resolving a racemic mixture if an enzyme is available that will react with the compound.

## 14.2     Polarimetric Instrumentation

Optical activity is measured with a polarimeter. This instrument, invented by Jean Baptiste Biot in the early nineteenth century, allows polarized light of a specific wavelength to shine through a sample contained in a long tube. The amount of rotation is measured in the analyzer. A schematic description of a polarimeter is given in Figure 14.1. All commercially available polarimeters have the same general features. The analyzer of older polarimeters is adjusted manually (Figure 14.2), whereas more recent instruments have all the components housed in a case and are completely automated,

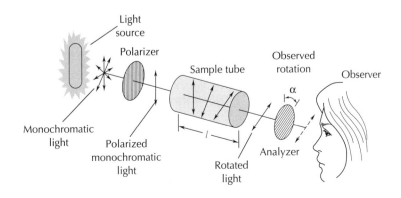

**FIGURE 14.1**
Schematic description
of a polarimeter.

FIGURE 14.2
Polarimetric
instrumentation.

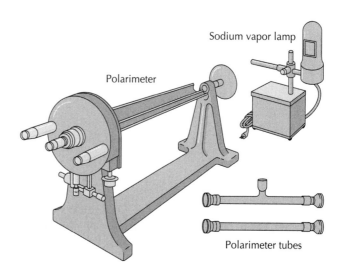

Sodium vapor lamp

Polarimeter

Polarimeter tubes

with a digital readout of the observed rotation. Both types of instruments use the same kinds of sample tubes (see Figure 14.2).

A normal light beam has wave oscillations in all the planes perpendicular to the direction in which the beam is traveling. When the light beam hits certain crystals or a piece of Polaroid plastic, which have ranks and files of molecules arranged in a highly ordered fashion, only the light whose oscillations are in certain planes is transmitted through the solid. The light that gets through is called plane-polarized light. The remaining waves are refracted away or absorbed by the polarizer. In a rough analogy, the light beam is made to pass through a solid whose atoms are ordered like the slats of wood in a picket fence. Only those light waves whose oscillations are parallel to the slats pass through.

The analyzer is a second polarizer whose files of molecules must also be lined up with the polarized light waves impinging on it in order to transmit the light. If the polarized light has been rotated by an optically active substance in the sample tube, the analyzer must be rotated the same amount to let it through. The rotation is measured in degrees. In practice, it is simpler to calibrate the analyzer to let a minimum of light through, that is, to set the analyzer at zero degrees rotation when its orientation is perpendicular to the polarizing crystal.

Monochromatic light is preferred in polarimetric measurements because the optical activity or rotatory power of chiral compounds depends on the wavelength of the light used. For example, the rotation at 431 nm (blue light) of a sample of sucrose is 2.8 times greater than it is at 687 nm (red light). The most common light source is a sodium lamp, which has two very intense emission lines at 589 and 589.6 nm. This closely spaced doublet is called the *sodium D line.* Another light source that is becoming more common is a mercury lamp, using the intense 546.1-nm emission line. The human eye is more sensitive to the mercury emission in the green region than to the sodium line in the yellow region of the visible spectrum.

A number of techniques are used to detect the rotation of polarized light by an optically active substance. The simplest way is to

**(a)** Off field inequality

**(b)** Zero position
(Rotation measured at
minimum light )

FIGURE 14.3
Representative images
in the light field of a
polarimeter.

rotate the analyzer until no light at all comes through. However, this method depends not only on the sensitivity of our eyes but also on our ability to remember quantitatively the amount of brightness we have just seen. In practice, this is difficult to do.

Various optical devices can be used to make the measurement of rotation easier. They depend on a sudden change of contrast when the minimum amount of light is transmitted by the analyzer crystal. What you would see in the analyzer of a typical manual polarimeter is shown in Figure 14.3.

## 14.3          Specific Rotation

The magnitude of optical rotation depends on the concentration of the solution, the length of the light path through the solution, the wavelength of the light, the nature of the solvent, and the temperature. A typical rotation, that of common table sugar, sucrose, is written in the following manner:

$$[\alpha]_D^{20} = +66.4°(H_2O)$$

The symbol $[\alpha]_\lambda^{T°}$ is called the *specific rotation* and is an inherent property of the optically active molecule. $T°$ signifies the temperature of the measurement in degrees Celsius, $\lambda$ the wavelength of light used. In the sucrose example, the sodium D line was used. The specific rotation is calculated from the observed angle of rotation:

$$[\alpha]_\lambda^{T°} = \frac{\alpha}{l \cdot c}$$

where $\alpha$ = the observed angle of rotation, $l$ = the length of the light path through the sample in decimeters, and $c$ = the concentration of the sample ($g \cdot mL^{-1}$ of solution).

The cell length is always given in decimeters (dm, $10^{-1}$ m) in the calculation of specific rotation. When a pure, optically active liquid is used as the sample, its concentration is simply the density of the liquid.

Sometimes a rotation of an optically active substance is given as a *molecular rotation:*

$$[M]_\lambda^{T°} = \frac{M}{100}[\alpha]$$

where M is the molecular weight of the compound.

The value of the specific rotation can change considerably from solvent to solvent. It is even possible for an enantiomer to have a different sign of rotation in two different solvents. Such solvent effects are due to specific solvent-solute interactions. The four most common solvents for polarimetry are water, methanol, ethanol, and chloroform.

The intrinsic specific rotation of a compound is generally considered to be a constant in dilute solutions at a particular temperature and wavelength. However, if you wish to compare the optical activity of a sample with that obtained by other workers, you should use the same concentration in the same solvent. Sucrose makes an excellent reference compound for polarimetry because its specific rotation in water is essentially independent of concentration up to 5–10% solutions.

A change in the specific rotation due to temperature variation may be caused by a number of factors, including changes in molecular association, dipole-dipole interactions, conformation, and solvation. When nonpolar solutes are dissolved in nonpolar solvents, variation in the specific rotation with temperature may not be large. But for polar compounds, there are cases where the specific rotation varies markedly with temperature. Near room temperature, the rotation of tartaric acid may vary by more than 10% per degree.

## 14.4     Using the Polarimeter

The control knobs on various models of polarimeters differ enough that no attempt will be made to discuss the details of operating a polarimeter. Nevertheless, certain techniques of polarimetry are general enough to merit discussion here. Polarimeter tubes are expensive and must be handled carefully. They come in different lengths, with 1-dm and 2-dm tubes being the most common. If the tube you are using closes by screwing on an end plate, be careful not to screw it on too tightly, because strain in an end window can produce an apparent optical rotation.

To obtain the rotation of an optically active sample, you should first clean out a polarimeter tube with some care. After the tube is clean, it should be rinsed out with the solvent you plan to use for the solution of your optically active compound. After the tube has been well drained, it should be rinsed with two or three small portions of your solution to ensure that the concentration of the solution in the polarimeter tube is the same as the solution that you prepared. You may want to save these optically active rinses, because your chiral compound can be recovered from them later.

When you fill a polarimeter tube with a solution, make sure that the tube has no air bubbles trapped in it, because bubbles will refract the light coming through. Also make sure that there are no suspended particles in a solution whose rotation you wish to measure, or you may get so little transmitted light that measurement of the rotation becomes very difficult. If you have a solution that you suspect is too turbid for polarimetry measurements, filter it by gravity through a small plug of glass wool.

A polarimeter can be standardized by filling a tube with an optically inactive solvent such as distilled water and adjusting the instrument to the zero position (see Figure 14.3). Always approach the setting of a manual analyzer from the same direction so that identical conditions are maintained in your measurements. If you are using a manual instrument, you should use the average of five to seven trials in setting the zero position.

Next you can try your hand at measuring the optical rotation of a sucrose-water solution whose concentration you know accurately. Use a 5.0% or a 10.0% solution. Check your ability to use the polarimeter by calculating the specific rotation of sucrose, using the equation for $[\alpha]$. Again, you should calculate the average of five to seven readings of the optical rotation.

## 14.5    Enantiomeric Excess (Optical Purity)

The results of polarimetric analysis of a stereoisomer are reported in terms of enantiomeric excess (or, less accurately, optical purity). Enantiomeric excess (% ee) is calculated from the expression

$$\% \text{ ee} = \left( \frac{[\alpha]_{\text{observed}}}{[\alpha]_{\text{pure}}} \right) \times 100\%$$

Thus, if we observe a rotation for 2-butanol of $+6.5°$, we can calculate the enantiomeric excess (% ee) of the sample if we know the specific rotation of pure 2-butanol ($[\alpha] = +13.00°$):

$$\% \text{ ee} = \left( \frac{6.5}{13.00} \right) \times 100\% = 50\%$$

It is instructive to examine the molecular composition of 100 molecules of the mixture with a % ee = 50%. We have an excess of 50 (+) molecules, which causes the optical activity. The remaining 50 molecules, because they have no net optical activity, are composed of 25 (+) molecules and 25 (−) molecules. Thus we have a total of 25 (−) molecules, and 50 + 25 = 75 (+) molecules.

## 14.6    Chiral Shift Reagents and Chiral Chromatography

After a resolution procedure has been completed, it is important to determine how successful the process has been by converting a mixture of enantiomers to a corresponding mixture of diastereomers. The composition of the diastereomers corresponds to the composition of the original mixture of enantiomers. Here we briefly describe two methods of measuring enantiomeric enrichment: use of lanthanide shift reagents that are chiral and use of chiral chromatography columns.

***Chiral Shift Reagents***   Derivatives of camphor provide shift reagents that are rich in chiral character; Eu(hfc)$_3$, called tris[3-heptafluoropropylhydroxymethylene)-(+)-camphorato] europium, is such a compound. This compound undergoes rapid and reversible coordination with Lewis bases, (B:), establishing the following equilibrium:

$$Eu(hfc)_3 + B: \rightleftharpoons B: Eu(hfc)_3$$

The complex B: Eu(hfc)$_3$ brings a paramagnetic ion, Eu$^{3+}$, into close proximity to the chiral organic compound (B:), which induces changes in the $^1$H NMR chemical shifts of B: in the complex. The shifts are different in each of the two coordinated enantiomers because the formation of the diastereomeric pair causes those protons to become diastereotopic and thus nonequivalent.

Therefore, we merely look for the peaks of the two diastereomeric protons and measure their areas to determine the composition of the B: Eu(hfc)$_3$ complex, which equals the enantiomeric composition of the original compound.

***Chiral Chromatography***   The same general principle is used in chiral chromatography. When a mixture of enantiomers passes through a chiral chromatography column, we observe different coordination strengths between each enantiomer and the stationary phase, differences that lead to separation of the enantiomers. A typical stationary phase that produces this effect is composed of a chiral substance such as albumin or a carbohydrate immobilized by bonding to silica gel. The separation of the enantiomers of the 3,5-dinitrobenzamide of phenylglycine, an amino acid, is an example of chiral chromatography. The less tightly coordinated enantiomer passes through the column more rapidly than the tightly coordinated enantiomers.

## References

1. Gordon, A. J.; Ford, R. A. *The Chemist's Companion: A Handbook of Practical Data, Techniques, and References;* Wiley-Interscience: New York, 1972.

2. Morrill, T. C. (Ed.) *Lanthanide Shift Reagents in Stereochemical Analysis;* VCH Publishers: New York, 1986.

## Questions

1. A sample of 2-butanol shows an optical rotation of +3.25°. Determine the % ee and the molecular composition of this sample. The optical rotation of pure (+)-2-butanol is +13.0°.
2. A sample of 2-butanol (see Question 1) shows an optical rotation of −9.75°. Determine the % ee and the molecular composition of this sample.
3. An optical rotation study gives a result of α = +140°. Suggest a dilution experiment to test whether the result is indeed +140°, not −220°.
4. The structures of strychnine (R = H) and brucine (R = CH$_3$O) are examples of alkaloid bases that can be used for resolutions. These molecules are rich sources of chirality (respectively, [α]$_D$ = −104° and −85° in absolute

ethanol). Assume that nitrogen inversion is slow and identify the eight chiral centers in each of these two nitrogen heterocyclic compounds.

R = H, strychnine
R = CH$_3$O, brucine

5. Only one of the two nitrogens in strychnine and brucine acts as the basic site for the necessary acid-base reaction for a resolution. Which nitrogen, and why?

# PART

# 2

# Chromatography

Few experimental techniques rival chromatographic analysis for versatility or usefulness in separating complex mixtures of compounds. First used in the early 1900s by the Russian botanist Mikhail Semenovich Tsvet, chromatography got its name because it was used to separate mixtures of different colored substances.

Although the development of chromatographic methods was very slow until the 1930s, progress accelerated once chemists realized that chromatography could be used to separate colorless substances as well as colored ones.

Three types of chromatography are used extensively in organic chemistry and are discussed in detail in Part 2: *thin-layer chromatography (TLC), gas-liquid chromatography (GC),* and *liquid (column) chromatography (LC). Flash chromatography* and *high-pressure liquid chromatography (HPLC)* are discussed briefly.

## Principles of Chromatography

All chromatographic methods depend on the distribution of the substances being separated between the two phases of the chromatographic system, a mobile phase and a stationary phase. The *mobile phase* consists of a liquid or gas that carries the sample through the solid or liquid that forms the *stationary phase.* For example, in both thin-layer and liquid chromatography, a finely ground solid forms the stationary phase. A liquid solvent provides the mobile phase.

The compounds in a mixture separate because of differences in their affinities for the stationary phase and their solubilities in the mobile phase. A dynamic equilibrium exists between the sample components attached to the stationary phase and those dissolved in the mobile phase. The compounds being separated move slowly along the stationary phase in the direction of the mobile phase flow. Separations occur by two processes: by adsorption onto the stationary phase followed by desorption into the mobile phase or by partition between the stationary phase and the mobile phase.

## Adsorption Chromatography

In *adsorption chromatography,* the compounds being separated adsorb onto and desorb from the stationary phase many, many times as the solvent passes through the stationary phase. The components of the sample separate because polar compounds rather than nonpolar compounds are preferentially adsorbed onto the stationary phase. Therefore, polar compounds remain on the stationary phase longer than less polar compounds and require greater amounts of solvent (mobile phase) to carry them through the stationary phase.

## Partition Chromatography

It is possible to create a liquid stationary phase by coating solid particles or a solid surface with a liquid that does not dissolve off the solid when a solvent passes through the chromatographic system. In this instance, the stationary phase is actually the coating liquid, not the solid support. For example, water and methanol may bind so tightly to a polar solid surface that they stay put. In such a case, the compounds to be separated partition themselves between this stationary liquid phase and the mobile liquid phase (the solvent traveling through the stationary phase). This partitioning of a substance occurs in the same way a solute partitions itself between the two immiscible solvents used for an extraction [see Technique 8]. This kind of chromatography is called *partition chromatography.*

# 15

# THIN-LAYER CHROMATOGRAPHY

*If Technique 15 is your introduction to chromatographic analysis, read the introduction to chromatography on pp. 151–152 before you read Technique 15.*

Thin-layer chromatography, which appeared in the late 1950s, has become a widely used analytical technique. It is simple, inexpensive, fast, efficient, reasonably sensitive, and requires only milligram quantities of material. Thin-layer chromatography, or TLC, is especially useful for determining the number of components in a mixture, for possibly establishing whether or not two compounds are identical, and for following the course of a reaction.

## 15.1 Principles of Thin-Layer Chromatography

In thin-layer chromatography, glass, metal, or plastic plates coated with a thin layer of adsorbent serve as the stationary phase. The mobile phase is a pure liquid compound or a solution of several liquid compounds; the composition of the mobile phase depends on the polarities of the compounds in the mixture being separated. Most nonvolatile solid organic compounds can be analyzed by thin-layer chromatography. However, TLC does not work well for liquid compounds because their volatility can often lead to loss of the sample by evaporation from the TLC plate during the analysis.

*Overview of TLC Analysis*

To carry out a TLC experiment, a small amount of the mixture being separated is applied or spotted on the adsorbent near one end of the plate. Then the TLC plate is placed in a closed chamber, with the edge nearest the applied spot immersed in a shallow layer of *developing solvent*, which is the mobile phase (Figure 15.1). The solvent rises through the stationary phase by capillary action, a process called *developing the chromatogram.*

As the solvent ascends the plate, the sample is distributed between the mobile liquid phase and the stationary solid phase. The separation during the development process occurs as a result of the many equilibrations taking place between the mobile and stationary phases and the compounds being separated. **The more tightly a compound binds to the adsorbent, the more slowly it moves on the TLC plate.** The developing solvent moves nonpolar substances up the plate most rapidly. Polar substances travel up the plate slowly or possibly not at all as the solvent ascends by capillary action.

The TLC plate is removed from the developing chamber when the *solvent front* (leading edge of the solvent) is about 1 cm from the top of the plate. The position of the solvent front is marked immediately with a pencil, before the solvent evaporates. The plate is then placed in a ventilating hood to dry.

Several methods are available to *visualize* the spots. If the TLC plate is impregnated with a fluorescent indicator, the plate can be

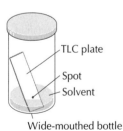

TLC plate

Spot

Solvent

Wide-mouthed bottle

**FIGURE 15.1**
Developing chamber containing a thin-layer plate.

illuminated by exposure to ultraviolet radiation. Alternatively, placing the plate in a jar containing a few iodine crystals may color the spots with the brown of an iodine charge-transfer complex. In another method, the TLC plate is dipped in a reagent that undergoes reaction with the compounds on the plate to produce a colored product when the plate is heated. The developed and visualized plate is then ready for analysis of the separation.

**Determination of the $R_f$**

Under a constant set of experimental conditions for a TLC chromatographic analysis, a given compound always travels a fixed distance relative to the distance traveled by the solvent front (Figure 15.2). This ratio of distances is called the $R_f$ *(ratio to the front)* and is expressed as a decimal fraction:

$$R_f = \frac{\text{Distance traveled by compound}}{\text{Distance traveled by developing solvent front}}$$

The $R_f$ value for a compound depends on its structure and is a physical characteristic of the compound, just as its melting point is a physical characteristic. Whenever a chromatogram is done, the $R_f$ value should be calculated for each substance and the experimental conditions recorded. The important data include the following:

1. Brand, type of backing, and adsorbent on the TLC plate
2. Developing solvent
3. Method used to visualize the compounds
4. $R_f$ value for each substance

To calculate the $R_f$ value for a given compound, measure the distance that the compound has traveled from where it was originally spotted and the distance that the solvent front has traveled (see Figure 15.2). The measurement is made from the center of a spot. The best data are obtained from chromatograms in which the spots are less than 5 mm in diameter. If spots show "tailing," measure from the densest point of the spot. The $R_f$ values for the two substances separated in the TLC plate in Figure 15.2 follow.

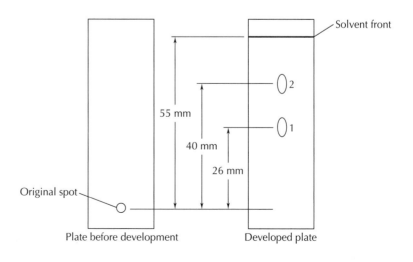

**FIGURE 15.2**
Calculating the $R_f$ value.

$$\text{Compound 1:} \qquad R_f = \frac{26 \text{ mm}}{55 \text{ mm}} = 0.47$$

$$\text{Compound 2:} \qquad R_f = \frac{40 \text{ mm}}{55 \text{ mm}} = 0.73$$

When two samples have identical $R_f$ values, you should not conclude that they are the same compound without doing further analysis. There are perhaps 100 distinct $R_f$ = values, whereas there are greater than $10^8$ known organic compounds. You could conclude that the samples are not the same if subsequent TLC analyses on the two samples with different developing solvents revealed different $R_f$ values for each sample. Further analysis by IR or NMR spectroscopy would be needed to provide definitive evidence as to whether the compounds were identical or not.

## 15.2     Plates for Thin-Layer Chromatography

**Adsorbents**

Three solid adsorbents—silica gel ($SiO_2 \cdot xH_2O$), aluminum oxide ($Al_2O_3$; also called alumina), and cellulose—are commonly used as stationary phases for thin-layer chromatography. A number of intermolecular forces cause organic molecules to bind to the solid stationary phase. Only weak van der Waals forces bind nonpolar compounds to an adsorbent, but polar molecules can also adsorb by dipole-dipole interactions, hydrogen bonding, and coordination to the highly polar metal oxide surfaces. The strength of the interaction varies for different compounds, but one generality can be stated: **The more polar the compound, the more strongly it binds to silica gel or alumina.**

Cellulose is used for the partition chromatography of water-soluble and quite polar organic compounds, such as sugars, amino acids, or nucleic acid derivatives. Cellulose can adsorb up to 20% of its weight in water; the substances being separated partition themselves between the developing solvent and the water molecules that are hydrogen bonded to the cellulose particles.

Silica gel and aluminum oxide are prepared from activated, finely ground powder. Activation usually involves heating the powder to remove adsorbed water. Silica gel is acidic, and it separates acidic and neutral compounds that are not too hydrophilic ("water-loving"). Aluminum oxide is available in acidic, basic, and neutral formulations.

**Backing for TLC Plates**

A number of manufacturers sell TLC plates that are precoated with a layer of adsorbent. Plates are available with plastic, glass, or aluminum backing. $R_f$ values for the same adsorbent and solvent may differ if plates of different backings are used.

*We suggest TLC plates*
*of 2.5 × 6.7 cm;*
*24 plates can be cut*
*from a standard*
*20 × 20 cm sheet.*

Plastic-backed silica gel plates are usually the least expensive. They can be cut to any desired size with scissors. The adsorbent surface is of uniform thickness, usually 0.10 mm. Results are quite reproducible, and sharp separation is normal. The plastic backing is generally a solvent-resistant polyester plastic. The adsorbent is bound to the plastic by solvent-resistant polyvinyl alcohol, which binds tightly to the adsorbent and to the plastic. Precoated plates impregnated with a fluorescent indicator are also available; these plates facilitate the visualization of many colorless compounds [see Technique 15.6].

TLC plates with a glass or aluminum backing are also available in the standard 20 × 20 cm sheets. Both types can be heated without melting the backing; this property is valuable if the plate is to be visualized with a reagent that requires heating [see Technique 15.6]. Aluminum sheets can be cut with scissors into convenient sizes for TLC plates.

If the plastic seal on the package containing the precoated sheets has been broken for some time, the TLC plates should be activated before use to remove adsorbed water. This practice enhances the reproducibility of $R_f$ values. Activation is done simply by heating the sheets in a clean oven at 100°C for 15–30 min.

## 15.3     Sample Application

*If you are analyzing a*
*solid, dissolve 10–20 mg*
*of it in 1 mL of the*
*solvent.*

The sample must be dissolved in a volatile organic solvent; a 1–2% solution works best. The solvent needs a high volatility so that it evaporates almost immediately. Acetone and dichloromethane are commonly used. If you are analyzing a solid, dissolve 10–20 mg of it in 1 mL of the solvent.

---

**SAFETY PRECAUTION**

Wear neoprene gloves and work in a hood while using $CH_2Cl_2$.

---

Tiny spots of the sample are carefully applied with a micropipet near one end of the plate. Keeping the spots small assures the cleanest separation. It is also important not to overload the plate with too much sample; overloading leads to large tailing spots and poor separation.

Commercial micropipets are available in 5- and 10-μL sizes and work well for applying samples on plastic-backed plates. Glass and aluminum-backed plates require micropipets of a smaller diameter. A micropipet can be made easily from an open-ended, thin-walled, melting-point capillary tube. The capillary tube is heated at its mid-point with a Bunsen burner. The softened glass tube is stretched and drawn into a narrower capillary.

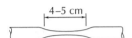

4–5 cm

**FIGURE 15.3**
Constricted capillary tube.

*No type of pen should be used for marking TLC plates because ink separates during development and may obscure the samples; the graphite and clay used in pencils do not dissolve in developing solvents.*

A microburner works best because only a small flame is required. While heating the tube, rotate it until it is soft on all sides over a length of 1–2 cm. When the tubing is soft, draw out the heated part until a constricted portion 4–5 cm long is formed (Figure 15.3). After cooling the tube for a minute or so, score it gently at the center with a file and break it into two capillary micropipets. The diameter at the end of a micropipet should be about 0.3 mm. The break must be a clean one, at right angles to the length of the tubing, so that when the tip of the micropipet is touched to the plate, liquid is pulled out by the adsorbent.

The micropipet is filled by dipping the constricted end into the solution to be analyzed. Only 1–5 µL of the sample solution are needed for most TLC analyses. Hold the micropipet vertically to the plate and apply the sample by touching the micropipet gently and *briefly* to the plate about 1 cm from the bottom edge (Figure 15.4). Mark the edge of the plate with a pencil at the same height as the center of the spot; this mark indicates the compound's starting point for your $R_f$ calculation. It is important to touch the micropipet to the plate very lightly so that no hole is gouged in the adsorbent. The spot delivered should be no more than 2–3 mm in diameter to avoid excessive broadening of the spot during the development. If you need to apply more sample, touch the micropipet to the plate a second time at exactly the same place. Let one spot dry before applying the next. The spotting procedure may be repeated numerous times, if necessary.

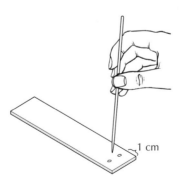

1 cm

**FIGURE 15.4**
Spotting a thin-layer plate.

You can quickly test for the proper amount of solution to spot on the plate by spotting two different amounts on the same slide; develop the plate and decide which spot gives better results.

If you are using 2.5 × 6.7 cm TLC plates, two side-by-side spots can be applied to one plate (see Figure 15.4). As a result of diffusion, the spots become larger during the development step. If the spots get too close to each other or to the edge of the plate, the chromatograms become difficult to interpret.

If available, an authentic standard should be included on the sample plate for comparison. If two compounds have the same $R_f$ value, they may be the same compound; if the $R_f$ values differ significantly, they most definitely are not the same compound. If $R_f$ values are quite close, it is best to run the chromatogram again, using a longer TLC plate or a different solvent.

| 15.4 | ## Development of a TLC Plate |

To ensure good chromatographic resolution, the developing chamber must be saturated with solvent vapors to prevent the evaporation of solvent as it rises on the TLC plate. Inserting a piece of filter paper halfway around the inside of the developing bottle helps to saturate the atmosphere with solvent vapor by wicking solvent into the upper region of the chamber. Not having constant evaporation of the solvent within the chamber is a major cause of nonreproducible $R_f$ values. Use enough developing solvent to allow a shallow layer (3–4 mm) to remain on the bottom after the closed chamber has been shaken to wet the filter paper. If the solvent level in the jar is too high, the spots on the plate may be below the solvent level. Under these conditions, the spots leach into the solvent, thereby ruining the chromatogram.

*The solvent depth in the developing chamber must be less than the height of the spots on the TLC plate.*

Uncap the developing chamber and place the TLC plate inside with a pair of tweezers (Figure 15.5). Recap the chamber, and let the solvent move up the plate. The adsorbent will become visibly moist. When the solvent front is about 1 cm from the top of the plate, remove the plate from the developing chamber with a pair of tweezers and *immediately* mark the adsorbent at the solvent front with a pencil. To get accurate $R_f$ values, the final position of the front must be marked before any evaporation occurs.

*Do not touch the adsorbent side of the TLC plate with your fingers. Hold the plate by the edges or with a pair of tweezers.*

*Do not lift or otherwise disturb the chamber while the TLC plate is developing.*

### SAFETY PRECAUTION

Evaporate the solvent from a developed chromatogram in a fume hood.

The development of a chromatogram is finished within 5–10 min. Let the developing solvent evaporate from the plate before visualizing the results.

**FIGURE 15.5**
Typical developing
chambers.

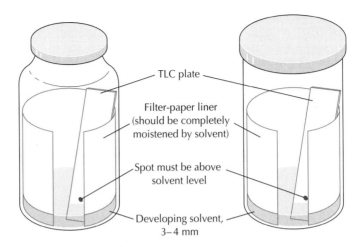

TLC plate

Filter-paper liner
(should be completely
moistened by solvent)

Spot must be above
solvent level

Developing solvent,
3–4 mm

---

## 15.5 Choice of Developing Solvent

One vital question remains. What developing solvent should you use? This question has no simple answer, but experience gives a number of helpful leads. In general, you should use a nonpolar developing solvent for nonpolar compounds and a polar developing solvent for polar compounds. Table 15.1 shows the relative polarity of common TLC developing solvents and organic compounds by functional-group class.

Chromatographic behavior is the result of competition between the compounds being separated and the developing solvent for the active surface sites of the adsorbent. If the solvent is adsorbed too

---

**TABLE 15.1 Relative polarities of common TLC solvents and organic compounds by functional group class**

| Common solvents | Increasing polarity | Organic compounds by functional-group class |
|---|---|---|
| Alkanes, cycloalkanes | | Alkanes |
| Toluene | | Alkenes |
| Dichloromethane | | Conjugated dienes, aromatic hydrocarbons |
| Diethyl ether | | Ethers, halocarbons |
| Chloroform | | Aldehydes, ketones, esters |
| Ethyl acetate | | Amines |
| Acetone | | Alcohols |
| Ethanol | | Carboxylic acids |
| Methanol | | |
| Acetonitrile | | |
| Water | | |

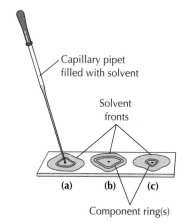

**FIGURE 15.6**
Rapid method for determining an
effective solvent: (a) good development;
(b) and (c) poor development.

(a)        (b)        (c)

Component ring(s)

well, the adsorbent is deactivated, $R_f$ values are increased, and separation may be incomplete. Therefore, a solvent that causes all the spotted material to move with the solvent front is too polar, whereas one that does not cause any compounds to move from the original spot is not polar enough. An appropriate solvent for a TLC analysis gives $R_f$ values of 0.20–0.70, with ideal values of 0.30–0.50.

With a silica gel plate, nonpolar hydrocarbons should be developed with hydrocarbon solvents, but a mixture containing an alcohol and an ester might be developed with a toluene-dichloromethane mixture. When a very polar solvent is required to move spots on a particular TLC adsorbent, better results are obtained by switching to a less active adsorbent and a less polar solvent.

As a rapid way to determine the best TLC developing solvent among several possibilities, three or four samples can be spotted along the length of the same plate (Figure 15.6). Fill a micropipet with the solvent to be tested and gently touch one of the spots. The solvent will diffuse outward in a circle, and the sample will move out with it. Mixtures of compounds will be partially separated and approximate $R_f$ values can be estimated.

Adding a small amount of a second solvent can dramatically affect the developing power of the resulting solution relative to that of the first solvent. Thus, the purity of the developing solvent is an important factor in the success of thin-layer chromatography. For example, chloroform is commonly stabilized with 1% ethanol. The presence of ethanol changes the developing power substantially from that of pure chloroform.

## 15.6    Visualization Techniques

*Fluorescence*

Chromatographic separations of colored compounds can usually be seen directly on the TLC plate, but colorless compounds require an indirect method. The simplest visualization technique involves the use of adsorbents that contain a fluorescent indicator. A common fluorescent indicator is calcium silicate, activated with lead and manganese. The insoluble inorganic indicator rarely interferes in any way with the chromatographic results and makes visualization straightforward. When the output from a short-wavelength ultraviolet lamp (254 nm) shines on the plate in a darkened room or dark box, the plate fluoresces visible light (Figure 15.7).

---

**SAFETY PRECAUTION**

Never look directly at an ultraviolet radiation source. Like the sun, it can cause eye damage.

---

The separated compounds appear as dark spots on the fluorescent field because the substances quench the fluoresence of the adsorbent as shown in Figure 15.7a. Sometimes substances are visible by their own fluorescence. Outline each spot with a pencil while the plate is under the UV source to give you a permanent record of the analysis and allow the calculation of $R_f$ values. Not all substances are visible on fluorescent silica gel, so visualization by one of the following methods should also be tried on any unknown sample.

***Visualization Reagents***

Another way to visualize colorless organic compounds uses their absorption of iodine ($I_2$) vapor. The TLC plate is put in a bath of iodine vapor prepared by placing 0.5 g of iodine crystals in a tightly capped bottle. Colored spots are gradually produced from the reaction of the substances with gaseous iodine. The spots are dark brown on a white to tan background. After 10–15 min, the plate can be removed from the bottle. Sometimes it is necessary to warm the bottle on a steam bath or hot plate to increase the amount of iodine vapor surrounding the TLC plate. The colored spots disappear in a short period of time and should be outlined with a

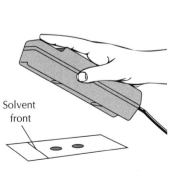

Solvent front

**FIGURE 15.7**
Visualization.

(a) Using an ultraviolet lamp

(b) Using an ultraviolet lamp with dark box

pencil *immediately* after the plate is removed from the iodine bath. The spots will reappear if the plate is again treated with iodine.

TLC plates can be dipped in visualizing solutions containing reagents that react to form colored compounds. Alternatively, the TLC plates can be sprayed with the visualizing solution. Visualization occurs by heating the dipped or sprayed TLC plates with a heat gun or on a hot plate. Many visualization reagents are specific for certain functional groups. Two common visualizing solutions are *p*-anisaldehyde and phosphomolybdic acid.[*] Consult the references at the end of this chapter for detailed discussions of visualization reagents.

**Quantitative Information**

The size and intensity of the spots can be used as a rough measure of the relative amounts of the substances. These parameters can be misleading, however, especially with fluorescent visualization. Some organic compounds interact much more intensely with ultraviolet radiation than do others. It is sometimes possible to obtain quantitative information by running mixtures of known composition alongside the mixture being studied. The spots can also be scraped off the plate and studied spectroscopically. It should be noted, however, that quantitative information is not one of the strengths of thin-layer chromatography.

## 15.7 Summary of TLC Procedure

1. Obtain a precoated TLC plate of the proper size for the developing chamber.
2. Spot the plate with a small amount of a 1–2% solution containing the materials to be separated and mark its position on the plate.
3. Add a filter-paper wick to the developing jar and cap it.
4. Develop the chromatogram with a suitable solvent.
5. Mark the solvent front immediately after removing the plate from the developing chamber.
6. Visualize the chromatogram and outline the separated spots.
7. Calculate the $R_f$ value for each compound.

## References

1. Touchstone, J. C.; Dobbins, M. F. *Practice of Thin Layer Chromatography;* 2nd ed.; Wiley: New York, 1983.
2. Fried, B.; Sherma, J. *Thin-Layer Chromatography: Techniques and Applications;* 3rd ed.; Chromatographic Science Series, Vol. 66, Marcel Dekker: New York, 1994.
3. Sherma, J.; Fried, B. (Eds.) *Handbook of Thin-Layer Chromatography;* 2nd ed.; Chromatographic Science Series, Vol. 71, Marcel Dekker: New York, 1996.

[*] *p*-Anisaldehyde developing solution: 2 mL of *p*-anisaldehyde in 2 mL of concentrated sulfuric acid, 36 mL of 95% ethanol, and 5 drops of acetic acid. Phosphomolybdic acid developing solution: 20% by weight in ethanol.

## Questions

1. The $R_f$ value of compound A is 0.34 when developed in hexane and 0.47 when developed in dichloromethane. Compound B has an $R_f$ value of 0.42 in hexane and 0.69 in dichloromethane. Which solvent would be better for separating a mixture of compounds A and B? Explain.

2. A student wishes to use TLC analysis on a mixture containing an alcohol and a ketone. After consulting Table 15.1, suggest a likely developing solvent.

3. When 2-propanol was used as the developing solvent, two substances moved with the solvent front ($R_f = 1$) during TLC analysis on a silica gel plate. Can you conclude that they are identical? If not, what additional experiment(s) would you perform?

4. If the structures of the two compounds in Question 2 are very similar except for the functional group, which would have the larger $R_f$ (a) on silica gel? (b) on alumina?

*If Technique 16 is your introduction to chromatographic analysis, read the introduction to chromatography on pp. 151–152 before you read Technique 16.*

**TECHNIQUE**

# 16   GAS-LIQUID CHROMATOGRAPHY

Few techniques have altered the analysis of volatile organic chemicals as much as gas-liquid chromatography (GC). Before GC became widely available, organic chemists usually looked for ways to convert liquid compounds to solids in order to analyze them. Gas-liquid chromatography provides a quick, easy way for both qualitative and quantitative analysis of volatile organic mixtures. In addition, GC has a truly fantastic ability to separate complex mixtures. Gas-liquid chromatography does, however, have several limitations. It is useful only for the analysis of compounds that have vapor pressures high enough to allow them to pass through a GC column, and, like TLC, gas-liquid chromatography does not identify compounds unless known samples are available. The current practice of coupling a gas-liquid chromatographic instrument, called a *chromatograph,* with an IR, NMR, or mass spectrometer combines the superb separation capabilities of GC with the superior identification methods of spectroscopy.

Introduction of practical gas-liquid chromatography, developed by the British biochemists Archer J. P. Martin and Richard M. L. Synge, occurred in 1952, the year in which they received the Nobel prize in chemistry for their development of paper chromatography. Gas-liquid chromatography is known by a number of names: vapor-phase chromatography (VPC), gas-liquid partition chromatography (GLPC), or simply gas chromatography. We use the abbreviation "GC" throughout this book. GC is a technique you will use on many occasions as you study organic chemistry.

FIGURE 16.1
Microview of a packed column.

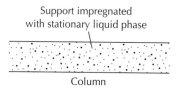

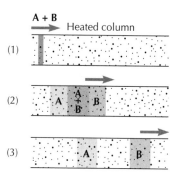

FIGURE 16.2
Stages in the separation of a two-component (**A, B**) mixture as it moves through the column.

***Overview***
***of Gas-Liquid***
***Chromatography***

In gas-liquid chromatography, the stationary phase consists of a nonvolatile liquid with a high boiling point. For packed column chromatographs, the liquid is coated on a porous, inert *solid support* that is then packed into a tube, which forms the column (Figure 16.1). For capillary column chromatographs, a uniform film of the liquid phase is applied to the interior wall of a capillary tube or to a thin layer of solid support lining the capillary tube, leaving a channel through the center for gas passage.

A flow of inert gas, such as helium or nitrogen, serves as the mobile phase. When the mixture being separated is injected into the flow of gas at the heated injection port, the components vaporize and pass into the column, where separation occurs. The components of the mixture *partition themselves between the gas and the liquid phases* in an equilibrium that depends on the temperature, the rate of gas flow through the column, and the solubility of the components in the liquid phase (Figure 16.2).

## 16.1    Instrumentation

The basic parts of a gas-liquid chromatograph are

> source of high-pressure pure *carrier gas*
> *flow controller*
> heated *sample inlet,* or *injection port*
> *column*
> *detector*
> *recorder*

These components are shown schematically in Figure 16.3.

A small hypodermic syringe is used to inject the sample through a sealed rubber septum or gasket into the stream of carrier gas flowing through the heated injection port (Figure 16.4). The sample vaporizes immediately and the carrier gas sweeps it into the column, a metal, glass, or fused-silica tube that contains the liquid phase (Figure 16.5). The column is enclosed in an oven whose temperature can be carefully regulated from just above room temperature to 250°C, even up to 400°C on some instruments. A GC column has thousands of theoretical plates as a result of the huge surface area on which the gas and liquid phases can interact [see Technique 11.4, p. 118, for a discussion of theoretical plates]. After the sample's components are separated by the column, they pass into a detector, where they produce signals that can be amplified and recorded. The three most crucial features of a gas chromatograph are the type of column, the liquid stationary phase, and the detection system.

**FIGURE 16.3**
Schematic diagram of a gas-liquid chromatograph.

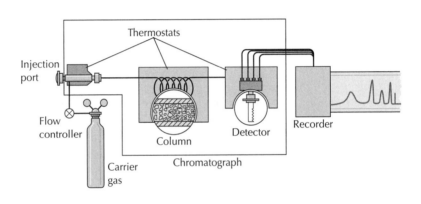

**FIGURE 16.4**
Injection port during sample injection.

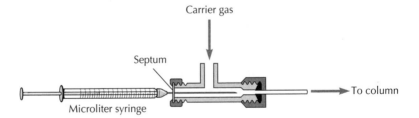

**FIGURE 16.5**
GC columns.

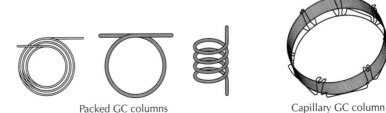

Packed GC columns

Capillary GC column

## 16.2    Columns and Liquid Stationary Phases

Mixtures separate during gas chromatography because their components interact in different ways with the liquid stationary phase. The partitioning of a substance between the liquid stationary phase and the vapor phase depends both on its relative attraction for the liquid phase and on its vapor pressure. All gases mix to form solutions, and the solubility of any given compound in the gas phase is a function of its vapor pressure alone. The greater a compound's vapor pressure, the greater its tendency to go from the liquid stationary phase to the gas phase. So, in the thousands of liquid-gas equilibria that take place as substances travel through a GC column, a more volatile compound spends less time in liquid solution and more time in the vapor state, thus traveling through the column faster than a less volatile compound. In general, **lower-boiling-point compounds with their higher vapor pressures travel through a GC column faster.**

What factors most influence the efficiency of separation? The answer to this question involves the surface area of the liquid stationary phase within the column, the nature of the liquid stationary phase, and the column temperature.

*Types of Columns*

A gas chromatograph may have either a packed column or a capillary column; the latter is also called an open tubular column. A packed column usually has an interior diameter of 2–4 mm, is 2–3 m in length, and contains the packed solid support coated evenly with a liquid phase of 0.05–1 μm in thickness. A capillary column has a much smaller interior diameter of only 0.25–0.50 mm, a length of 10–100 meters, with coating thicknesses ranging from 0.1 to 5 μm.

Several types of capillary columns are available. In a *wall-coated open tubular column (WCOT)*, the liquid phase coats the interior surface of the tube, leaving the center open; in a *support-coated open tubular column (SCOT)*, the liquid phase coats a thin layer of solid support that is bonded to the capillary wall. *Fused-silica open tubular columns (FSOT)* represent a recent improvement in column design. These columns have better physical strength, lower reactivity toward sample components, and better flexibility for bending into coils than do glass SCOT and WCOT columns. The characteristics of typical GC columns are summarized in Table 16.1.

Capillary columns give much better resolution (separation) than do packed columns. The greater length of capillary columns plus the higher gas flow rate provide more theoretical plates where equilibration between the components of the sample with the liquid stationary phase and the gas phase can occur than a packed column does.

*Solid Support
Within a Column*

The solid support in packed columns or SCOT capillary columns consists of a porous, inert material that has a very high surface area. The most commonly used substance is calcined diatomaceous

| TABLE 16.1 | **Characteristics of typical GC columns** | | | | | | | |
|---|---|---|---|---|---|---|---|---|
| Type of column | Length, m | Inside diameter, mm | Efficiency, plates/m | Sample size, ng | Relative pressure | Relative speed | Chemical inertness | Flexible? |
| FSOT[a] | 10–100 | 0.1–0.53 | 2000–4000 | 10–75 | Low | Fast | Best | Yes |
| WCOT[b] | 10–100 | 0.25–0.75 | 1000–4000 | 10–1000 | Low | Fast | | No |
| SCOT[c] | 10–100 | 0.5 | 600–1200 | 10–1000 | Low | Fast | | No |
| Packed | 1–6 | 2–4 | 500–1000 | $10-10^6$ | High | Slow | Poorest | No |

SOURCE: D. A. Skoog and J. J. Leary, *Principles of Instrumental Analysis,* 4th ed., 1992, p. 615, by permission of the publishers, Harcourt, Brace and Co., Orlando, FL.
a. FSOT, fused-silica open tubular column.
b. WCOT, wall-coated open tubular column.
c. SCOT, support-coated open tubular column.

earth, which contains the crushed skeletons of algae, especially diatoms. Its major component is silica. The efficiency of separation increases with decreasing particle size as a consequence of the expanded surface area available for the liquid coating, although there is a practical lower limit to the particle size imposed by the gas pressure needed to push the mobile phase through the column.

***Nature of the Liquid Stationary Phase***

The liquid stationary phase interacts with the substances being separated by a number of intermolecular forces: dipolar interactions, van der Waals forces, hydrogen bonding, and specific functional group interactions. These intermolecular forces determine the relative volatility of the adsorbed compounds and play dominant roles in the separation process.

As a general rule, the liquid phase is most effective if it is chemically similar to the material being separated. Nonpolar liquid coatings are used to separate nonpolar compounds, whereas polar compounds are best separated with polar liquid phases. In part, this rule is simply a manifestation of the old adage "Like dissolves like." Unless the sample dissolves well in the liquid phase, little or no separation occurs as the substance passes through the column. Table 16.2 lists commonly used liquid stationary phases for both packed and capillary columns and the type of compounds that can be separated on them.

The polysiloxanes have the same silicon-oxygen backbone, with some variation in the —R substituent:

$$R\diagdown\underset{\underset{R}{|}}{\overset{\overset{R}{|}}{Si}}\diagup O\left[\diagdown\underset{\underset{R}{|}}{\overset{\overset{R}{|}}{Si}}\diagup O\right]_n \underset{\underset{R}{|}}{\overset{\overset{R}{|}}{Si}}\diagdown R$$

Polydimethyl siloxane has —CH$_3$ for all —R groups, making it a nonpolar stationary phase. The other polysiloxanes have various

**TABLE 16.2   Common stationary phases for GC**

| Stationary phase | Common trade name | Maximum temperature, °C | Common applications |
|---|---|---|---|
| Polydimethyl siloxane | OV-1, SE-30 | 350 | General purpose nonpolar phase; hydrocarbons; polynuclear aromatics; drugs; steroids; PCBs |
| Poly(phenylmethyl-dimethyl) siloxane (10% phenyl) | OV-3, SE-52 | 350 | Fatty acid methyl esters; alkaloids; drugs, halogenated compounds |
| Poly(phenylmethyl) siloxane (50% phenyl) | OV-17 | 250 | Drugs; steroids; pesticides; glycols |
| Poly(trifluoropropyl-dimethyl) siloxane | OV-210 | 200 | Chlorinated aromatics; nitroaromatics; alkyl-substituted benzenes |
| Polyethylene glycol | Carbowax 20M | 250 | Free acids; alcohols; ethers; essential oils; glycols |
| Poly(dicyanoallyl-dimethyl) siloxane | OV-275 | 240 | Polyunsaturated fatty acids; rosin acids; free acids; alcohols |

SOURCE: D. A. Skoog and J. J. Leary, *Principles of Instrumental Analysis*; 4th ed., 1992, p. 617, by permission of the publishers, Harcourt, Brace and Co., Orlando, FL.

numbers of methyl groups replaced by other functional groups such as phenyl ($-C_6H_5$), 3,3,3-trifluoropropyl ($-CH_2CH_2-CF_3$), and dicyanoallyl ($-CH=CH-CH(CN)_2$). These functional groups increase the polarity of the liquid and allow the design of a wide variety of stationary phases suited to almost any application.

Polyethylene glycol, commonly used for separating polar compounds, has the structure

$$HO-CH_2-CH_2-(O-CH_2-CH_2)_n-OH$$

*Column Temperature*     Another important characteristic of the liquid phase is its useful temperature range. A stationary phase cannot be used under conditions in which it decomposes or in which its vapor pressure is high. Therefore, most columns have a specified temperature maximum. Table 16.2 includes the temperature limits for several common stationary-phase liquids. All liquid stationary phases evaporate or "bleed" if they are heated to a high enough temperature; this vaporized material then fouls the detector.

The proper choice of a liquid stationary phase is a trial-and-error process. Published experimental procedures usually specify the type of GC column used, but eventually you will need to make your own choices. Tables of appropriate liquid phases for specific classes of compounds can be found in the references at the end of the chapter.

## 16.3          Detectors

Two kinds of detectors are most often used in gas-liquid chromatography: *thermal conductivity detectors* and *flame ionization detectors.* The function of a detector is to "sense" a material and convert that sensing into an electrical signal.

*Thermal Conductivity Detectors*

Thermal conductivity detectors operate on the principle that heat can be conducted away from a hot body at a rate that depends on the composition of the gas surrounding the body. In other words, heat loss is related to gas composition. The electrical component of the detector is a hot wire or filament. Most of the heat loss from the hot wire of the detector occurs by conduction through the gas and depends on the rate at which gas molecules can diffuse to and from the metal surface. Helium, the carrier gas most often used with thermal conductivity detectors, has an extremely high thermal conductivity. However, organic molecules are less efficient heat conductors because they diffuse more slowly. With only carrier gas flowing, a constant heat loss is maintained and there is constant electrical output. As a sample reaches the detector, the gas composition changes, causing the hot filament to heat up and its electrical resistance to increase. The change in electrical resistance creates an imbalance in the electrical circuit that can be recorded.

As the separated materials leave the column, they enter the detector cell. In practice, the filament of the detector, a tungsten-rhenium or platinum wire, operates at temperatures from 200°C to over 400°C. An enlarged view of the most common thermal conductivity detector is shown in Figure 16.6.

Thermal conductivity detectors have the advantages of stability, simplicity, and recovery of the separated materials but the disadvantage of low sensitivity. Because of their low sensitivity, they are unsuitable for use with capillary columns.

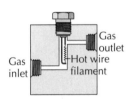

**FIGURE 16.6**
Thermal conductivity detector.

*Flame Ionization Detectors*

Flame ionization is much more sensitive than thermal conductivity and is used with capillary columns, where the amount of sample reaching the detector is substantially less than that emanating from a packed column. Noncombustible gases such as water, carbon dioxide, sulfur dioxide, and various nitrogen oxides produce no response with a flame ionization detector, thus allowing the analysis of samples contaminated with water or these other oxides. However, the destruction of the sample is one disadvantage of a flame ionization detector. An instrument equipped with a splitter that detours some of the column effluent before it reaches the detector eliminates this disadvantage.

In a flame ionization detector, the carrier gas leaving the column is mixed with hydrogen and air and burned. The combustion process produces ions that alter the current output of the detector (Figure 16.7). A typical flame ionization detector is shown in Figure 16.8. In the chromatograph, the electrical output of the flame is fed to an electrometer, whose response can be recorded.

**FIGURE 16.7**
Chemical reactions
in a flame ionization
detector.

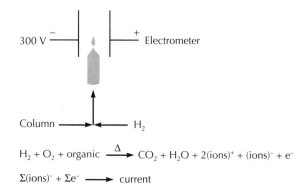

$$H_2 + O_2 + organic \xrightarrow{\Delta} CO_2 + H_2O + 2(ions)^+ + (ions)^- + e^-$$

$$\Sigma(ions)^- + \Sigma e^- \longrightarrow current$$

**FIGURE 16.8**
Flame ionization
detector.

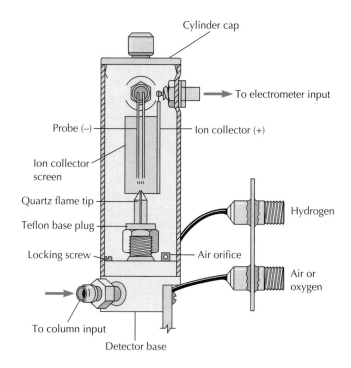

16.4          **Interpreting Chromatograms**

The recorded response of the detector's electrical signal as the sample passes through it is called a chromatogram. A typical chromatogram for a mixture of alcohols is shown in Figure 16.9. The line at the left shows when the sample was injected into the column. The pen responds to the changes in the electrical signal as each component of the mixture passes through the detector. You will notice that the later peaks are broader. This pattern is typical; the longer a compound remains on the column, the broader its peak will be when it passes through the detector.

**FIGURE 16.9**
Typical chromatogram.

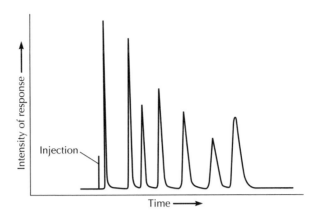

*Retention Time*

Under a definite set of experimental conditions, a compound always travels through a GC column in a fixed amount of time, called the *retention time*. The retention time for a compound, like the $R_f$ value in thin-layer chromatography, is an important number, and it is reproducible if the same set of instrumental parameters is maintained from one analysis to another.

In Figure 16.10 you can see how retention times are calculated from a GC chart record. The distance from the time of injection to the time the pen reaches the peak maximum is the retention time for that compound. You can determine the retention time by measuring the distance on the chromatogram with a ruler and dividing the distance by the recorder chart speed. Chromatographs equipped with digital integrators also record the retention times of peaks.

The retention time depends on many factors. Of course, the compound's structure is one factor. Beyond that, the kind and amount of stationary liquid phase used on the column, the length of the column, the carrier gas flow, the column temperature, the solid support, and the column diameter are most important. To

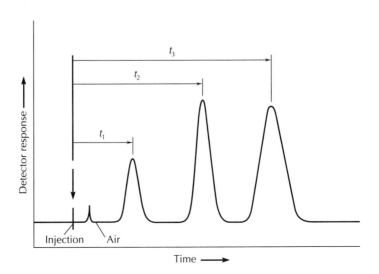

**FIGURE 16.10**
Measuring
retention times.

some extent, the sample size can also affect the retention time. Always record these experimental variables when you note a retention time in your lab notebook.

---

## 16.5    Practical GC Operating Procedures

The procedure for a GC analysis involves several steps. Decide whether a polar or nonpolar column is needed to separate the sample being analyzed. Make sure that the chromatograph and the recorder are warmed up and ready to go and that the GC column oven, detector, and injector port are at the correct temperatures. Temperature equilibration usually requires 20–30 min for a given set of parameters.

Ascertain that the carrier-gas pressure is properly set. The necessary pressure depends on the instrument and columns you are using, so you should check with your instructor before changing the pressure setting. Smaller-diameter packed columns are normally operated at lower flow rates. For example, with a packed column 6 ft long and 1/8 inch in diameter, a flow rate of 20–30 mL·min$^{-1}$ is common. For a 1/4-inch column of the same length, 60–70 mL·min$^{-1}$ is usual. Flow rates for capillary columns generally range from 60 to 70 mL·min$^{-1}$.

A convenient measure of the carrier-gas flow rate in a packed column chromatograph is made at the exit port by using a soap-film (bubble) flowmeter (Figure 16.11). The rate at which a soap film moves up the calibrated tube is measured by using a sweep second hand or stopwatch. Capillary column chromatographs have built-in flowmeters.

*Turning on the Detector*

A thin filament with electricity running through it can oxidize and burn in the presence of oxygen. Only after the inert carrier-gas flow has been on for 2–3 min should the current be turned on in a thermal conductivity detector. Within a few minutes, the current stabilizes. The flame ionization detector is not ignited until the injector port, column oven, and detector have reached the temperatures needed for the analysis. Before the sample is injected, the detector circuit must be balanced and the proper sensitivity (attenuation) chosen for

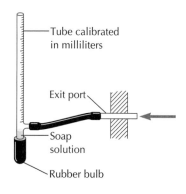

**FIGURE 16.11**
Bubble flowmeter.

the analysis. Your instructor will show you how to do these operations, because the techniques differ for different instruments.

*Sample Size to Inject*

As soon as the instrument is ready, you can inject your sample. Gas-liquid chromatographs take very small samples; if too much sample is injected, poor separation will occur from overloading of the column. Normally, for a packed column GC, 1–3 μL of a volatile mixture are injected through the rubber septum with a special microliter hypodermic syringe. For capillary columns, the sample must be extremely dilute, usually one or two drops diluted with a milliliter of a volatile solvent such as diethyl ether; 0.5–1.0 μL of the diluted sample solution is injected. Consult your instructor about sample preparation and sample size for the chromatographs in your laboratory.

*Injection Technique*

Proper injection technique is important if you want to get well-formed peaks on the chromatogram. The needle should be pushed into the injection port and the injection made immediately with a smooth, rapid motion (Figure 16.12). Withdraw the syringe needle quickly after completing the injection. This procedure ensures that the entire sample reaches the column at one time.

*Marking injection time.* The time of the injection can be recorded on the chromatogram in several ways. A mark can be made on the recorder base line just after the sample has been injected, but this is hard to do reproducibly. With a packed column, a better way is to include several microliters of air in your syringe. The air is injected at the same time as the sample, and it comes through the column very quickly. The first tiny peak in Figure 16.9 is due to air. Retention times are then calculated, using this air peak as the injection time. A third way to mark the injection time works well with some packed column instruments. As you inject the sample, turn the base line

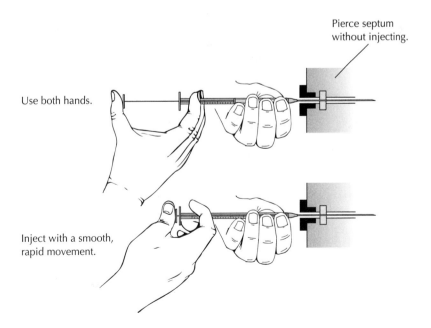

Pierce septum without injecting.

Use both hands.

Inject with a smooth, rapid movement.

**FIGURE 16.12**
Injecting the sample into the column.

knob on the recorder back and forth to make a mark on the chart paper. For an instrument equipped with an automatic digital integrator, simply press the start button at the time of sample injection.

***Microliter syringes.*** A microliter syringe has a tiny bore that can easily become clogged if it is not rinsed after use. If viscous organic liquids or solutions containing acidic residues are allowed to remain in the syringe, you may find that it is almost impossible to move the plunger in and out. For this reason, a small bottle of acetone is often kept beside each instrument. One or two fillings of the syringe with acetone will normally suffice, if done directly after its use. However, it is unnecessary to rinse the syringe with acetone after each injection in a series of analyses. This practice may even cause confusion if traces of acetone show up on the chromatogram. For multiple analyses, it is best to rinse the syringe several times with the next sample to be analyzed before filling the syringe with the injection sample.

After injection, wait for the peaks to appear on the moving chart. Sometimes it is difficult to know exactly how long to wait before injecting another sample, because components with unexpectedly long retention times may be present in the first sample. The analysis of unknown mixtures is a matter of trial and error. However, when you are analyzing mixtures with a known number of components, you need wait only until the last component has come through the column.

*Record the following experimental parameters: injection port temperature, column temperature, detector temperature, carrier gas flow rate, injection sample size, and length of column and nature of its stationary phase.*

When you have finished your analyses, thoroughly rinse out the microliter syringe with acetone and tear off the chart paper record. You will want to attach your GC traces firmly in your notebook, along with a notation of the experimental conditions under which the chromatograms were run.

**Poor Separations**

If the components of your mixture are not well separated, a number of factors can be adjusted. You may have injected too much sample into the column, the column temperature may be too high, the wrong stationary liquid phase may have been used, or the packed column may be too short.

---

## 16.6     Identification of Components Shown on a Chromatogram

GC analysis can quickly assess the purity of a compound, but as with thin-layer chromatography, a compound cannot be identified by gas-liquid chromatography unless a known sample is available. Comparison of retention times, peak enhancement, and spectroscopy are among the methods used to identify the components of a mixture.

**Comparison of Retention Times**

One method of identification compares the retention time of a known compound with the peaks on the chromatogram of the sample. If the operating conditions of the instrument are unchanged, a match of the known's retention time to one of the sample peaks

**FIGURE 16.13**
Identification by the
peak enhancement
method.

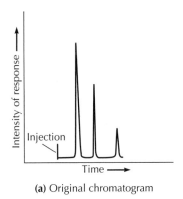

(a) Original chromatogram

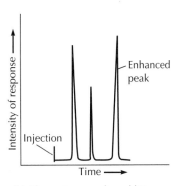

(b) Chromatogram after addition
of a known compound identical
to a compound in the sample

serves to identify it. This method will not work for a mixture in which the identity of the components is totally unknown, because several compounds could have identical retention times.

*Peak Enhancement*

When mixtures containing known compounds are being analyzed, *peak enhancement* serves as a method for identifying a peak in a chromatogram. The sample being analyzed is "spiked" with a drop of the known compound and the mixture injected into the chromatograph. If the known that is added is identical to one of the compounds in the mixture, its peak area is enhanced relative to the other peaks (Figure 16.13).

*Spectroscopic
Methods*

Positive identification of the compounds in a completely unknown mixture requires the pairing of gas chromatographic methods for separation with a spectroscopic method such as IR, NMR, or mass spectrometry. The separated components from the gas chromatograph can be collected for analysis with one of these instruments. More frequently, the instruments are interfaced with the chromatograph and the separated components pass directly from the chromatograph to the spectrometer; a gas chromatograph is most often paired with a mass spectrometer.

## 16.7     Quantitative Analysis

Gas-liquid chromatography is particularly useful for quantitative analysis of the components in volatile mixtures. A comparison of relative peak areas on the chromatogram often gives a good approximation of relative amounts of compounds in a mixture. For accurate quantification of a GC analysis, however, the response of each component to the detector must be determined from known samples.

*Determination
of Peak Areas*

One great advantage of GC over other chromatographic methods is that approximate quantitative data are almost as easy to obtain as information on the number of components in a mixture. If we assume

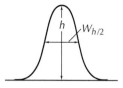

**FIGURE 16.14**
Determining peak area:
$h$, height; $W_{h/2}$, width
at half-height.

equal response by the detector to each compound, then the relative amounts of compounds in a mixture are proportional to their peak areas. Most peaks are approximately the shape of either an isosceles or a right triangle, whose areas are simply $A = 1/2 \times$ base $\times$ height. Measuring the base of most GC peaks is difficult because abnormalities in the shape usually occur in this region. A more accurate estimate of peak area is $A =$ height $\times$ width at half-height (Figure 16.14).

The easiest method for calculating the percentage composition of a mixture is through internal normalization. *The percentage of a compound in a mixture is its peak area divided by the sum of all peak areas.*

$$\% A = \frac{\text{area}_A}{\text{area}_A + \text{area}_B} \times 100$$

$$\% B = \frac{\text{area}_B}{\text{area}_A + \text{area}_B} \times 100$$

Mechanical or electronic digital integrators, common on most modern chromatographs, determine peak areas. Chromatograms produced by these recorders include a table of data that lists retention times and relative peak areas, making the calculations just described unnecessary.

***Relative Response Factors***

Each compound has a unique response in the detector, but detector response varies more between classes of compounds than it does within a homologous series of compounds. For accurate quantitative interpretation of a chromatogram, analysis of standard mixtures of known concentration must be carried out and a correction factor, called a ***response factor,*** $f$, determined for each compound. The area under a chromatographic peak, $A$, is proportional to the concentration, $C$, of the sample producing it; the response factor is the proportionality constant.

$$A = fC \tag{1}$$

Response factors can be determined as either ***weight factors*** or ***mole factors*** depending on the units of concentration used for the standard sample.

In chromatographic analyses, the samples being analyzed usually have more than one component; therefore the ***relative response factors*** of one compound to the other compounds in the sample are usually determined. For a two-component system, the response-factor equation for each component is

$$A_1 = f_1 C_1 \tag{2}$$

$$A_2 = f_2 C_2 \tag{3}$$

The relative response factor of compound 1 to compound 2 can be determined by dividing equation 2 by equation 3:

$$\frac{A_1}{A_2} = \frac{f_1}{f_2} \times \frac{C_1}{C_2} \tag{4}$$

**FIGURE 16.15**
Chromatogram of a standard mixture containing known concentrations of both components.

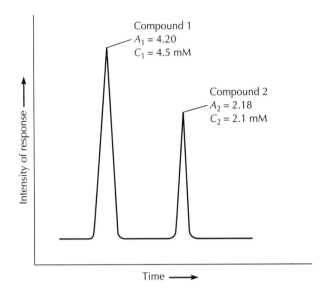

Rearranging equation 4 gives the ratio of response factors, $f_1/f_2$, the relative response factor of compound 1 to compound 2:

$$\frac{f_1}{f_2} = \frac{A_1}{A_2} \times \frac{C_2}{C_1} \tag{5}$$

Using data from the chromatogram shown in Figure 16.15 as an example, equation 5 can be used to calculate the molar response factor of compound 1 relative to compound 2, which is arbitrarily assigned a response factor of 1.00.

$$\frac{f_1}{f_2} = \frac{4.20}{2.18} \times \frac{2.1}{4.5}$$

Therefore, the molar response factor for compound 1 is 0.90 relative to 1.00 for compound 2.

Once relative molar response factors have been determined, the composition of a mixture can be calculated from the areas of the peaks on a chromatogram. Table 16.3 shows how molar response factors (designated $M_f$) can be used to determine the corrected mole percentage composition of a sample containing heptane and ethyl acetate, and it compares the results to the uncorrected composition. The substantial differences between the corrected and uncorrected compositions illustrate the necessity of using response-factor corrections for accurate quantitative analysis.

**TABLE 16.3  Molar percentage composition data for a heptane/ethyl acetate mixture corrected and uncorrected for molar response factors, $M_f$**

| Compound | Area (A) (arbitrary units) | Uncorrected % $((A/75.8) \times 100)$ | $M_f$ | $A \times M_f$ | Corrected mol % $((A \times M_f/54.2) \times 100)$ |
|---|---|---|---|---|---|
| Heptane | 50.2 | 66.2 | 0.57 | 28.6 | 52.8 |
| Ethyl acetate | 25.6 | 33.8 | 1.00 | 25.6 | 47.2 |
| Total | 75.8 | 100 | — | 54.2 | 100 |

## References

1.  Miller, J. M. *Chromatography: Concepts and Contrasts;* Wiley-Interscience: New York, 1988.
2.  Skoog, D. A.; Leary, J. J. *Principles of Instrumental Analysis;* 5th ed.; Saunders: New York, 1998.
3.  Ravindranath, B. *Principles and Practice of Chromatography;* Wiley: New York, 1989.
4.  Grob, R. L. *Modern Practice of Gas Chromatography;* 2nd ed.; Wiley: New York, 1985.

## Questions

1.  Why is a GC separation more efficient than a fractional distillation?
2.  What characteristics must a carrier gas have to be useful for GC? What must the characteristics of the column packing be?
3.  How do (a) the flow rate of the carrier gas and (b) the column temperature affect the retention time of a compound on a GC column?
4.  Describe a method whereby a compound can be identified by using GC analysis.
5.  Describe two methods by which a compound separated by and collected from a gas chromatograph can be identified.

6.  If the resolution of two components in a GC analysis is mediocre but shows some peak separation, what are two adjustments in the operating parameters that can be made to improve the resolution (without changing columns or instruments)?
7.  Suggest a suitable liquid stationary phase for the separation of (a) ethanol and water; (b) cyclopentanone (bp 130°C) and 2-hexanone (bp 128°C); (c) phenol (bp 182°C) and pentanoic acid (bp 186°C); (d) ethyl acetate (bp 77°C) and carbon tetrachloride (bp 77°C).

**TECHNIQUE**

# 17

# LIQUID CHROMATOGRAPHY

*If Technique 17 is your introduction to chromatographic analysis, read the introduction to chromatography on pp. 151–152 before you read Technique 17.*

Liquid chromatography (LC), also called column chromatography (and the related methods of flash chromatography and high-performance liquid chromatography—HPLC), completes the triad of chromatographic methods so important in experimental organic chemistry. Like thin-layer chromatography, liquid chromatography can be done under adsorption or partition conditions. Liquid chromatography is generally used to separate mixtures of low volatility, whereas gas chromatography works only for volatile mixtures. Unlike TLC, liquid chromatography can be carried out with a wide range of sample quantities, ranging from a few nanograms for HPLC up to 100 g for column chromatography.

| **17.1** | **Adsorbents** |

Two adsorbents are commonly used for the stationary phase in liquid chromatography: silica gel ($SiO_2 \cdot xH_2O$) and aluminum oxide (alumina, $Al_2O_3$). Both adsorbents produce a polar stationary phase (aluminum oxide is more polar), and both are generally used with nonpolar to moderately polar elution solvents as the mobile phase. Most chromatographic separations today use silica gel, because it allows the separation of compounds with a wide range of polarities. Silica gel also has the advantage of being less likely than alumina to cause a chemical reaction with the substances being separated. Alumina may be used for separations of compounds of low to medium polarity. Polar organic compounds, such as carboxylic acids, amines, phenols, and carbohydrates, can adsorb so tightly on an alumina surface that highly polar solvents are needed to dislodge them.

*Silica Gel*

Chromatographic silica gel has 10–20% adsorbed water by weight and acts as the solid support for water under the conditions of partition chromatography. Compounds separate by partitioning themselves between the elution solvent and the water that is strongly adsorbed on the silica surface. The partition equilibria depend on the relative solubilities of the compounds being separated in the two liquid phases. The adsorptive properties of silica gel may vary greatly from one manufacturer to another or even within different lots of the same grade from one manufacturer. Therefore, the solvent system previously used for a particular analysis may not work for another separation of the same sample mixture when silica gel from another manufacturer or from a different lot from the same manufacturer is substituted for the stationary phase.

*Alumina*

Activated alumina, made explicitly for chromatography, is available commercially as a finely ground powder in neutral (pH 7), basic (pH 10), and acidic (pH 4) grades. Different brands and grades vary enormously in adsorptive properties, mainly because of the amount of water adsorbed on the surface. Heating alumina at 400°–450°C until no more water is lost produces the most active grade, called Activity I in the classification system of Brockmann. Addition of 3% water gives alumina of Activity II, 6% water content gives Activity III, and 10% water yields Activity IV.

The strength of the adsorption holding a substance on aluminum oxide depends on the strength of the bonding forces between the substance and the polar surface of the adsorbent. Dipole-dipole and van der Waals forces, as well as stronger hydrogen-bonding and coordination interactions, are important in the adsorption of compounds to aluminum oxide particles.

*Adsorbents for Reverse-Phase Chromatography*

The separation of very polar compounds may require an adsorbent of lesser polarity than either silica gel or alumina. In this situation, *reverse-phase chromatography*, which uses a stationary phase that

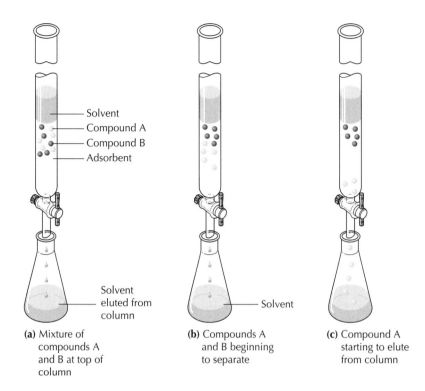

— Solvent
— Compound A
— Compound B
— Adsorbent

Solvent
eluted from
column

**(a)** Mixture of
compounds A
and B at top of
column

— Solvent

**(b)** Compounds A
and B beginning
to separate

**(c)** Compound A
starting to elute
from column

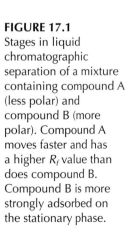

**FIGURE 17.1**
Stages in liquid
chromatographic
separation of a mixture
containing compound A
(less polar) and
compound B (more
polar). Compound A
moves faster and has
a higher $R_f$ value than
does compound B.
Compound B is more
strongly adsorbed on
the stationary phase.

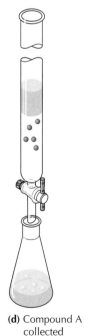

**(d)** Compound A
collected

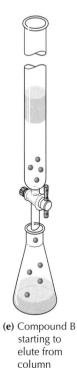

**(e)** Compound B
starting to
elute from
column

**(f)** Compound B
collected

is *less* polar than the mobile phase, may be useful in separating the compounds. Under reverse-phase conditions, elution of the more polar compounds occurs first, with the less polar compounds adsorbed more tightly to the stationary phase. Reverse-phase columns are especially useful in high-pressure liquid chromatography [see Technique 17.9]. The surface of silica particles is rendered less polar by replacing the hydrogen of terminal Si—OH groups with trialkyl silyl groups such as methyl and long-chain alkyl groups ($C_{12}$—$C_{18}$).

$$\text{www}\,O\!-\!\underset{\underset{O}{|}}{\overset{\overset{O}{|}}{Si}}\!-\!O\!-\!\underset{\underset{CH_3}{|}}{\overset{\overset{CH_3}{|}}{Si}}\!-\!(CH_2)_{17}CH_3$$

www = Siloxane bonds $\left(\!\!-\!O\!-\!Si\!-\!\right)_n$

***Separation Process on Adsorbents***

Figure 17.1 illustrates the separation of two compounds, A and B, on a chromatographic column. With a polar adsorbent such as alumina or silica gel, the compound represented by A would be less polar than compound B. In reverse-phase chromatography, a relatively nonpolar adsorbent would be used, and the compound represented by A would be more polar than compound B.

---

## 17.2    Elution Solvents

In column chromatography, the liquids used to dislodge the compounds adsorbed on the column, called *eluents,* are made progressively more polar as the separation progresses. Nonpolar compounds bind less tightly than polar compounds on a solid adsorbent and dislodge easily with nonpolar solvents. Therefore, the nonpolar compounds in a mixture will exit the column first. The more polar compounds must be *eluted,* or washed out of the column, with polar solvents.

For example, consider an activated alumina column with a band of cholesterol adsorbed at the top of the column. Cholesterol is an alcohol with a large surface area and medium polarity. Let us compare the action of a nonpolar elution solvent, such as hexane, to that of diethyl ether, a solvent of medium polarity. The primary forces between alumina and hexane molecules are weak van der Waals interactions. Hexane is not bound strongly to the column. The adsorption-desorption equilibria that occur as hexane travels down the column do little to dislodge cholesterol from the alumina surface. Diethyl ether, however, binds more strongly with alumina by dipole-dipole and coordination interactions. It can readily displace cholesterol as it travels down the column. In time, a band of cholesterol, dissolved in ether, can be collected from the bottom of the column.

HO
Cholesterol

**FIGURE 17.2**
Relative eluting
power of common
elution solvents.

INCREASING
ELUTING POWER
ON ALUMINA

Alkanes (petroleum ether, hexane, cyclohexane)
Carbon tetrachloride
Toluene
Dichloromethane, diethyl ether
Chloroform
Acetone
Ethyl acetate
Ethanol
Methanol

*Rate of Elution*

The balance between the activity of the adsorbent and the polarity of the solvent controls the rate at which materials elute from a column. If compounds elute too rapidly and poor separation occurs, either the adsorbent should be stronger or the solvent should be less polar. If elution is so slow that only polar solvents are effective, a milder adsorbent is needed. Figure 17.2 gives the usual order of elution, from the least potent (nonpolar) to the most powerful (polar) elution solvent. It should be pointed out that there is no universal series of eluting strengths, because this property depends not only on the activity of the adsorbent but also on the compounds being separated.

*Purity of*
*Elution Solvents*

Elution solvents for column chromatography must be rigorously purified and dried for best results. Small quantities of polar impurities can radically alter the eluting properties of a solvent. The presence of water in a solvent increases its eluting power. For example, wet acetone may have an eluting power greater than dry ethanol.

## 17.3 Column Dimensions

After deciding which adsorbent to use for a separation, you need to decide how much adsorbent to use. The choice naturally depends on the amount of material you wish to separate, but it also depends on the diameter of the cylindrical column to be used. A long, thin adsorbent column retains compounds more tenaciously than does a short, fat one. In general, for a moderately challenging separation, you should use about 20 times (by weight) as much adsorbent as the material to be separated (Table 17.1). More adsorbent should be used for a difficult separation, less for an easy one.

**TABLE 17.1 Representative columns**

| Sample, g | Alumina, g | Column dimensions, diameter × height in cm |
|-----------|------------|--------------------------------------------|
| 10 | 250 | 3.4327 |
| 2–3 | 60 | 2.0319 |

Silica gel has a bulk density of about $0.3 \ \text{g} \cdot \text{cm}^{-3}$, much lower than that of alumina. You would need to use a larger-diameter column for the same amount of silica gel. An 8:1 or 10:1 ratio of the height of the adsorbent to the column diameter is normal. Alumina has a bulk density of about $1 \, g \cdot cm^{-3}$. A calculation using the volume of a cylinder ($\pi r^2 h$) shows, for example, that 20 g of alumina would fill a column with a 1.5-cm inner diameter to a height of about 11 cm. Therefore, if you are using 20 g of alumina, choose a column with an inner diameter of about 1.5 cm.

Generally we use a chromatography column with a stopcock near the bottom (Figure 17.3a). However, a glass tube with a tapered end, finished with a small piece of rubber tubing fitted with a screw clamp and a small piece of glass tubing at the bottom (Figure 17.3b) works just as well. The results of a chromatographic separation are best when the flow of solvent is not stopped until the separation is complete.

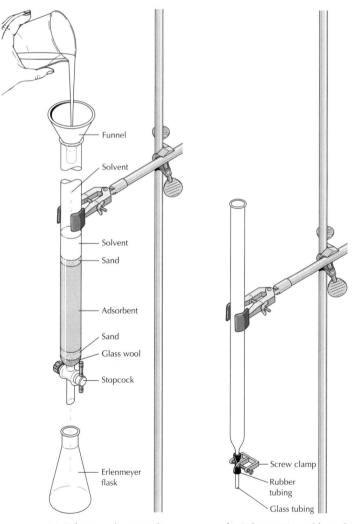

**FIGURE 17.3**
Chromatographic
columns.

(a) Column with stopcock

(b) Column using rubber tubing and
screw clamp instead of stopcock

| 17.4 | Preparing a Miniscale Column |

After you have selected a chromatography column and weighed the requisite amount of adsorbent, you are ready to prepare the column. A chromatographic column in operation is shown in Figure 17.3a. The construction of such a column is just as crucial to the success of the chromatography as is the choice of adsorbent and elution solvents. If the column of adsorbent has cracks or channels or if the top surface is not flat, separation will be poor.

Clamp the chromatography tube in an upright position on a ring stand or vertical support rod and fill it approximately one-half full either with the first developing solvent you plan to use or with a less polar solvent. Add a small piece of glass wool as a plug, and push it to the bottom of the column with a glass rod, making sure all the air bubbles are out of the glass wool. Cover the plug with approximately 6 mm of white sand. The glass wool plug and sand serve as a level support base to keep the adsorbent in the column and prevent it from clogging the stopcock or tip.

*Dry Adsorbent Method*

Pour the adsorbent slowly into the solvent-filled column. The stopcock should be closed. Take care that the adsorbent falls uniformly to the bottom. Do not add the adsorbent too quickly or clumping may occur. The adsorbent column should be firm, but if it is packed too tightly the flow of elution solvents becomes too slow.

The top of the adsorbent must always be horizontal. Gentle tapping on the side of the column as the adsorbent falls through the solvent prevents the appearance of bubbles in the adsorbent. If large bubbles or channels develop in the column, the adsorbent should be discarded, and the column should be repacked. Any irregularities in the adsorbent column may cause poor separation, because part of the advancing material will move faster than the rest. The time consumed in repacking will be much less than the time wasted trying to make a poor column function efficiently.

After all the adsorbent has been added, carefully pour approximately 4 mm of white sand on top. The layer of sand protects the adsorbent from mechanical disturbances when new solvents are poured into the column during the separation process.

*Slurry Method*

If you use a liquid more polar than alkanes or dichloromethane in making a column, you may need to mix, or slurry, the adsorbent in some of the liquid before pouring it into the column. This precaution prevents the formation of clumps or bubbles, which can form when heat is produced by the interaction between polar solvents and the surface of the metal oxide.

**The column must never be allowed to dry out once it is made,** and the solvent level should never be allowed to fall below the upper

*Be sure that the adsorbent is covered with solvent throughout the chromatographic procedure.*

level of the adsorbent. On drying, the adsorbent may pull away from the walls of the column and form channels. Once you begin a chromatographic separation, finish it without interruption. When you are finished, the chromatography tube can be emptied by opening the stopcock, pointing the upper end of the column into a beaker, and using gentle air pressure at the tip to push out the adsorbent.

---

## 17.5        Elution Techniques

***Application of Sample***

A mixture of solids is always poured onto the column dissolved in a solvent; liquid samples can be applied. The solvent used in packing the column or a solvent very similar in polarity is preferred for the sample solution. The sample solution should be as concentrated as possible, preferably no more than 5 mL in volume. Alternatively, if the sample's components are not soluble in the first elution solvent, a small amount of a more polar solvent can be used to dissolve them.

Before the sample solution is applied to the column, the liquid used in packing the column should be allowed to drain out until the liquid level is just at the top of the upper sand layer. Then close the stopcock and pour the mixture to be separated into the column. After reopening the stopcock and allowing the upper level of the solution to reach the top of the sand, stop the flow, fill the column with elution solvent, and proceed to develop the chromatogram. Follow the same procedure when changing elution solvents.

Another alternative is to preabsorb the sample by placing a small amount of adsorbent into a solution of the sample. The solvent is removed on a rotary evaporator and the dry powder is added to the top of the column.

***Choice of Elution Solvents***

The proper choice of elution solvents and the amount to use are, in part, a trial-and-error process. Polar compounds always require more polar elution solvents than do nonpolar compounds. For example, the separation of 1-decene from 2-chlorodecane requires developing solvents of low polarity, such as alkanes. However, the separation of the alcohol 2-decanol from its oxidation product, 2-decanone, requires more polar solvents, such as a hexane/diethyl ether mixture.

Thin-layer chromatography [see Technique 15] is one method that provides an estimate of how effective a certain solvent will be as an eluent for a liquid chromatographic separation. The TLC plates should have the same adsorbent that will be used for the column. A useful elution solvent for liquid chromatography is one in which the highest $R_f$ value of any compound in the mixture is about 0.3.

***Elution of the Sample Components***

Elution of the compounds in the sample is done by using a series of increasingly polar elution solvents. The less polar compounds elute first with the less polar solvents. Polar compounds usually come out of a column only after a switch to a more polar solvent. As the elution proceeds, the compounds in the mixture separate into a series of bands along the column (Figure 17.4).

A mixture of two solvents is also commonly used for elution. Addition of small amounts of a polar solvent to a less polar one increases the eluting power in a gentle fashion. For example, the development of the column can begin with hexane and, if nothing elutes from the column with this solvent, a 10% or 20% solution of diethyl ether in hexane can be used next, followed by increasing concentrations of ether, then pure diethyl ether for the most polar compounds.

If the change of solvent is made too abruptly, particularly with alumina as the adsorbent, enough heat may be generated from adsorbent/solvent bonding to cause cracking or channeling of the adsorbent column. In some cases, a low-boiling-point elution solvent may actually boil on the column. The bubbles that form will degrade the efficiency of the column.

*Flow Rate of Eluent*    A greater height of developing solvent above the adsorbent layer provides a faster flow rate through the column. An optimum flow rate is about 2–3 mL · min$^{-1}$. If the flow is too slow, poor separation may result from diffusion of the bands. A reservoir at the top of a column can be used to maintain a proper height of elution solvent above the adsorbent so that an adequate flow rate is maintained. A separatory funnel makes a good reservoir. It can be filled with the necessary amount of solvent and clamped directly above the column. The stopcock of the separatory funnel can be adjusted so that liquid flows into the column as fast as it is flowing out the bottom.

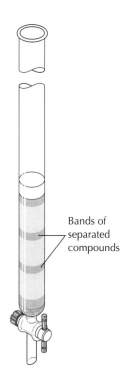

Bands of separated compounds

**FIGURE 17.4**
Chromatography column during elution.

***Size of Eluent Fractions***

The size of the eluent fractions collected at the bottom of the column depends on the particular experiment. Common fraction sizes range from 25 to 150 mL.

If the separated compounds are colored, it is a simple matter to tell when the different fractions should be collected. However, column chromatography is not limited to colored materials. With an efficient adsorbent column, each compound in the mixture being separated is eluted separately. After one compound has come through the column, there is a time lag before the next one appears. Hence, there are times when only solvent drips out of the column. To ascertain when you should collect a new fraction of eluent, either note the presence of crystals forming on the tip of the column as the solvent evaporates or collect a few drops of liquid on a watch glass and evaporate the solvent in a hood. Any relatively nonvolatile compounds that are being separated will remain on the watch glass.

***Recovery of Separated Components***

The purity of each fraction can be ascertained by GC or TLC analysis. The fractions containing the same component should be combined before the pure components are recovered by evaporation of the solvent. Evaporation methods include blowing off the solvent with a stream of nitrogen or air, evaporation with the aid of a partial vacuum, and use of a rotary evaporator [see Technique 8.9].

## 17.6    Microscale Column Chromatography

Column chromatography is an effective way to remove small amounts of impurities from the product of a microscale synthesis. Silica gel usually works as the adsorbent for separating most organic compounds. Thin-layer chromatography on silica gel plates can be used to determine the best solvent system for separating the mixture [see Technique 15.5]; a solvent that moves the desired compound to an $R_f$ of approximately 0.3 should be a good elution solvent. The separation on a silica gel thin-layer plate with a particular solvent reflects the separation that the mixture will undergo with a silica gel column and the same solvent. A column suitable for separating 50–100 mg of sample can be prepared in a large-volume Pasteur pipet.* Regular-size Pasteur pipets can be used for 10–30 mg of sample.

*Assemble all equipment and reagents for the entire chromatographic procedure before you begin to prepare the column.*

Obtain about 50 mL of the elution solvent in a flask fitted with a cork. Dissolve the mixture being separated in a small test tube by adding approximately 1 mL of the elution solvent, or some other solvent that is less polar than the elution solvent, to dissolve the sample. Spot a silica gel TLC plate with the sample solution; develop the plate with the elution solvent that you propose to use in the column, then visualize the plate [see Technique 15]. Cork the

*Available from Fisher-Scientific, catalog no. 13678-8; the pipets have a capacity of 4 mL.

test tube containing the sample solution while you analyze the TLC plate. When TLC shows that you have chosen a good eluent, prepare the column. Label a series of 10 test tubes (13 × 100 mm) for fraction collection. Pour 4 mL of elution solvent into one test tube and mark the liquid level on the outside of the test tube. Place a corresponding mark on the outside of the other test tubes.

Pack a small plug of glass wool into the stem of a large-volume Pasteur pipet, using a wooden applicator stick or a thin stirring rod (Figure 17.5). Clamp the pipet in a vertical position and place a 25-mL Erlenmeyer flask underneath it to collect the elution solvent that you will be adding to the top of the column. Weigh approximately 3 g of silica gel adsorbent in a tared 50-mL beaker; add enough elution solvent to make a thin slurry. Transfer the adsorbent slurry to the column, using a regular-sized Pasteur pipet. Continue adding slurry until the column is two-thirds full of adsorbent. Fill the column four to five times with the elution solvent to pack the adsorbent well. This eluted solvent can be reused.

*Keep the adsorbent covered with liquid throughout the chromatographic procedure.*

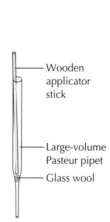

Wooden applicator stick

Large-volume Pasteur pipet

Glass wool

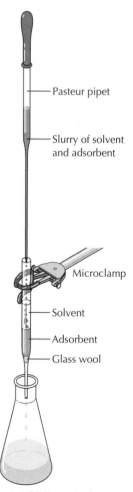

Pasteur pipet

Slurry of solvent and adsorbent

Microclamp

Solvent

Adsorbent

Glass wool

**FIGURE 17.5**
Setting up a microscale column.

**1.** Pack glass wool plug in large-volume Pasteur pipet.

**2.** Add slurry of solvent and adsorbent.

Allow the solvent level to almost reach the top of the adsorbent and place the test tube labeled "Fraction 1" under the column. Begin transferring the sample mixture to the column, using a Pasteur pipet. When all the sample is on the column, begin adding the elution solvent. Collect fractions of approximately 4 mL in your labeled test tubes. Determine the composition of each fraction by TLC analysis. Combine the fractions with the same composition and evaporate the solvent to recover each substance.

## 17.7  Summary of Column Chromatography Procedure

1. Prepare a properly packed column of adsorbent.
2. Add the sample mixture to the column in a small volume of solution.
3. Elute the adsorbed compounds with progressively more polar liquids.
4. Collect the eluted compounds in fractions from the bottom of the column.
5. Evaporate the liquids to recover the separated compounds.

## 17.8  Flash Chromatography

Simple gravity column chromatography [see Techniques 17.1–17.6] can be extremely time consuming. In the research laboratory, gravity chromatography has been largely replaced by flash chromatography, in which gas pressure is used to push the eluent through the adsorbent. The flash technique is not only much faster but also more efficient, because silica gel with a very small pore size (23–40 μm, 230–400 mesh compared to 63–200 μm, 70–230 mesh for gravity columns) is used. The total time to prepare and completely elute the column is often less than 15 min. The small particle size of the stationary phase requires pressures up to 20 pounds per square inch (psi), thus necessitating a special glass apparatus with a chromatographic column that does not leak and a source of nitrogen gas or compressed air. Although it is desirable to have an $R_f$ difference of $\geq 0.35$ for the compounds being separated, it is possible to separate components with a difference of $\sim 0.15$.

The pressure of the gas controls the flow rate of eluent through the column. One type of apparatus consists of a glass column topped by a variable bleed device (Figure 17.6). The bleed device has at its top a Teflon needle valve that controls the pressure applied to the top of the solvent in the column. Table 17.2 provides column and solvent dimensions for preparation of a silica gel column of 5–6 inches in height. Table 17.2 shows that a smaller

**FIGURE 17.6**
Apparatus for flash
chromatography.

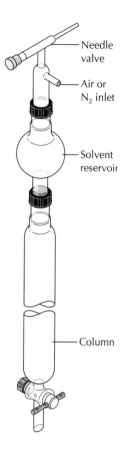

Needle
valve

Air or
N$_2$ inlet

Solvent
reservoir

Column

column diameter requires that the collected fraction sizes be correspondingly smaller. In addition, the smaller the difference in
$R_f$ values, the smaller the size of the sample that can be placed on
the column. Eluent fractions must be analyzed by TLC or GC.

Before running a flash column, the TLC characteristics of
the sample components should be determined. Ideally a solvent
system that provides an $R_f$ difference of $\geq 0.35$ should be used.
Systems that have been found useful include petroleum ether
($30°-60°C$) mixed with one of the following: dichloromethane,
ethyl acetate, or acetone.

**TABLE 17.2** **Column dimensions and solvent volumes for flash chromatography**

| Column diameter, mm | Volume of eluent, mL | Typical sample size, mg, $\Delta R_f > 0.2$ | $\Delta R_f \approx 0.1$ | Recommended fraction size, mL |
|---|---|---|---|---|
| 10 | 100 | 100 | 40 | 5 |
| 20 | 200 | 400 | 160 | 10 |
| 30 | 400 | 900 | 360 | 20 |
| 40 | 600 | 1600 | 600 | 30 |
| 50 | 1000 | 2500 | 1000 | 50 |

SOURCE: Still, W. C.; Kahn, M.; Mitra, A. *J. Org. Chem.* **1978**, *43*, 2923–2925.

Consult Table 17.2 when assembling the column. Either the available column sets the range of sample sizes that can be accommodated, or the size of the sample to be separated indicates the column size needed. To prepare the column, put a glass wool plug at the bottom of the column (a smaller glass tube may be used to insert the plug) and cover it with a thin layer (about 1/8 inch thick) of sand, 50–100 mesh. With the stopcock open, add, with tapping, 5–6 inches of 23–40 μm silica gel,* or the adsorbent can be packed by the slurry method. Add a second layer of sand (about 1/8 inch thick) at the top of the silica gel and level it with gentle tapping.

Fill the column to the top with the solvent system being used for elution. Insert the connected flow controller and, with the bleed valve open, gently turn on the flow of pressurized gas. Control the pressure by placing your finger (wear gloves) over the exit tube (located around the needle valve), and manipulate the pressure such that the column is packed tightly by forcing all entrapped air from the bottom. With practice, you should reach an equilibrium pressure that causes the level of solvent to drop at a rate of 2 inches · min$^{-1}$. **Never let the column run dry** (the solvent should not go below the level of the top sand layer). When the solvent has just reached the level of the sand, turn off the stopcock and remove the flow controller. Introduce a 20–25% solution of your sample in the elution solvent to the column with a pipet. Reinsert the flow controller and adjust the pressure until the level of solvent at the top of the column drops at about 2 inches · min$^{-1}$. Collect the proper fraction volumes of eluent solution (see Table 17.2) until all the solvent you planned to use is in the column or until fraction monitoring indicates that the desired components have been eluted.

*If the sample is not very soluble, use a small amount of polar solvent to get the sample on the column.*

---

## 17.9    High-Performance Liquid Chromatography

Improved techniques and technology for gas chromatography have led to another advance in liquid chromatography, namely, the development of high-performance liquid chromatography (HPLC). High-performance liquid chromatography allows separations and analyses to be completed in minutes instead of hours. The principles of separation are the same, but HPLC gives better separation and detection of compounds using other liquid chromatographic techniques. The stationary phase in an HPLC column has a particle size of only 3–10 μm. The enhanced separation and sensitivity of the column comes from the increased surface area provided by these very small particles. However, particles of this small size pack very tightly, a condition that severely restricts the flow of

---

*Aldrich and other suppliers indicate whether the silica gel is suitable for flash chromatography.

solvent through the column. Pressures of 1000–6000 psi are required to force the solvent through the column at a rate of 1–2 mm · min$^{-1}$.

The instrumentation for high-pressure liquid chromatography consists of a column, a sample injection valve, a solvent reservoir, a pump, a detector, and a recorder or computer. Figure 17.7 shows a block diagram of a typical high-pressure liquid chromatograph. The column can range from 5 to 30 cm long, with an inner diameter of 1–5 mm for analytical HPLC of 0.1–1 mg samples. Preparative HPLC columns range from 5 to 50 cm long with an interior diameter of 2.5–5.0 cm; these columns separate samples of 0.5–1.0 g.

HPLC columns commonly have a liquid covalently bonded to microporous silica ($SiO_2$) particles as the stationary phase.

$$\text{Silica particle}-Si-OH + (CH_3CH_2O)_3SiR \longrightarrow \text{silica particle}-Si-O-\underset{\underset{OCH_2CH_3}{|}}{\overset{\overset{OCH_2CH_3}{|}}{Si}}-R + CH_3CH_2OH$$

Columns suitable for **normal-phase** chromatography have a stationary phase with polar substituents such as $-(CH_2)_3NH_2$ or $-(CH_2)_3CN$ for the $-R$ group in the foregoing equation. In normal-phase HPLC, the stationary liquid phase is more polar than the solvent (liquid) phase; therefore, the less polar compounds in a mixture being analyzed elute first. **Reverse-phase** columns with long-chain alkyl groups such as $-(CH_2)_{17}CH_3$ for the $-R$ group are used to separate moderately polar to polar compounds, for example, amino acids. In this case, the solvent is more polar than the hydrophobic stationary phase, so the more polar compounds elute first. Methanol and acetonitrile, often mixed with water, are common eluents for reverse-phase chromatography.

The solvent is stored in a reservoir and passes through a filtration system before being pumped through the injector port, the column, and the detector. The solvents used for HPLC must be of

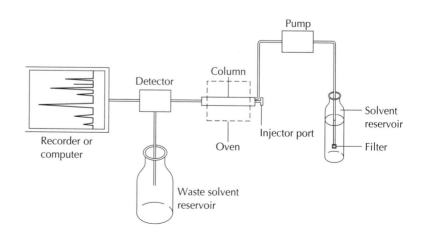

**FIGURE 17.7**
Schematic representation of a typical high-performance liquid chromatograph.

very high purity because impurities degrade the column by irreversible adsorption on the particles of the solid phase. More sophisticated instruments have several solvent reservoirs and a gradient elution system that allows the composition of solvent mixtures to be changed during the separation.

The detector used for HPLC has a high sensitivity, usually in the microgram to nanogram range. The two most common types are ultraviolet (UV) and refractometer detectors. A UV detector is relatively inexpensive and can be used with gradient elution. It detects any organic compound that absorbs in the UV region. The limitations of a UV detector preclude its use with solvents that absorb in the UV region themselves or with samples that do not have a chromophore for UV absorption. Refractometer detectors measure changes in the refractive index of the eluent as solutes move off the column in the solvent stream. A refractometer detector cannot be used with gradient elution because the base line would change as the solvent composition changes.

## References

1. Miller, J. M. *Chromatography: Concepts and Contrasts;* Wiley-Interscience: New York, 1988.
2. Still, W. C.; Kahn, M.; Mitra, A. "Rapid Chromatographic Technique for Preparative Separations with Moderate Resolution"; *J. Org. Chem.* **1978,** *43,* 2923–2925.
3. Kirkland, J. J.; Snyder, L. R. *Introduction to Modern Liquid Chromatography;* Wiley-Interscience: New York, 1974.
4. Snyder, L. R.; Glajch, J. L.; Kirkland, J. J. *Practical HPLC Method Development;* Wiley: New York, 1988.

## Questions

1. Once the adsorbent is packed in the column, it is important that the elution solvent level not drop below the top of the adsorbent. Why?
2. What precautions must be taken when you introduce the mixture of compounds to be separated onto the adsorbent column?
3. What effect will the following factors have on a chromatographic separation: (a) too strong an adsorbent; (b) collection of large elution fractions; (c) solvent level below the top of column adsorbent?
4. Arrange the following compounds in order of decreasing ease of elution from a column of silica gel: (a) 2-octanol; (b) 1,3-dichlorobenzene; (c) *tert*-butylcyclohexane; (d) benzoic acid.
5. Compare the advantages and limitations of liquid chromatography with those of thin-layer and gas-liquid chromatography.

# PART

# 3

# Spectroscopic Methods

Part 3 introduces the three spectroscopic methods most frequently used today by organic chemists to determine the structure of organic compounds: **infrared (IR) spectroscopy, nuclear magnetic resonance (NMR) spectroscopy,** and **mass spectroscopy (MS).** Throughout the study of organic chemistry, you are asked to think in terms of structure, because structure determines the properties of molecules. The experienced organic chemist can anticipate many of the physical and chemical properties of various compounds by a quick glance at their structures. Fifty and more years ago, structure was determined largely by very time-consuming chemical methods. Modern organic spectroscopic methods have produced a revolution in the determination of the structures of complex organic molecules. What used to take months can now often be done in a day. For organic molecules having molecular weights of 300 or less, the job can often be done within an hour or so. The new techniques are based in large part on the absorption of radiation from various portions of the electromagnetic spectrum. In effect, spectroscopic techniques provide "snapshots" of a molecule.

Infrared spectroscopy is the oldest of the three spectroscopic methods, and in some ways it has been outshone by the newer NMR and mass spectroscopy. However, IR spectroscopy provides quick and valuable information on functional groups present in a molecule (for example, $-C\!\!=\!\!O$, $-OH$, and $-COOH$). NMR spectroscopy reveals many details of an organic compound's structure, particularly the interrelated connectivity of its hydrogen and carbon atoms. Chemical shifts, spin-spin coupling patterns, and integration can be invaluable in structure determination, and integration of mixtures can determine product ratios from chemical reactions. Today NMR is arguably the most powerful spectroscopic method in organic chemistry. It is no surprise that it is the major focus of this spectroscopic methods section. Mass spectroscopy, which uses highly energetic electrons to ionize molecules, allows chemists to determine the molecular weight of a compound, and the fragmentation pattern of the ionized molecule provides data that can assist in the identification of the compound.

Integrating the data obtained from the three different spectroscopic methods discussed in Part 3 is important in the characterization of an organic compound. One spectral method may reveal features about a compound that may not be clear from another method, and in some cases one spectral method may confirm the existence of a structural unit suggested by another method. At the end of Part 3 are integrated problems with data about specific compounds obtained by several spectroscopic techniques. Solving these problems requires you to consider how the data from one spectroscopic technique assist in interpreting the data obtained by another. The mass spectroscopy data are usually a good starting point because they include the molecular weight of the compound. Next, consider the IR spectrum, which provides data for the identification of the functional groups present in the compound. Interpretation of the $^1$H NMR spectrum allows you to complete the structural analysis.

# 18

# INFRARED SPECTROSCOPY

One of the most useful spectroscopic methods in organic chemistry is *infrared (IR) spectroscopy*, also called *vibrational spectroscopy* because the photon energy of IR radiation absorbed by a covalent molecule corresponds to the energy it takes to stretch or bend covalent bonds. The energy of infrared light is on the order of 8–40 kJ/mole (2–10 kcal/mole). This amount is not enough energy to break a covalent bond, but it is enough to make the bond vibrate. When infrared radiation is absorbed, the sample becomes warm as its molecules increase in kinetic energy.

IR absorption is generally used to identify organic functional groups. The detailed analysis of the carbon skeleton and any attached hydrogen atoms is usually carried out by other techniques, especially nuclear magnetic resonance spectroscopy [see Technique 19].

## 18.1 Theory of Molecular Vibration

Molecules selectively absorb specific frequencies of IR radiation that correspond to the frequencies of the vibrational oscillations of its atoms. When a molecule absorbs IR radiation, the amplitudes of these vibrations increase. The absorption corresponding to these oscillations appears in certain definite regions of the IR spectrum, regardless of the particular compound in which the group of atoms is contained. For example, the stretching regions of oxygen-hydrogen bonds in all alcohols appear at nearly the same frequency. In the same way, the C=O vibrations of all carbonyl compounds appear within a narrow frequency range.

The frequency at which a characteristic IR absorption occurs depends on the mass of the atoms involved in the vibration and the strength of the bond connecting them. Consider, for example, the molecule HCl. When it absorbs IR radiation at the frequency for stretching the HCl bond, the motion within the molecule becomes more vigorous and the average displacement (amplitude) of the atoms from the equilibrium bond length becomes greater.

Water has three fundamental vibrational frequencies, two related to stretching vibrations and one to a bending vibration. They are drawn in Figure 18.1. Vibrations (a) and (b) correspond to the stretching of the O—H bonds. Figure 18.1a shows the *symmetric vibration* in which the hydrogen atoms stretch in and out simultaneously. In the *asymmetric vibration* in Figure 18.1b, as one hydrogen atom springs out, the other springs in, and the oxygen atom, moving slowly because of its large mass, shifts a little to one side. The last of the three vibrations (Figure 18.1c) involves the bending of the O—H bond, a kind of scissoring motion in which the H—O—H bond angle changes back and forth.

**FIGURE 18.1**
Fundamental
vibrations of water.
The oxygen moves to
maintain a constant
center of gravity.

| (a) | (b) | (c) |
|---|---|---|
| Symmetric | Asymmetric | Bending |
| coupled | coupled | vibration |
| stretch | stretch | |

---

## 18.2    IR Spectrum

When IR radiation is absorbed by a sample, the changes in bond vibrations are picked up electrically and passed on to a detector. The raw data are run through a computer and eventually recorded as a two-dimensional trace—called an IR spectrum—with the frequency or wavelength of the absorption on the horizontal axis and intensity of the absorption on the vertical axis. Figures 18.2 and 18.3 are examples of IR spectra for cyclopentanone recorded on two different types of spectrometers.

The equation that relates the energy $(E)$ of the absorbed radiation to its frequency $(v)$ is

$$E = hv$$

where $h$ = Planck's constant. Frequency is inversely proportional to the wavelength $(\lambda)$ of the absorbed radiation, so

$$E = hc(1/\lambda)$$

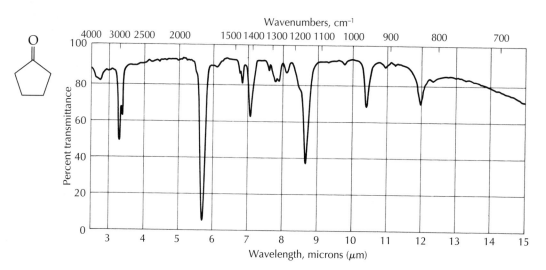

**FIGURE 18.2** IR spectrum of cyclopentanone. The sample was prepared as a liquid film (thin film), and the spectrum was recorded on a dispersive instrument with prism optics. Note that the wavelength scale is linear; consequently, the wavenumber scale is not.

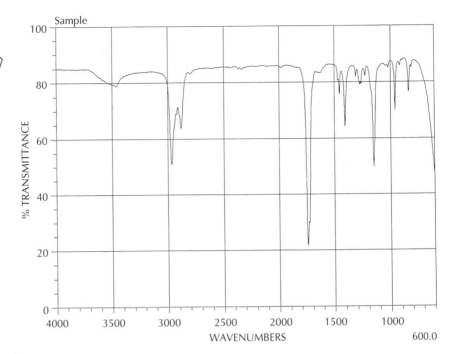

**FIGURE 18.3** IR spectrum of cyclopentanone recorded from a capillary film between salt plates (thin film) on an FTIR instrument. In this spectrum, the linear scale is in wavenumbers, not wavelength. Reproduced from R. A. Tomasi, *A Spectrum of Spectra,* Sunbelt R & T Pub., Tulsa, OK, 1992, with permission.

The quantity $c$ is the speed of light; $(1/\lambda)$, which is proportional to the frequency, is called the *wavenumber* $(\bar{v})$ and is usually expressed in units of reciprocal centimeters (cm$^{-1}$). The wavenumber is directly proportional to the energy of the IR absorption.

$$E = hc\bar{v}$$

An IR absorption band is often called a peak, even though it is really an inverse peak, or trough. In any case, the peak maximum is defined as the position of maximum absorption in wavenumber units (cm$^{-1}$). The IR region of interest to organic chemists is the *mid infrared,* extending from 4000 to 600 cm$^{-1}$, where absorptions due to most molecular vibrations in a typical organic compound appear. (The lower limit may vary from one instrument to another.) Both Figures 18.2 and 18.3 show the most intense peak for cyclopentanone between 1700 and 1800 cm$^{-1}$. These two IR spectra of cyclopentanone reveal the same peak positions, but their shapes are different because they have been measured on different types of spectrometers.

Organic compounds, which contain 10–20 atoms or more, can give rise to a substantial number of IR bands; thus the spectrum of such a compound can be quite complex. Fortunately, many of the peaks can often be ignored. The large number of fundamental vibrations, their overtones, and combinations of vibrations make it far too difficult to understand quantitatively the entire IR spectra of most organic compounds, but as you will see, they can easily yield a great deal of qualitative information about the functional groups the compounds contain.

The intensity (peak size) of an IR absorption can be reported in terms of either transmittance ($T$) or absorbance ($A$). **Transmittance** is the ratio of the amount of infrared radiation transmitted by the sample ($I$) to the intensity of the incident beam ($I_0$). Percent transmittance is $T \times 100$.

$$T = I/I_0$$

$$\% \, T = (I/I_0) \times 100$$

Peak intensity can also be reported as absorbance ($A$), where absorbance is the log of the inverse of transmittance. In practice, peak intensities are reported in a less quantitative fashion.

It is desirable to record a spectrum in which the most intense peak nearly fills the vertical height of the chart. Peaks of that order of magnitude are termed **strong** (s); smaller peaks are called either **medium** (m) or **weak** (w). Peaks can also be described as **broad** (br) or sharp. It is important that the most intense peak in a spectrum be above 0% transmittance (5–10% is good) so that its peak maximum can be measured accurately. Infrared spectra are commonly calibrated by using a polystyrene standard. Often its peak at 1600 cm$^{-1}$ is printed out separately on the spectrum to show that this has been done.

What determines the position and intensity of IR bands? The most important factors are as follows:

- Whether the vibration involves stretching or bending
- Bond order
- Masses of the atoms attached by covalent bonding
- Electronegativity difference between the two atoms or groups in a bond

In general, stretching covalent bonds takes more energy than bending vibrations, so stretching vibrations in the infrared appear at higher frequencies. Bond order is simply the amount of bonding between two atoms. For example, the bond order between carbon atoms increases from one to two to three for ethane ($CH_3$—$CH_3$), ethene (ethylene, $CH_2$=$CH_2$), and ethyne (acetylene, HC≡CH), respectively. In general, the higher the bond order, the greater the energy required to stretch the bond. Higher bond order produces a higher frequency IR absorption:

| Bond order | Absorption frequency (cm$^{-1}$) |
|---|---|
| C≡C, C≡N | 2300–2000 |
| C=C, C=O, C=N | 1900–1500 |
| C—C, C—O, C—N | 1300–800 |

The frequency of the IR absorption also relates to the masses of the vibrating atoms. Covalent bonds to hydrogen occur at high frequencies compared to bonds between heavier atoms—a light weight on a spring tends to oscillate faster than a heavy weight.

| Bond order | Absorption frequency (cm$^{-1}$) |
|---|---|
| C—H, O—H, N—H | 3800–2700 |

Bond polarity doesn't much affect the position of IR absorption, but it greatly influences the intensity of IR bands. If a vibration (stretch or bend) induces a significant change in the dipole moment, an intense IR band will result. Thus it is not surprising to find that stretching C—O, C=O, and O—H bonds produces intense IR peaks. In each of these bonds, there are large electronegativity differences between the atoms. Stretching these polar bonds causes a large separation of charge and strong IR peaks.

## 18.3        IR Instrumentation

Two major classes of instruments are used to measure IR absorption: dispersive and Fourier transform (FT) spectrometers. Dispersive spectrometers were developed first and for a long time were the standard research instruments. The advent of computers allowed the development of Fourier transform spectrometers in the 1960s. In recent years, instruments incorporating powerful and relatively inexpensive microcomputers have allowed many laboratories to convert to FTIR instruments.

*Dispersive Spectrometer*

In a *dispersive IR spectrometer* the source of radiation (for example, a heated filament) provides a beam of IR radiation that is focused onto either a prism or a grating, which incrementally separates the beam into the various infrared energies that are available from the source. The beam is then directed by mirrors through both sample and reference cells. This process eventually allows only a narrow region of energy to reach the detector. Collecting scans over the entire infrared region gives the complete spectrum. The reference cell allows the instrument to produce IR signals that represent the difference between absorptions due to the sample and those of the pure solvent. The spectrum in Figure 18.2 is a dispersive spectrum using prism optics. Note that the wavelength scale is linear in microns (μm), that is, micrometers.

*FTIR Spectrometer*

An *FTIR spectrometer* uses an interferometer rather than a prism or a grating. The IR radiation is focused from a source filament onto a beam splitter, which directs half the radiation to a stationary mirror and half to a movable scan mirror. Both mirrors reflect radiation back to the beam splitter, where they combine, pass through the sample, and reach the detector. Bursts of radiation containing the entire spectrum of IR frequencies are used. After this radiation has passed through the sample, Fourier transform mathematics is used to sort out the components of the spectrum. Each burst of radiation provides a complete IR spectrum. This approach offers the advantage of providing the entire spectral scan within one burst without a reference cell, although multiple scans are often used to improve the signal-to-noise ratio. Typically, a spectrum recorded by an FTIR instrument, such as the one in Figure 18.3, has a linear wavenumber scale.

The FTIR method has a number of advantages. Because the monochromator used in a dispersive IR spectrometer is not used in FTIR, the entire range of radiation is passed through the sample simultaneously, with a great savings in time. A spectrum with a high signal-to-noise ratio and excellent resolution is rapidly obtained. Results of a number of scans are combined to average out random artifacts, and excellent spectra can be obtained from very small samples. Moreover, computers easily manipulate the IR data. FTIR scans may show undesirable peaks due to atmospheric water and $CO_2$, but computer subtraction of an atmospheric scan can often greatly improve the quality of a spectrum.

Because the original dispersive IR spectrometers recorded data as a linear function of wavelength, many collections of IR spectra appear in this format. With FTIR instruments, the horizontal axis of the spectrum is usually plotted with a linear wavenumber scale. However, because the data obtained on an FTIR spectrometer are digitized, it is very easy to convert the data from one scale to the other for comparison purposes.

Wavelengths in micrometers ($\mu$m) and the frequency in wavenumbers ($cm^{-1}$) can be interconverted by the following relationship:

$$\bar{v} = 10,000/\lambda$$

The frequency, $\bar{v}$, is again in units of $cm^{-1}$, and the wavelength, $\lambda$, is in micrometers ($\mu$m, $10^{-6}$ meters, simply called $v$ or microns in the older literature).

---

## 18.4    Techniques of Sample Preparation

To obtain an IR absorption spectrum, the sample must be either a liquid, dissolved in a solution, or ground to a very fine dispersion. Otherwise too much of the IR radiation is reflected away rather than absorbed by the sample. The sample could also be a gas, requiring a special gas cell for sampling. This situation is only rarely encountered in organic chemistry.

*Liquid Samples*     If the compound of interest is a liquid, IR spectra are often obtained by using one of the devices shown in Figure 18.4a and b. These devices allow films of "neat" samples (pure liquids with no added

**FIGURE 18.4**
Three sample holders.

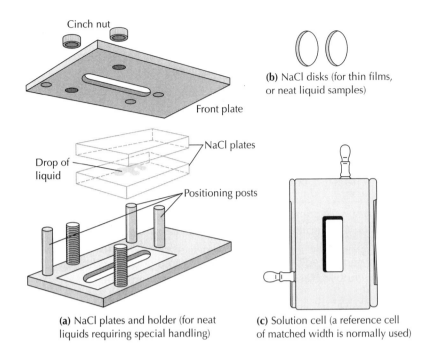

Cinch nut

Front plate

**(b)** NaCl disks (for thin films, or neat liquid samples)

NaCl plates

Drop of liquid

Positioning posts

**(a)** NaCl plates and holder (for neat liquids requiring special handling)

**(c)** Solution cell (a reference cell of matched width is normally used)

*When the sample is "neat" (that is, simply a thin film of the pure liquid), the reference is merely air and no reference cell is used.*

solvents) to be analyzed. When the sample has a low viscosity, the holder shown in Figure 18.4a is a better choice because it keeps the sample film tightly in contact with the plates. When the liquid is fairly viscous, it can be smeared between two disks as shown in Figure 18.4b; the disks are then placed in the sample beam.

Naturally, the sample cells themselves must be transparent to IR radiation. Unfortunately, glass absorbs far too much IR radiation to be a useful cell material. Most sample cells for IR spectroscopy are made from alkali halides; polished sodium chloride is by far the most common. Because the cells used in IR spectroscopy are usually made of sodium chloride disks or plates, all samples must be completely dry. Water makes sodium chloride disks cloudy and virtually useless.

**Solution Cells**

Solid or liquid compounds can be dissolved in a solvent and analyzed in a solution cell such as the one shown in Figure 18.4c. This type of solution cell can be used in most IR spectrometers. A solvent should be selected from Table 18.1. It is, of course, desirable to use a solvent that has a minimum number of IR absorptions in the regions of interest. When solvents do absorb, you should pick one that does not absorb in regions important to your sample. Dispersive instruments are double-beam instruments; thus the use of a solution cell requires a matched reference cell. Here the reference cell should have exactly the same path length as the sample cell and should be filled with pure solvent. The solvent peaks are subtracted, and the resulting spectrum should show only peaks due to the compound of interest. In practice, although well-matched cells

| TABLE 18.1 | Absorption regions of common solvents and mulling compounds |
|---|---|
| Carrier | Absorption region (cm$^{-1}$) |
| *Solvents* | |
| Chloroform[a] | 3125–2940 |
| | 1250–1190 |
| | 835–625 |
| Dichloromethane[a,b] | 3300–2850 |
| | 2450–2300 |
| | 1550–1150 |
| | <930 |
| *Mulling compounds* | |
| Fluorolube | 1300–1080 |
| | 1000–920 |
| | 910–870 |
| | <670 |
| Nujol | 0000–2700 |
| | 1500–1450 |
| | 1430–1360 |
| | 720–750 |
| Potassium bromide | Transparent |

a. Toxicity hazard.
b. Highly flammable.

may remove the solvent peaks, weak sample peaks that occur in the region where the solvent absorbs are often difficult to observe.

FTIR spectrometers are single-beam instruments. To obtain the spectrum of a compound in solution with this type of instrument, the spectrum of the solution and the spectrum of the solvent are obtained separately using the same sample cell. Because the data are digitized, the spectrum of the solvent can be subtracted from the spectrum of the solution using software provided with the instrument. The result of the subtraction is a spectrum of the compound devoid of any contributions from the solvent. In any frequency regions where the solvent has strong absorptions, signals are not reliable and should be ignored.

Chloroform is a simple, symmetrical molecule with few intense IR absorptions, and therein lies its attraction as an IR solvent. However, it does have intense absorptions due to the carbon-chlorine stretching vibrations as well as absorptions due to carbon-hydrogen stretching and bending. Such a small amount of IR radiation passes through $CHCl_3$ in these frequency regions that the spectrometer may simply not respond to any other absorptions that the sample may have. These absorbance regions are given in Table 18.1.

**Thin Films for Solid Samples**

Three other techniques used for preparing solid samples are thin films, mulls, and pressed KBr pellets. Perhaps the most useful is the thin film, which is prepared by placing a drop of a concentrated

**FIGURE 18.5**
Aids in the preparation
of KBr pellets.

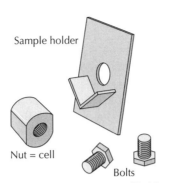

Sample holder

Nut = cell

Bolts

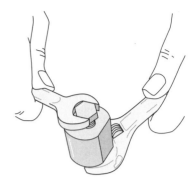

(a) Minipress components and holder          (b) Squeezing the KBr pellet in the press

solution of the compound in the center of a clean sodium chloride plate. For best results the salt plate must have a smooth, polished surface; scratched and pitted plates lead to uneven distribution of the sample. The solvent is allowed to evaporate. The NaCl disk is then placed on a sample holder like that shown in Figure 18.5a.

*Mulls*

Mulls are not true solutions but are fine dispersions of organic compounds either in a solid, such as anhydrous potassium bromide (KBr), or in a liquid, such as Nujol (a brand of mineral oil, which is a mixture of long-chain alkanes) or Fluorolube (a mixture of completely fluorinated alkanes). Mulls are used for IR samples of solid compounds. The fluorinated mulling substances are often used for more polar compounds. To make a mull, 15–20 mg of the solid should be ground using a small agate or nonporous ceramic mortar and pestle until the sample is highly dispersed and has a caked, glassy appearance. One drop of mulling liquid is then added to the ground solid. This mixture is ground together thoroughly to make a uniform paste with the consistency of toothpaste. The paste is transferred to plates such as those shown in Figure 18.4a and b, and the plates are gently pressed together and put in a sample holder, which is placed in the beam of the spectrometer. Nujol and Fluorolube display IR peaks that, of course, may cover peaks due to the compound dispersed in these liquids (see Table 18.1). The preparation of a good mull requires care and practice.

*KBr Pellets*

The technique for solids that provides the highest-quality IR spectra is to press them into a transparent disk made from potassium bromide, a compound that does not absorb IR radiation. A pellet is prepared by carefully grinding a small quantity of the solid sample (0.5–2.0 mg) first alone and then together with 100 mg of completely dry potassium bromide (KBr). The grinding operation is done with a polished mortar and pestle made of agate or some other nonporous material or by vibrating the mixture in a small ball mill similar to that used by dentists to mix amalgam fillings. The mixture is pressed into transparent disks with special dies (see Figure 18.5).

The KBr/compound mixture is subjected either to 14,000–16,000 psi in a high-pressure disk press or to hand pressure in the minipress shown in Figure 18.5b. Although KBr disks are excellent for IR analysis, their preparation is challenging and requires great care. For example, the smallest traces of water in the disk can disrupt homogeneous sample preparation and can also produce spurious O—H peaks in the spectrum. Some modern FTIR instruments make it unnecessary to prepare KBr disks. IR instruments now available can detect 1 mg of sample simply sprinkled on the surface of dry KBr.

*Sampling Cards*

A relatively recent innovation in IR sampling is a disposable sampling card. The sample is applied to an inert, microporous substrate in the middle of the card. Liquids are applied neat. Solids are applied in solution and the solvent is allowed to evaporate. The card is placed in the sample beam and scanned. A card usually has one of two varieties of substrate, polyethylene or polytetrafluoroethylene. The polyethylene substrate has strong absorptions in the regions 2918–2849 cm$^{-1}$, 1480–1430 cm$^{-1}$, and 740–700 cm$^{-1}$. The polytetrafluoroethylene substrate has strong absorptions in the regions 1270–1100 cm$^{-1}$ and 660–460 cm$^{-1}$.

*Reflectance Methods*

Since modern FTIR instruments are extremely sensitive, techniques have been developed that can obtain an infrared spectrum by bouncing a beam of infrared energy directly off the surface of a liquid or a finely ground solid. Techniques such as diffuse reflectance and attentuated total reflectance (ATR) make it unnecessary to prepare KBr disks, but they require specialized sampling accessories.

## 18.5    Interpreting IR Spectra

Confirming the identity of a compound is one of the most important uses of IR spectroscopy. A compound thought to have a certain structure will have an IR spectrum superimposable on that of an authentic sample of the known compound. No two different compounds are known to have identical IR spectra. However, all too often, one cannot count on knowing enough about the identity of a compound to choose an authentic sample for comparison. Therefore, it becomes necessary to interpret the IR spectrum.

Interpreting IR spectra is an acquired skill. If the spectrum is adequately resolved, the peaks are of the proper intensity, and the spectrometer is calibrated so that the position of the peaks is recorded accurately, the spectrum can be very useful in determining the compound's proper structure. Since a precise analysis of all the vibrations and modes of molecular motion is not usually feasible except in the simplest compounds, the spectra must be analyzed empirically.

IR absorption frequencies are characterized according to fundamental stretching absorptions of bond types (Figure 18.6). Because

**FIGURE 18.6**
Approximate regions of
chemical bond stretches
in an infrared spectrum.

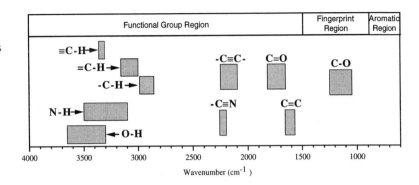

the absorptions of some bonds vary a good deal and are frequently of weak intensity, an absorption frequency table, such as Table 18.2, is useful only within restricted guidelines. One has to learn which regions of the IR spectrum are most useful and interpret the spectral information accordingly.

The IR spectrum can be broken down into three regions:

| IR region | Wavenumber (cm$^{-1}$) |
|---|---|
| Functional group region | 4000–1500 |
| Fingerprint region | 1500–900 |
| Aromatic region | 900–600 |

The functional group region (4000–1500 cm$^{-1}$) provides unambiguous, reasonably strong peaks for most major functional groups. The fingerprint region (1500–900 cm$^{-1}$) is normally so complex that except for a few intense absorptions it can be used for little more than fingerprint identification. That is, the identity of an unknown compound can be established by matching its IR spectrum with the spectrum of a known compound. The aromatic region (900–600 cm$^{-1}$) provides useful information for most benzenes and other aromatic compounds.

Keep in mind that the absence of IR bands is also useful information. The absence of substantial bands in the 900–600 cm$^{-1}$ region rules out aromatic compounds. The absence of appropriate IR bands in the 4000–1500 cm$^{-1}$ region argues against the presence of many functional groups. C=C, C≡C, C≡N, N—H, O—H, and C=O bonds show IR bands in the 4000–1500 cm$^{-1}$ region.

***Where to Begin?***

An efficient approach to interpreting an IR spectrum usually starts with a survey of the 4000–1500 cm$^{-1}$ functional group region, creating an inventory of bond types present in the molecule. This inventory allows you to get a good idea of what functional groups are and are not in the compound. The functional group region can be subdivided into smaller regions that are characteristic of certain bond

**TABLE 18.2  Characteristic infrared absorption peaks of functional groups**

| Vibration | Position (cm$^{-1}$) | Intensity[a] | Comment |
|---|---|---|---|
| *Alkanes* | | | |
| C—H stretch | 2962–2853 | m to s | |
| C—H bend | 1470–1430 and 1375–1365 | m to w | |
| *Alkenes* | | | |
| =C—H stretch | 3100–3010 | m | |
| C=C stretch | 1680–1640 (sat.), 1630–1590 (conj.)[b] | w to m | |
| C—H bend | 995–685 | s | See Table 18.3 for detail. |
| *Alkynes* | | | |
| ≡C—H stretch | 3300–3200 | s | |
| C≡C stretch | 2250–2100 | m to w | |
| *Aromatic* | | | |
| C—H stretch | 3080–3020 | m to w | |
| C=C stretch | 1650–1580 | m to w | |
| C—H bend | 900–680 | s | See Table 18.3 for detail. |
| *Alcohols* | | | |
| O—H stretch | 3650–3550 | m | Free |
| | 3550–3300 | br, s | Hydrogen bonded |
| C—O stretch | 1260–1000 | s | |
| *Amines* | | | |
| N—H stretch | 3500–3100 | br, m | 1° (two bands), 2° (one band) |
| *Nitriles* | | | |
| C≡N stretch | 2280–2200 | s | |
| *Aldehydes* | | | |
| C—H stretch | 2900–2820 and 2775–2700 | w | H—C=O |
| C=O stretch | 1740–1720 (sat.), 1705–1680 (conj.) | s | |
| *Ketones* | | | |
| C=O stretch | 1725–1705 (sat.), 1685–1665 (conj.) | s | |
| *Esters* | | | |
| C=O stretch | 1750–1735 (sat.), 1730–1715 (conj.) | s | |
| C—O stretch | 1250–1160 | s | |
| *Acids* | | | |
| O—H stretch | 3300–2500 | m to w | |
| C=O stretch | 1725–1700 (sat.), 1715–1680 (conj.) | s | |
| *Amides* | | | |
| N—H stretch | 3500–3150 | m | 1° (two bands), 2° (one band) |
| C=O stretch | 1680–1630 | s | |
| *Anhydrides* | | | |
| C=O stretch | 1850–1800 and 1790–1740 | s | |
| C—O stretch | 1050–900 | s | |
| *Acid chlorides* | | | |
| C=O stretch | 1820–1770 | s | |
| *Nitro* | | | |
| NO$_2$ stretch | 1660–1490 and 1390–1260 | s | |

a. s = strong, m = medium, w = weak, br = broad
b. sat. = saturated; conj. = conjugated

types. We will describe each important region and give examples of spectra illustrating the fundamental stretching bands. Besides correlating a stretching vibration with a wavenumber, it is important to get a good idea of the general appearance of the signal. Is it sharp? Is it broad? Is it weak? Is it strong?

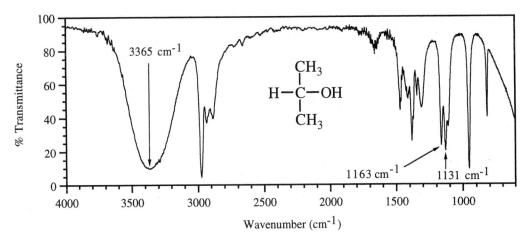

**FIGURE 18.7** IR spectrum of 2-propanol (thin film).

***O—H and N—H Stretch (3700–3100 cm⁻¹)***

$O$—$H$ and $N$—$H$ Stretch $(3700{-}3100\ cm^{-1})$

There are strong IR bands for oxygen-hydrogen bonds in alcohols, phenols, and carboxylic acids, as well as for nitrogen-hydrogen bonds in primary or secondary amines. The appearance of absorptions in the 3700–3100 cm⁻¹ region is highly varied, which can actually add to their usefulness.

The IR spectrum of 2-propanol shown in Figure 18.7 exhibits a broad, strong O—H stretching absorption at 3365 cm⁻¹. The broad shape of this signal is due to hydrogen bonding and is typical of O—H stretches in concentrated or neat alcohol samples. In dilute samples of alcohols and phenols, the signal due to the O—H stretch can be much sharper and shifted to higher wavenumbers, approximately 3600 cm⁻¹.

Primary and secondary amines also display N—H stretching vibrations in this region. The number of signals depends on the number of hydrogen atoms attached to the nitrogen atom. Primary amines show two peaks while secondary amines show only one. The IR spectrum of 4-methylphenylamine is shown in Figure 18.8. Because it is a primary amine, there are two absorptions

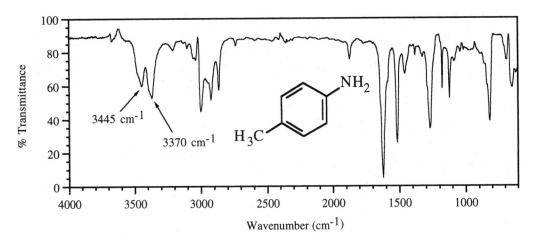

**FIGURE 18.8** IR spectrum of 4-methylphenylamine (CHCl₃ solution).

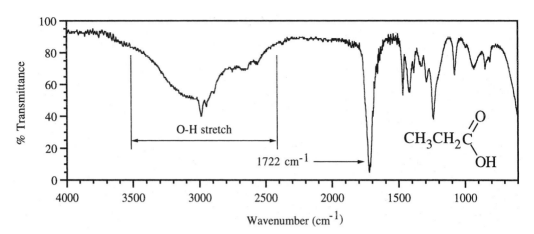

**FIGURE 18.9** IR spectrum of propanoic acid (thin film).

(at 3445 and 3370 cm$^{-1}$) due to symmetric and asymmetric N—H stretching vibrations. Amines are also capable of hydrogen bonding, so the position and shape of the absorption can vary. The higher the concentration of the compound and the better the compound can hydrogen bond, the broader the absorption. Hydrogen bonding also causes N—H stretching absorptions to appear at lower wavenumbers.

Carboxylic acids also show an O—H stretching absorption, but the band often tails from about 3500 cm$^{-1}$ all the way down to 2500 cm$^{-1}$ due to extensive hydrogen bonding. The spectrum of propanoic acid shown in Figure 18.9 illustrates this behavior.

$$CH_3CH_2C \underset{O-H------O}{\overset{O------H-O}{\phantom{CH_3CH_2C}}} CCH_2CH_3$$

In this spectrum the O—H stretching band is so broad that the sharper C—H stretch at approximately 3000 cm$^{-1}$ appears on top of it. This is not uncommon with the O—H stretching vibrations of carboxylic acids.

**C—H Stretch (3300–2850 cm$^{-1}$)**

Because most organic compounds contain hydrogen atoms, you can expect to find signals in the 3300–2850 cm$^{-1}$ region. The position of the C—H stretch depends on the hybridization of the carbon atom to which the hydrogen is bound. If the carbon atom is sp hybridized, the absorption appears near 3300 cm$^{-1}$. A good example is found in the spectrum of phenylacetylene shown in Figure 18.10. The C—H stretch of the acetylene appears at 3277 cm$^{-1}$. This band could be confused with a signal due to an O—H or N—H stretch; however, the band is much sharper than the typical O—H or N—H stretch in this region.

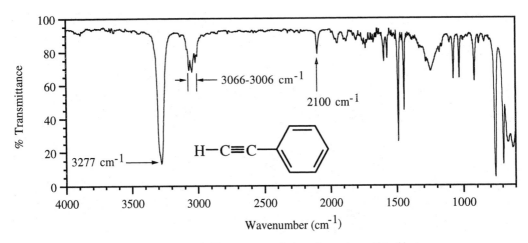

**FIGURE 18.10** IR spectrum of phenylacetylene (thin film).

Peaks due to hydrogens attached to $sp^2$ hybridized carbons of alkenes and aromatic compounds appear in the region from 3150 to 3000 cm$^{-1}$. In the spectrum of phenylacetylene (see Figure 18.10), the aromatic-hydrogen stretching vibrations appear from 3066 to 3006 cm$^{-1}$. In the spectrum of 1-hexene shown in Figure 18.11, the vinyl-hydrogen stretch appears at 3084 cm$^{-1}$.

Hydrogens attached to $sp^3$ hybridized carbons exhibit absorption bands in the region 3000–2850 cm$^{-1}$. In the spectrum of 1-hexene, the peaks due to C—H stretching vibrations of hydrogens attached to saturated carbon atoms appear from 2966 to 2880 cm$^{-1}$.

**$C\equiv C$ and**
**$C\equiv N$ Stretch**
**$(2280{-}2100 \text{ cm}^{-1})$**

Only the triple bonds of nitriles and alkynes have absorptions in the 2280–2100 cm$^{-1}$ region. If it is a strong absorption, it is likely to be that of a nitrile, such as the signal at 2231 cm$^{-1}$ in the spectrum

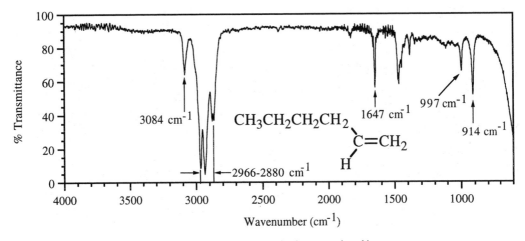

**FIGURE 18.11** IR spectrum of 1-hexene (thin film).

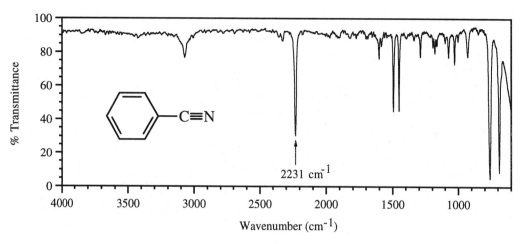

**FIGURE 18.12** IR spectrum of benzonitrile (thin film).

of benzonitrile (Figure 18.12). The absorption is strong because the C≡N bond is polarized due to the electronegativity difference between carbon and nitrogen. Alkynes have weak absorption bands in this region because the C≡C bond is not very polar. The C≡C stretch in phenylacetylene is the small peak at 2100 cm$^{-1}$ (see Figure 18.10).

*C=O Stretch (1820–1660 cm$^{-1}$)*
If there is a C=O present in the molecule, there will be a strong, sharp absorption band in the 1820–1660 cm$^{-1}$ region. A good example is the strong band at 1722 cm$^{-1}$ in the spectrum of propanoic acid (see Figure 18.9). If there is no strong band in this region, there is no C=O in the molecule. The position of the signal within this region depends on what type of functional group contains the C=O group.

| Functional group | Example | C=O stretch |
|---|---|---|
| Amides | Acetamide | 1660 cm$^{-1}$ |
| Ketones | Acetone | 1715 cm$^{-1}$ |
| Carboxylic acids | Propanoic acid | 1722 cm$^{-1}$ |
| Aldehydes | Acetaldehyde | 1730 cm$^{-1}$ |
| Esters | Methyl acetate | 1755 cm$^{-1}$ |
| Acid chlorides | Acetyl chloride | 1800 cm$^{-1}$ |
| Acid anhydrides | Propanoic anhydride | 1815 and 1745 cm$^{-1}$ |

These base values for saturated acyclic compounds are affected by several factors, such as ring strain and conjugation. Ring strain causes the position of the absorption band to move to higher wavenumbers indicating that the strength of the bond has increased.

| Ketone | C=O stretch | Ester | C=O stretch |
|--------|-------------|-------|-------------|
| Cyclopropanone | 1815 cm$^{-1}$ | — | — |
| Cyclobutanone | 1775 cm$^{-1}$ | Propanolactone | 1840 cm$^{-1}$ |
| Cyclopentanone | 1750 cm$^{-1}$ | 4-Butanolactone | 1770 cm$^{-1}$ |
| Cyclohexanone | 1715 cm$^{-1}$ | Methyl acetate | 1755 cm$^{-1}$ |
| Acetone | 1715 cm$^{-1}$ | 5-Pentanolactone | 1735 cm$^{-1}$ |
| Cycloheptanone | 1705 cm$^{-1}$ | 6-Hexanolactone | 1735 cm$^{-1}$ |

Conjugation with a double bond or with an aromatic ring decreases the bond order of the C=O slightly and causes the position of absorption to move to lower wavenumbers. Compare the positions of the C=O absorption bands of 4-methylpentan-2-one (Figure 18.13) and 4-methyl-3-penten-2-one (Figure 18.14). The absorption band is shifted from 1719 cm$^{-1}$ to 1695 cm$^{-1}$. A similar

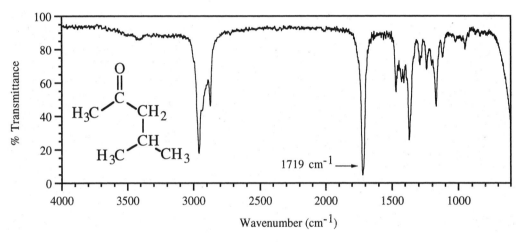

FIGURE 18.13 IR spectrum of 4-methylpentan-2-one (thin film).

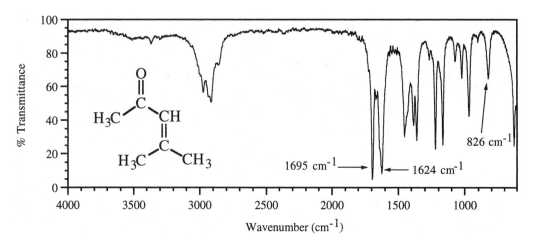

FIGURE 18.14 IR spectrum of 4-methyl-3-penten-2-one (thin film).

shift is observed in acetophenone where the C=O absorption band appears at 1690 cm$^{-1}$.

*C=C Stretch*
*(1680–1580 cm$^{-1}$)*

Absorptions in the 1680–1580 cm$^{-1}$ region are due to C=C bonds in alkenes as well as C=C bonds in aromatic compounds. Their intensities vary from weak to medium. Typical of an absorption of this type is the band at 1647 cm$^{-1}$ in the spectrum of 1-hexene (see Figure 18.11). Notice how weak the bands in this region are in the spectra of phenylacetylene (see Figure 18.10) and benzonitrile (see Figure 18.12). The position of the band and its intensity are affected by conjugation. The position of the C=C absorption in 4-methyl-3-penten-2-one (see Figure 18.13) appears at 1624 cm$^{-1}$ and the intensity is significantly stronger than the intensity of the band in 1-hexene.

*C—O Stretch*
*(1260–1000 cm$^{-1}$)*

Because the C—O bond is highly polarized, the absorption bands for the C—O bond are generally very strong. However, the assignment can sometimes be ambiguous because the peaks occur in the fingerprint region (1500–900 cm$^{-1}$), which is cluttered with many absorption bands due to bending vibrations, overtone bands, and combination bands. Esters, ethers, and alcohols show bands in this region. In the spectrum of 2-propanol (see Figure 18.7), the signals at 1165 and 1131 cm$^{-1}$ are attributed to the C—O stretching vibration. Strong absorptions in the region 1260–1000 cm$^{-1}$ have been correlated with the degree and type of substitution of an alcohol.

| Type of alcohol | C—O stretch |
|---|---|
| RCH$_2$—OH <br> Primary | 1075–1000 cm$^{-1}$ |
| R <br> \ <br> HC—OH <br> / <br> R′ <br> Secondary | 1125–1000 cm$^{-1}$ |
| R″ <br> \| <br> R—C—OH <br> \| <br> R′ <br> Tertiary | 1210–1100 cm$^{-1}$ |
| ⬡—OH <br> Phenol | 1260–1180 cm$^{-1}$ |

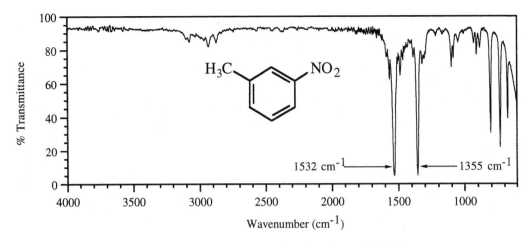

**FIGURE 18.15** IR spectrum of 3-nitrotoluene (thin film).

**NO₂ Stretches (1570–1500 cm⁻¹ and 1380–1300 cm⁻¹)**

Nitro groups have two very distinctive absorptions due to symmetric and asymmetric O—N—O stretches. They are usually the most intense bands in the spectrum. In the spectrum of 3-nitrotoluene, shown in Figure 18.15, the signals appear at 1532 and 1355 cm⁻¹.

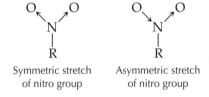

Symmetric stretch          Asymmetric stretch
of nitro group                of nitro group

**Other Useful Correlations**

An infrared spectrum can be highly cluttered with peaks, and not every one can be easily or directly correlated to a specific vibration. However, there are some absorption bands, in addition to the basic fundamental IR stretching vibrations, that can provide structural information. For example, aldehydes exhibit two weak absorptions between 2700 and 2900 cm⁻¹ due to C—H stretches coupled with the C=O stretch. These characteristic bands are evident at 2815 and 2743 cm⁻¹ in the spectrum of cinnamaldehyde shown in Figure 18.16.

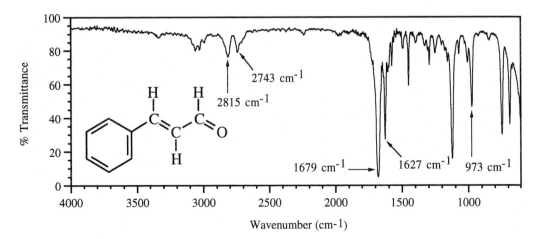

**FIGURE 18.16** IR spectrum of cinnamaldehyde (thin film).

Table 18.3 summarizes the diagnostic peaks in the region from 1000 to 600 cm$^{-1}$. Absorptions at 997 and 914 cm$^{-1}$ in the spectrum of 1-hexene (see Figure 18.11) are characteristic of a monosubstituted alkene. In the spectrum of cinnamaldehyde (see Figure 18.16), the *trans*-disubstituted C=C bond is indicated by the absorption at

| TABLE 18.3 | Out-of-plane C—H bending vibrations of alkenes and aromatic compounds |
|---|---|
| Structure | Position (cm$^{-1}$) |
| R\C=CH$_2$ /H | 997–985 915–905 |
| R\C=C/H H\R | 980–960 |
| R\C=C/R H\H | 730–665 |
| R\C=CH$_2$ /R | 895–885 |
| R\C=C/R R\H | 840–790 |
| ⬡—R | 770–730 710–690 |
| ⬡ R R (ortho) | 770–735 |
| R⬡R (meta) | 900–860 810–750 725–680 |
| R—⬡—R (para) | 860–800 |

973 cm$^{-1}$. The trisubstitued alkene in 4-methyl-3-penten-2-one is indicated by the absorption appearing at 826 cm$^{-1}$ in its IR spectrum (see Figure 18.14).

## 18.6    Sources of Confusion

Careful sample preparation is essential to producing a useful IR spectrum. If the sample is not scrupulously dry, the suspended or dissolved water in the sample will result in large and broad bands in the O—H stretching region (3500–3200 cm$^{-1}$). Besides producing a spectrum with misleading absorptions, the water will also etch the sodium chloride cells used to contain the sample.

If the sample is too concentrated or too thick in the case of thin films, the large bands will "bottom out" at 0% transmittance, producing wide signals from which it is impossible to determine an exact absorption position. Also, small signals will appear to have larger significance than they deserve, often leading to erroneous assignments. The remedy is to prepare a less concentrated solution or a thinner film.

Broad, indistinct signals throughout the spectrum indicate a problem with sample preparation. Usually the problem is one of the following:

- Water is present in the sample.
- In thin films, the sample has evaporated or migrated away from the sampling region of the infrared beam.
- In mulls and KBr pellets, the solid has not been ground finely enough.

A common mistake made when preparing Nujol mulls is to add too much of the oil, leading to a spectrum that is virtually indistinguishable from the spectrum of Nujol itself, shown in Figure 18.17. A problem that is difficult to avoid with Nujol mulls is a drifting

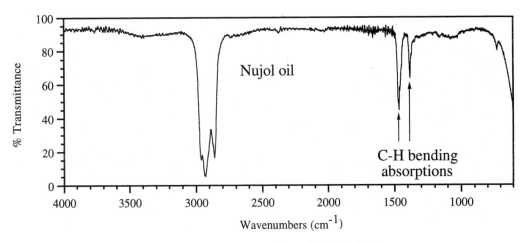

**FIGURE 18.17** IR spectrum of Nujol oil (thin film).

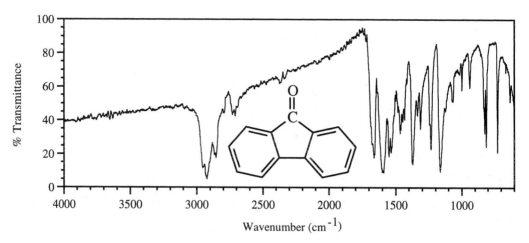

**FIGURE 18.18** IR spectrum of fluorenone (Nujol mull).

base line, as in the spectrum of fluorenone shown in Figure 18.18. Often a severely sloping baseline is the result of a poorly ground solid, but even with careful grinding, some samples still produce a spectrum with a sloping baseline.

Extra peaks in unexpected positions can lead to confusion. In most cases, the extraneous signals are overtones of very strong peaks in the spectrum. A good example of such signals is shown in the spectrum of methyl acetate (Figure 18.19). The signal at $3490 \text{ cm}^{-1}$ is in the region where O—H stretch absorptions appear, but this is clearly not an O—H stretch because of its weak intensity and the shape of the absorption. It is an overtone of the extremely strong C=O peak at $1745 \text{ cm}^{-1}$.

Sometimes you may encounter a spectrum that seems internally inconsistent. Certain absorptions argue for the presence of a functional group, but other characteristic absorptions of the functional group are not present. An example of this situation is seen in the spectrum of cyclohexene (Figure 18.20). The absorption at $3028 \text{ cm}^{-1}$ indicates that there are hydrogens attached to $sp^2$-hybridized carbon atoms in the compound. However, there is no characteristic C=C

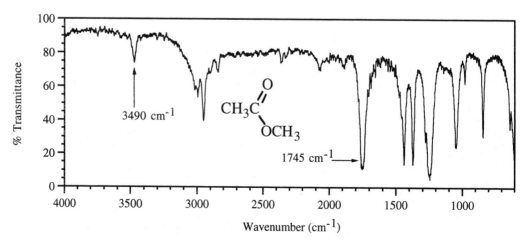

**FIGURE 18.19** IR spectrum of methyl acetate (thin film).

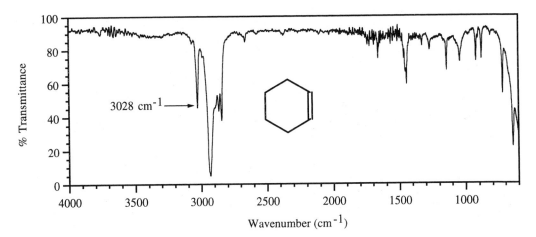

**FIGURE 18.20** IR spectrum of cyclohexene (thin film).

stretch at 1645 cm$^{-1}$. The absence of the C=C absorption is the result of symmetry; the C=C bond does not have a dipole bond moment.

In summary, IR analysis is a way to identify organic functional groups. In addition, some structural information can be obtained from IR spectroscopy, although, as we will see in Chapter 19, NMR is usually better for obtaining structural information. Finally, IR spectra are useful fingerprints for confirming the identity of organic compounds. Comparing an IR spectrum obtained in the laboratory to the spectra available in a compendium of IR spectra, such as Ref. 1, is an excellent way to identify a compound.

## References

1. *The Aldrich Library of FT IR Spectra*; 2nd ed.; Aldrich Chemical Co.: Milwaukee, WI, 1992; 3 volumes.
2. Silverstein, R. M.; Webster, F. X. *Spectrometric Identification of Organic Compounds*; 6th ed.; Wiley: New York, 1998.
3. Colthup, N. B.; Daly, L. H.; Wiberly, S. E. *Infrared Absorption Spectroscopy—Practical*; 3rd ed.; Academic: New York and London, 1990.

## Questions

1. Comment on how the sample preparation techniques used in each situation can affect the appearance of the high wavenumber end of the IR spectrum.

   a. The IR spectrum of a thin film of 1-pentanol reveals a single broad and strong band centered at 3300 cm$^{-1}$, whereas the spectrum of a dilute chloroform solution of the same compound shows both the same broad band and a sharp spike at 3650 cm$^{-1}$.

   b. IR analysis of *o*-hydroxyacetophenone shows the same appearance in a wide range of concentrations in chloroform: a broad band centered at 3080 cm$^{-1}$.

   c. When simple alcohols are subjected to analysis in any kind of mulling compound (Nujol, KBr, Fluorolube), the IR spectrum shows essentially the same broad, strong band centered at approximately 3300 cm$^{-1}$, whereas dilute solutions in a chlorinated hydrocarbon solvent might show the same 3300 cm$^{-1}$ band and very often show a new, sharp peak at approximately 3650 cm$^{-1}$.

2. In each of the following sets, match each compound with the appropriate set of IR bands.

   a. dodecane, 1-decene, 1-hexyne, 1,2-dimethylbenzene

   3311(s), 2961(s), 2119(m) cm$^{-1}$

   3020(s), 2940(s), 1606(s), 1495(s), 741(s) cm$^{-1}$

   3049(w), 2951(m), 1642(m) cm$^{-1}$

   2924(s), 1467(m) cm$^{-1}$

   b. phenol, benzyl alcohol, methoxybenzene

   3060(m), 2835(m), 1498(s), 1247(s), 1040(s) cm$^{-1}$

   3370(s), 3045(m), 1595(s), 1224(s) cm$^{-1}$

   3330(bs), 3030(m), 2980(m), 1454(m), 1023(s) cm$^{-1}$

   c. 2-pentanone, acetophenone, 2-phenylpropanal, heptanoic acid, 2-methylpropanamide, phenyl acetate, octyl amine

   3070(m), 2978(m), 2825(s), 2720(m), 1724(s) cm$^{-1}$

   3372(m), 3290(m), 2925(s) cm$^{-1}$

   3070(w), 1765(s), 1215(s), 1193(s) cm$^{-1}$

   3300–2500(bs), 2950(m), 1711(s) cm$^{-1}$

   3060(m), 2985(w), 1685(s) cm$^{-1}$

   3352(s), 3170(s), 2960(m), 1640(s) cm$^{-1}$

   2964(s), 1717(s) cm$^{-1}$

3. In an attempt to prepare diphenylacetylene, 1,2-dibromo-1,2-diphenylethane is refluxed with potassium hydroxide. A hydrocarbon with the chemical formula $C_{14}H_{10}$ is isolated. The infrared spectrum exhibits signals at 3100–3000 cm$^{-1}$ but no signals in the region 2300–2100 cm$^{-1}$. Is this spectrum consistent with a compound containing a carbon-carbon triple bond? Explain the absence of a signal in the 2300–2100 cm$^{-1}$ region.

4. Treatment of cyclohexanone with sodium borohydride results in a product that can be isolated by distillation. The IR spectrum of this product is shown in Figure 18.21. Identify the product and assign the major IR bands.

5. When benzene is treated with chloroethane in the presence of aluminum chloride, the product is expected to be ethylbenzene (bp 136°C). During the isolation of this product by distillation, some liquid of bp 80°C is obtained. Identify this product using its boiling point and the IR spectrum in Figure 18.22.

6. In principle, matched solution cells should remove IR bands due to the solvent and reveal peaks due to solute (sample) in the region where solvents absorb (see Table 18.1). In practice, although solvent band removal is easily achieved, it is often difficult to discern the peaks due to the sample in these regions. Explain.

7. Consider the IR spectra shown in Figures 18.23–18.30 and match them to the following compounds: biphenyl, 4-isopropylmethylbenzene, 1-butanol, phenol, 4-methylbenzaldehyde, ethyl propionate, benzophenone, acetanilide.

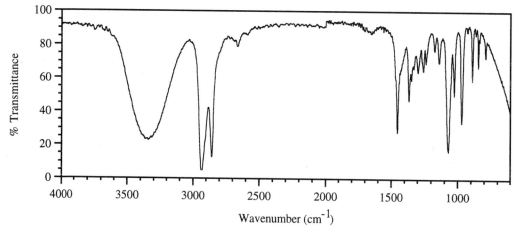

**FIGURE 18.21** IR spectrum for Question 4 (thin film).

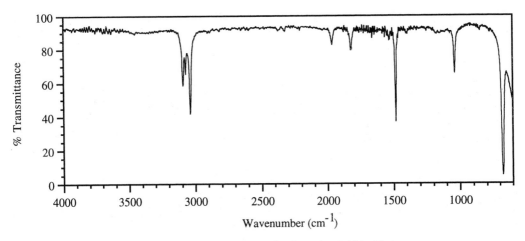

**FIGURE 18.22** IR spectrum for Question 5 (thin film).

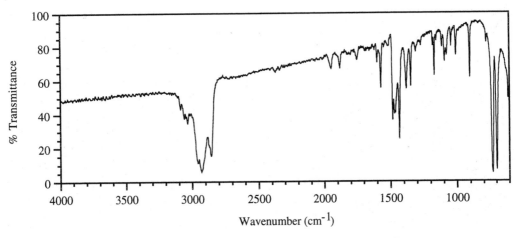

**FIGURE 18.23** IR spectrum for Question 7 (Nujol mull).

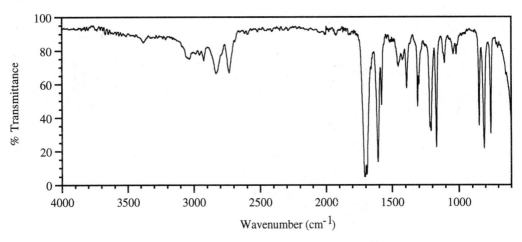

**FIGURE 18.24** IR spectrum for Question 7 (thin film).

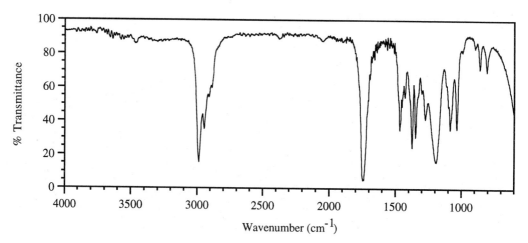

**FIGURE 18.25** IR spectrum for Question 7 (thin film).

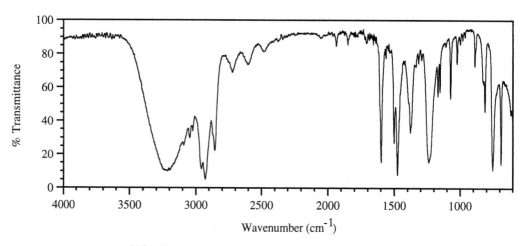

**FIGURE 18.26** IR spectrum for Question 7 (Nujol mull).

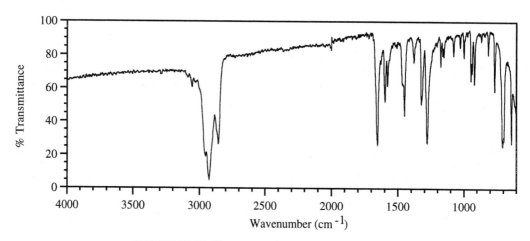

**FIGURE 18.27** IR spectrum for Question 7 (Nujol mull).

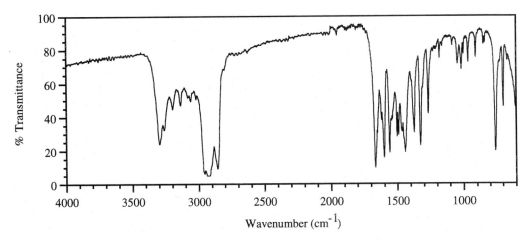

**FIGURE 18.28** IR spectrum for Question 7 (Nujol mull).

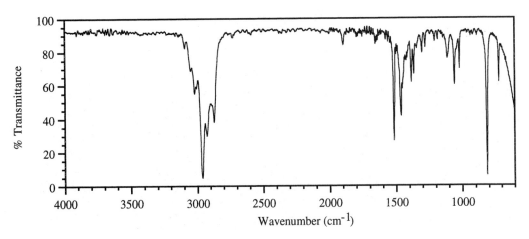

**FIGURE 18.29** IR spectrum for Question 7 (thin film).

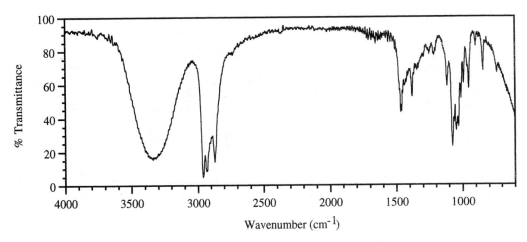

**FIGURE 18.30** IR spectrum for Question 7 (thin film).

# 19

# NUCLEAR MAGNETIC RESONANCE SPECTROSCOPY

Nuclear magnetic resonance (NMR) spectroscopy is one of the most important modern instrumental techniques used in the determination of molecular structure. The first measurements of the absorption of energy by atomic nuclei in a varying magnetic field were reported independently by the American physicists Felix Bloch (Stanford University) and Edward Purcell (Harvard University) in 1946. In 1952 they shared the Nobel prize in physics for this work. Although the initial absorptions measured were from $^{15}N$ nuclei, the technique was soon extended to hydrogen nuclei, which are major components of organic compounds. Since that time, nuclear magnetic resonance has been in the forefront of the spectroscopic techniques that have completely revolutionized organic structure determination. Like other spectroscopic techniques, NMR depends on the quantized energy changes that can be induced in simple molecules when they interact with electromagnetic radiation. The energy needed for NMR is in the radiofrequency range and is much lower energy than that of other spectroscopic techniques.

Any isotope whose nucleus has a nonzero magnetic moment is in theory detectable by NMR spectroscopy. Readily observable nuclei include $^{1}H$, $^{2}H$, $^{13}C$, $^{15}N$, $^{19}F$, and $^{31}P$, among others. Here we will focus on $^{1}H$ NMR, with a shorter description of $^{13}C$ NMR.

The physical model that explains the behavior of a spinning nucleus in an applied magnetic field comes directly from quantum mechanics (Figure 19.1). The precession of a child's top about a vertical axis as it spins can be used as a mechanical model for this

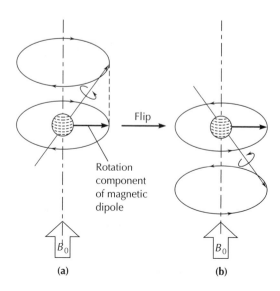

**FIGURE 19.1**
Resonance condition.

process, and this precession describes the upper circle in Figure 19.1a. The vertical axis is defined by the instrument's magnetic field, and nuclei with nonzero magnetic moments precess about this axis. In Figure 19.1a the vector at a 45° upward angle is aligned with the applied magnetic field $B_0$. The orientation of the 45° vector in Figure 19.1b is aligned against the field.

Precession of a nucleus about the applied magnetic field is a quantized phenomenon, and the absorption of the excitation frequency of radio waves can cause a transition to a higher energy level. If the applied frequency ($v$) is precisely tuned to the rotational frequency of the vector, the system is said to be in resonance, and the nucleus absorbs energy and flips to a position opposed to the magnetic field (Figure 19.1b). When a nucleus in the higher-energy state returns to the lower-energy state, it gives up a quantum of energy. This emitted energy gives rise to an NMR signal. The energy of this transition is extremely small by chemical standards— only about $10^{-6}$ kJ · mol$^{-1}$—which corresponds to energy in the radio-frequency region of the electromagnetic spectrum. Different nuclei give rise to signals at different frequencies because of their structural environments in an organic compound.

## 19.1    NMR Instrumentation

The first NMR spectrometers were continuous wave (cw) instruments. They used a radio-frequency transmitter to continuously sweep a sample with energy until a match was obtained between this energy and the magnitude of the energy difference between nuclei in the two states shown in Figure 19.1. When an exact match was obtained, resonance caused a signal to occur at a frequency indicative of the environment of the nucleus (usually $^1$H or $^{13}$C). A radio-frequency receiver was used to monitor the energy changes.

More recent instruments use a technique known as pulsed Fourier transform NMR (FT NMR). In this technique, a broad pulse of electromagnetic radiation excites all the nuclei simultaneously, resulting in a continuously decreasing oscillation caused by the decay of excited nuclei back to a stable Boltzmann energy distribution. This stable distribution contains a slightly greater population of the lower energy state, the state to the left in Figure 19.1. The oscillating, or decaying, sine curve is called a *free-induction decay* (FID). The FID, often referred to as a time domain signal, is converted to a set of frequencies, or "normal" spectrum, by the mathematical treatment of a Fourier transform. The relatively simple FID of a compound with only a single frequency is shown in Figure 19.2a. In Figure 19.2b, the FID from a compound with two frequencies is more complex. Constructive combinations of the two frequencies give rise to enhanced signals and destructive combinations give little or no signal. The Fourier transform of the FID in Figure 19.2b produces two signals.

**FIGURE 19.2**
FID and Fourier
transform of FID of
(a) one signal and
(b) two signals.

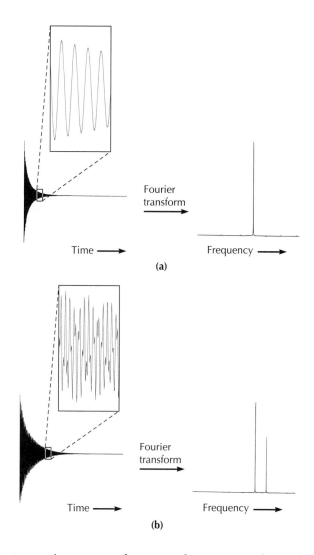

Most organic compounds are much more complex and the FID is made up of contributions of hundreds of frequencies. A computer program using Fourier transform mathematics is required to convert the FID to the "normal" spectrum. When the sample size is small, the acquisition of the signal from more than one pulse (or "scan") is necessary to obtain good NMR signals with the required signal-to-noise ratio. The NMR computer programs the multiple pulses and collects the data from these pulses. The computer is an indispensable part of the pulsed NMR spectrometer.

There are many different kinds of NMR spectrometers, and it is common practice to refer to them by their nominal operating frequency. In earlier instruments, equipped with permanent magnets or standard electromagnets, the typical magnet field ($B_0$) strength was 1.4092 T (T = Tesla), and hydrogens resonated at approximately 60 megahertz (MHz). Now it is common to use NMR instruments of substantially higher magnetic field strengths. Research instruments routinely operate at 300 to 500 MHz, and many laboratories have instruments with operating frequencies of 600 and even

800 MHz. The high magnetic fields necessary for these instruments can be achieved only by using superconducting electromagnets. Since the materials used to build the magnets are superconducting only at very low temperatures, these magnets are maintained in double-jacketed Dewar vessels cooled by liquid helium and liquid nitrogen.

There are numerous benefits of higher frequency and field strength. The higher field means larger energy differences between different nuclei and thus greater signal separation. The advantage of greater signal separation is evident from the spectra of cinnamyl alcohol, $C_9H_{10}O$, shown in Figure 19.3. The spectrum in Figure 19.3a

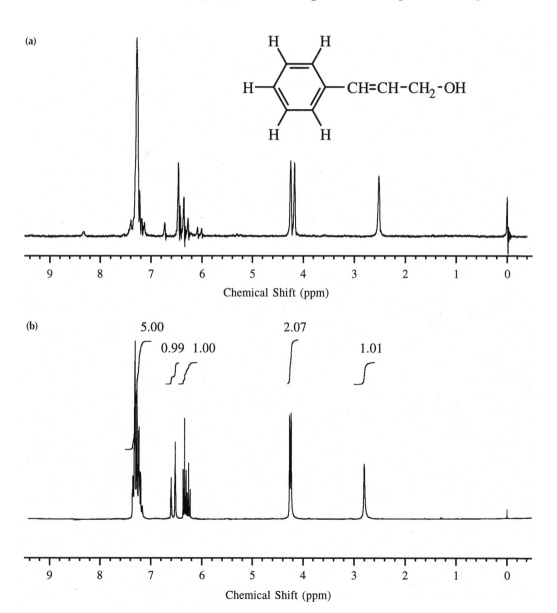

**FIGURE 19.3** $^1H$ NMR spectra of cinnamyl alcohol at (a) 60 MHz and (b) 200 MHz.

was obtained with a 60-MHz continuous wave (cw) NMR spectrometer. The spectrum in Figure 19.3b was obtained with a 200-MHz pulsed Fourier transform NMR spectrometer. In the 60-MHz spectrum is a complex set of signals between 6.0 and 7.0 ppm, which in the higher-frequency spectrum breaks into two groups of signals, one centered at 6.3 ppm and one centered at 6.6 ppm. These spectra are discussed further in the following sections. Additionally, higher-field instruments have a greater sensitivity, which translates into a stronger signal relative to background noise.

## 19.2    Preparing Samples for NMR Analysis

Almost all NMR analysis is done on dilute solutions, whether the sample is a solid or a liquid. Extremely concentrated solutions, undiluted liquids, and solids usually exhibit broad peaks that are not easy to interpret. Broad peaks occur because the nuclear-spin energy is efficiently transferred from one nucleus to another and quickly spreads out among many nuclei. The spectrometer sees many different magnetic environments for each kind of proton. Spectra with sharp, well-differentiated signals are obtained only with dissolved samples. NMR is a nondestructive analytical technique. Because none of the sample is destroyed, it can be recovered. For example, a solid can be recovered by evaporating the $CDCl_3$ solvent.

The choice of solvent for your NMR sample is important. Most of the sample in an NMR tube is solvent, so ideally we want a chemically unreactive solvent that does not absorb energy in the magnetic field. (A magnetically transparent solvent is a solvent with no protons.) Most of the solvents used for preparation of NMR samples are deuterated forms of common solvents such as chloroform ($CHCl_3$) and water. There are two major reasons why deuterated solvents are particularly useful:

1.  No large signals are due to the solvent. Although deuterium does have a magnetic moment, its signal is well removed from the region where protons absorb.
2.  The signal from the deuterium nuclei in the solvent is used by pulsed Fourier transform NMR spectrometers as a reference signal.

Modern NMR spectrometers "lock" on the deuterium signal, assuring that multiple acquisition scans are synchronized. The scans must be synchronized so that identical signals add together constructively, improving the signal-to-noise ratio. Deuterated chloroform ($CDCl_3$) is the most commonly used NMR solvent because it dissolves a wide range of organic compounds and is not prohibitively expensive. Table 19.1 lists several standard NMR solvents. These deuterated solvents are never 100% deuterated. For example, commercial $CDCl_3$ is commonly available as 99.8% (or more)

| TABLE 19.1 | Deuterated solvents for NMR spectroscopy | |
|---|---|---|
| Solvent | Residual $^1H$ signal (ppm) | $^{13}C$ Chemical shift (ppm) |
| $CDCl_3$ | 7.24 (singlet) | 77.0 |
| $CD_3(C=O)CD_3$ | 2.04 (quintet) | 29.8, 206.5 |
| $D_2O$ | 4.6 (broad singlet) | — |
| $CD_3(S=O)CD_3$ | 2.49 (quintet) | 39.7 |

deuterium and only 0.2% protium in its molecules. The residual protons give a small peak ($CHCl_3$) at 7.24 ppm. The residual proton signals for the various solvents are also shown in Table 19.1. It is important to be aware of the position of these residual signals since you do not want to confuse solvent signals with sample signals.

Most solvents used for preparing NMR samples have a small amount of a standard reference substance, tetramethylsilane (TMS), or $(CH_3)_4Si$, dissolved in them. The position of the signal from TMS is defined as the zero point of the spectrum. TMS was chosen because all its protons are equivalent, and they absorb at a magnetic field where very few other protons in typical organic compounds absorb. TMS is also chemically inert and is soluble in most organic solvents. The amount of TMS in the sample depends on the type of instrument being used. For a cw NMR spectrometer, the typical concentration of TMS is 1–2%. With a modern FT NMR instrument, the typical concentration of TMS is 0.1% or less. NMR solvents are commercially available with or without TMS. When you prepare a sample, you must check to make sure you are using the appropriate solvent. In fact, a modern FT NMR instrument does not actually require TMS in the solution; instead, it uses the small residual proton peak of the solvent as the reference peak for calibrating the spectrum (see Table 19.1).

Tetramethylsilane is not soluble in deuterium oxide ($D_2O$), so it cannot be used as a standard with this solvent. The reference substance used for $D_2O$ solutions is sodium 2,2-dimethyl-2-silapentane-5-sulfonate (DSS), $(CH_3)_3SiCH_2CH_2CH_2SO_3Na$. Its major peak appears at nearly the same position as the TMS absorption.

The appropriate concentration of the sample solution will depend on the type of NMR instrumentation available. Only about 10–20 mg of compound dissolved in approximately 0.5–0.7 mL of solvent is necessary for a modern high-frequency FT NMR sample. The cw 60-MHz NMR spectrometer is not as sensitive, and more concentrated solutions are necessary to obtain good spectra. A typical NMR sample for this instrument is prepared by dissolving 0.15 mL of a liquid sample or 150 mg of a solid sample in approximately 0.5 mL of solvent.

Before using an expensive deuterated solvent for an NMR analysis, be sure that your compound dissolves in the common, deuterium-free solvent. Prepare this preliminary sample using the necessary amount of solvent in a small vial or test tube. If the preliminary test is satisfactory, place the sample in a separate vial or

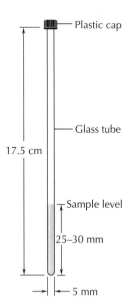

**FIGURE 19.4**
NMR sample tube filled to the correct height.

small test tube and add approximately 0.7 mL of deuterated solvent. Agitate the solution to facilitate dissolution of the sample. If a clear, homogeneous solution is obtained, transfer the sample to the NMR tube with a disposable glass pipet.

If a clear, homogeneous solution is not obtained, the particulate material (dust and/or undissolved sample) must be removed before the sample is transferred to the NMR tube. A convenient filter can be prepared by inserting a small wad of glass wool into the neck of a disposable glass pipet. The narrow end of the pipet filter is placed in the NMR tube and the sample to be filtered is transferred into the filter with a second disposable pipet. Pressure from a pipet bulb is used to force any solution trapped in the filter into the NMR tube.

Only a small part of an NMR tube is in the effective probe area of the instrument. Typically, the height of the sample in the tube should be 25–30 mm (Figure 19.4). However, the required height can vary from one instrument to another, and you should check the required minimum for your instrument. Often a gauge is available for checking the solution height for a given NMR spectrometer. About 0.5–0.7 mL of solution in the tube provides the required sample height. If the sample is slightly short, add a few drops of solvent to bring the solution to the required level and agitate the NMR tube to thoroughly mix the sample. Excessive amounts of sample in an NMR tube can adversely affect the quality of a spectrum, so it is important to use the required amount but not much more than the minimum amount.

NMR tubes are delicate precision equipment. The most commonly used NMR tubes are thin-walled glass and are easily chipped at the top if not handled carefully. Chipping occurs most often during pipetting of the sample into the tube and when trying to remove the plastic cap from the top of the tube.

After the spectrum has been obtained, the NMR tube should be cleaned, usually by rinsing with a solvent such as acetone or chloroform, and allowed to dry. Solvents cling tenaciously to the inside surface and a long drying period or baking in an oven is required to remove all the solvent. If NMR tubes are not cleaned soon after use, the solvent usually evaporates and leaves a caked or gummy residue that is difficult to clean.

To obtain spectra with sharp, well-defined peaks, the tube is usually spun in the NMR instrument to average the magnetic environments throughout the sample. In fact, NMR tubes are selected for uniform wall thickness and minimum wobble. Too much sample in the tube is a waste of material and tends to make it topheavy, often resulting in poor spinning performance and thus poor-quality spectra.

Polar chemicals, such as carboxylic acids and polyhydroxyl compounds, are usually not soluble in $CDCl_3$. However, in most cases these compounds are soluble in water. Deuterium oxide ($D_2O$) should then be used as the NMR solvent. If a carboxylic acid is not soluble in $D_2O$, it will probably be soluble in $D_2O$ containing sodium hydroxide. Adding a drop or two of concentrated sodium hydroxide solution to the sample in $D_2O$ is usually enough to dissolve the sample.

The use of $D_2O$ presents some problems. There will always be a broad peak at approximately 4.6 ppm due to a small amount of HOD present in the original $D_2O$ solvent. This solvent peak can cover important peaks from your compound. Also, $D_2O$ may exchange with protons in your sample compound, producing HOD. Consider, for example, what happens when a carboxylic acid or alcohol dissolves in $D_2O$:

$$R-\overset{\overset{\displaystyle O}{\|}}{C}-O-H + D_2O \rightleftharpoons R-\overset{\overset{\displaystyle O}{\|}}{C}-O-D + H-OD$$

Carboxylic acid

$$R-O-H + D_2O \rightleftharpoons R-O-D + H-OD$$

Alcohol

Deuterium nuclei are "invisible" in $^1H$ NMR spectra, and in an NMR solution there are many more molecules of solvent $D_2O$ than of the sample's acid or alcohol protons. The equilibrium positions lie well to the right, and the hydroxyl protons do not appear as separate signals but instead merge into the HOD signal.

## 19.3    Summary of Steps for Preparing an NMR Sample

1. Test solubility of the sample in ordinary, nondeuterated solvents. Select a solvent that dissolves the sample completely.
2. Place 10–25 mg of the sample in a clean, small vial or test tube.
3. Add 0.5–0.7 mL of the appropriate deuterated solvent.
4. Agitate the mixture in the vial to produce dissolution of the sample.
5. Transfer the sample solution into the NMR tube using a disposable glass pipet. If there are any solids in the sample, filter the sample through a small plug of glass wool.
6. Check the level of the sample in the tube. If needed, add drops of solvent to bring the solution to the recommended level for the instrument and agitate the mixture to produce a homogeneous solution.
7. Cap the NMR tube.
8. Wipe the outside of the tube to remove any material that may impede smooth spinning of the sample in the NMR instrument.

## 19.4    Interpreting NMR Spectra

Typically, four types of information are extracted from an NMR spectrum:

1. Number of different kinds of protons in the molecules of the sample, given by the number of groups of signals
2. Relative number of protons contributing to each group of signals in the spectrum, called the *integration*
3. Positions of the groups of signals along the horizontal axis, called the *chemical shift*
4. Pattern within a group of signals, called *spin-spin coupling*

Each of these four types of information is important in determining the chemical structure of the sample, and all are discussed in the following sections.

## 19.5    How Many Types of Hydrogens Are Present?

The first step in analyzing an NMR spectrum is to take a broad overview. A common mistake of novice analysts is to focus on some detail in the spectrum, often a prominent signal, and develop an analysis based entirely on an assumption that is consistent with only that detail. Sometimes this method works, but many times it does not. A general method for analysis starts by looking at the entire spectrum and counting the number of groups of signals. A structure consistent with the spectrum is required to have at least this many different kinds of protons. This number is a minimum requirement, and often, as the analysis is refined, it is possible to divide a group of signals into subsets of protons that are subtly different from each other. If you examine the 200-MHz NMR spectrum of cinnamyl alcohol (Figure 19.3b), you will see that there are five groups of signals, centered at 2.8, 4.2, 6.3, 6.6, and 7.3 ppm along the horizontal scale. Note that the horizontal scale is read from right to left, with 0.0 at the far right.

## 19.6          Counting Hydrogens (Integration)

Above each group of signals in the 200-MHz NMR spectrum of cinnamyl alcohol (see Figure 19.3b) is what looks like a step with a number over it. The height of each step corresponds to the total signal intensity encompassed by the step. The numbers correspond to the heights normalized to one of the signals. Software on modern digital NMR spectrometers makes this normalization an easy task. Reading from right to left, the normalized values for the groups of signals are 1.0, 2.1, 1.0, 1.0, and 5.3 ppm, respectively. The values are usually not neat, whole-number ratios. Deviations from whole numbers can be as much as 10% and are usually attributed to differences in the amount of time it takes different types of excited hydrogen nuclei to give up energy (relaxation time). In acquiring NMR data, it is necessary to allow enough time for the nuclei to relax. Otherwise, the measured integrals do not accurately reflect the numbers of protons responsible for the signals.

The integrals for this spectrum are interpreted as 1:2:1:1:5. The group of signals at 4.2 ppm is due to two equivalent or nearly equivalent hydrogen nuclei, and the group of signals at 7.3 ppm is due to five equivalent or nearly equivalent hydrogen nuclei. Further examination permits us to complete the analysis. Referring to the structure of cinnamyl alcohol, identify the protons responsible for the five-proton signal appearing at 7.3 ppm and the protons responsible for the two-proton signal appearing at 4.2 ppm.

If a sample is a mixture of compounds and unique signals can be determined for each of the components, integrals of the unique signals can be used to determine the exact composition of the sample. Quantitative analysis of a mixture using the integration is discussed in Technique 19.9.

## 19.7      Chemical Shift

An NMR spectrum is a plot of the intensity of the NMR signals versus magnetic field or frequency. Nuclei that are chemically equivalent, such as the four protons in methane ($CH_4$) or the two protons in dichloromethane ($CH_2Cl_2$), show only one absorption in the NMR spectrum. However, protons that are not chemically equivalent absorb at different frequencies. The local magnetic field experienced by the different protons in a chemical structure varies with different magnetic environments within the molecule. Therefore, the signals from the protons can appear over a range of frequencies. The positions of the signals along the horizontal scale can be correlated with the molecule's structural features.

*Chemical Shift Units*
*(Parts per Million)*

Because it is difficult to reproduce magnetic fields exact enough for NMR on an absolute basis, an internal standard is used as a reference point. The position of an NMR signal, called the *chemical shift*, is measured relative to the absorption of a reference standard. The standard for $^1H$ and $^{13}C$ NMR is tetramethylsilane (TMS), or $(CH_3)_4Si$. Chemical shifts can be measured at a frequency (Hz) corresponding to the sample signal's position relative to TMS. It is conventional, however, to convert frequency to a value $\delta$ *(ppm)* by dividing the chemical shift frequency by the frequency of the spectrometer:

$$\delta \text{ (ppm)} = \frac{\text{frequency of signal of interest (in Hz, from TMS)}}{\text{applied frequency (instrument field strength, MHz)}} \times 10^6$$

The resulting chemical shift scale is independent of the frequency of the spectrometer. On the chemical shift scale, the position of the TMS absorption at the far right is 0.0 ppm. A sample with a signal at 1.00 ppm on a 60-MHz spectrometer will also exhibit that signal at 1.00 ppm when measured on a 200-MHz spectrometer, even though the frequencies of the signal relative to TMS on the respective instruments are 60 Hz and 200 Hz.

There are two absorption peaks (in addition to the TMS standard) in the NMR spectrum of *tert*-butyl acetate (Figure 19.5). The nine equivalent methyl protons (a) in the *tert*-butyl group absorb at $\delta$ 1.45, whereas the three methyl protons (b) appear at $\delta$ 1.97; these numerical values of $\delta$ apply no matter the frequency of the instrument.

**The power of NMR spectroscopy comes from the correlation of molecular structure with positions of signals along the chemical shift scale.** In other words, the chemical shift of an NMR signal provides some of the most important information in an NMR spectrum. For example, the type of bonding and the proximity of nuclei other than carbon influence the chemical shift position of protons attached to carbon atoms. Signals for different types of protons attached to carbons appear in well-defined regions of the chemical shift scale. Charts and tables cataloging these relationships have been constructed by compiling large numbers of NMR signals from many

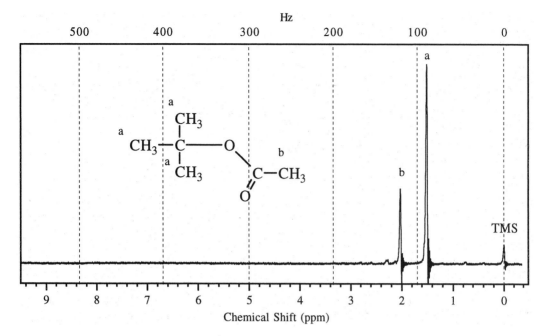

**FIGURE 19.5** ¹H NMR spectrum (60 MHz) of *tert*-butyl acetate. The relative areas (3:1) of peaks a and b correspond to the ratio of protons (9:3) causing the peaks.

organic compounds. These tables usually contain too much data to memorize, but it is crucial that you be able to read these charts and tables for the data you need to interpret an NMR spectrum.

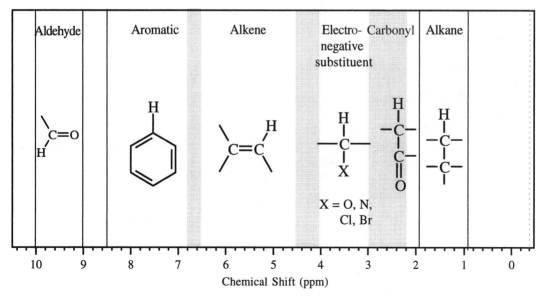

**FIGURE 19.6** Approximate regions of chemical shifts for different types of protons attached to carbon. Shaded areas indicate regions where chemical shifts of two types of protons overlap.

**TABLE 19.2   Characteristic $^1$H NMR chemical shifts**

| Compound | Chemical shift (δ, ppm) |
| --- | --- |
| TMS | 0.0 |
| Alkanes (CH—C) | 0.9–1.9 |
| Ketones, aldehydes, carboxylic acids, esters, and other acid derivatives (CH—C=O) | 1.9–3.0 |
| Halides (CH—X) | 2.1–4.52 |
| Alcohols, esters, ethers (CH—O) | 3.0–4.5 |
| Alkenes (CH=C) | 4.0–6.8 |
| Aromatic (Ar—H) | 6.5–8.5 |
| Aldehydes [(C=O)H] | 9.0–10.0 |
| Amines (—NH) | 1.0–3.0 |
| Alcohols (—OH) | 1.0–5.5 |
| Phenols (ArOH) | 4.0–8.0 |
| Amides [(C=O)NH] | 5.0–8.0 |
| Carboxylic acids [(C=O)OH] | 10.0–13.0 |

Figure 19.6 shows the approximate regions of signals for different types of protons attached to carbon atoms. Signals from protons attached to oxygen and nitrogen can occur over broader regions in a $^1$H NMR spectrum because of exchangeability and H-bonding. A list of chemical shifts for different types of protons is also found in Table 19.2.

Within a region, further correlations can be made based on the bonding (hybridization) at the attached carbon. Referring to the NMR spectrum of *tert*-butyl acetate in Figure 19.5, the signal due to the methyl group attached to the carbonyl group ($sp^2$) appears at 1.97 ppm. The signal due to the methyl groups attached to a saturated carbon ($sp^3$) appears at 1.45 ppm.

*Electronegativity Effects*

The position of the signal due to a proton attached to a carbon atom will also depend on the relative electronegativity ($\chi$) of the nuclei attached to carbon. If a methyl group is attached to silicon ($\chi = 1.8$), as in TMS, the signal appears at 0 ppm; if a methyl group is attached to carbon ($\chi = 2.5$), as in ethane, the signal appears at 0.9 ppm; if the methyl group is attached to oxygen ($\chi = 3.5$), as in methanol, the signal appears at 3.4 ppm; if the methyl group is attached to fluorine ($\chi = 4.0$), as in methyl fluoride, the signal appears at 4.3 ppm.

| Element | Electronegativity ($\chi$) | Example | δ of methyl group (ppm) |
| --- | --- | --- | --- |
| Si | 1.8 | $CH_3$—$Si(CH_3)_3$ | 0 |
| C | 2.5 | $CH_3$—$CH_3$ | 0.9 |
| O | 3.5 | $CH_3$—OH | 3.4 |
| F | 4.0 | $CH_3$—F | 4.3 |

| TABLE 19.3 | Chemical shifts (δ, ppm) of protons in various environments near functional groups | | |
|---|---|---|---|
| Group (Y) | Methyl, CH₃Y | Methylene, RCH₂Y | Methine, R₂CHY |
| —H | 0.2 | 0.9 | 1.3 |
| —CH=CH₂ | 1.7 | 2.0 | 2.2 |
| —C≡CH | 1.8 | 2.2 | 2.6 |
| —(C=O)CH₃ | 2.1 | 2.5 | 2.5 |
| —(C=O)OR | 2.0 | 2.3 | 2.5 |
| —(C=O)OH | 2.1 | 2.4 | 2.6 |
| —(C=O)H | 2.2 | 2.5 | 2.4 |
| —I | 2.2 | 3.2 | 4.2 |
| —C₆H₅ | 2.4 | 2.6 | 2.9 |
| —NH₂ | 2.5 | 2.7 | 3.1 |
| —(C=O)C₆H₅ | 2.6 | 2.9 | 3.6 |
| —NH(C=O)CH₃ | 2.7 | 3.2 | 4.0 |
| —Br | 2.7 | 3.4 | 4.2 |
| —Cl | 3.1 | 3.5 | 4.1 |
| —OR | 3.2 | 3.3 | 3.6 |
| —OH | 3.4 | 3.5 | 3.9 |
| —O(C=O)CH₃ | 3.7 | 4.0 | 4.9 |
| —OC₆H₅ | 3.7 | 4.0 | 4.5 |
| —O(C=O)C₆H₅ | 3.9 | 4.4 | 5.2 |
| —F | 4.3 | 4.4 | 4.8 |

**Substitution Effects**

*In NMR spectroscopy, primary hydrogens are called **methyl**, secondary hydrogens are called **methylene**, and tertiary hydrogens are called **methine**.*

As the carbon atom bearing the proton becomes more highly substituted, the chemical shift of the signal increases. You can see this by comparing the values of the chemical shifts within a row in Table 19.3. Thus, a tertiary hydrogen has a greater chemical shift than a primary hydrogen with the same electronegative groups nearby. In NMR spectroscopy, primary hydrogens are called *methyl*, secondary hydrogens are called *methylene*, and tertiary hydrogens are called *methine*.

Within the aromatic hydrogen region, the position of a signal depends on the substituents attached to the aromatic ring. Table 19.4 (page 238) details the chemical shift positions for the *ortho, meta,* and *para* protons of various monosubstituted benzenes.

**Estimation of Chemical Shifts**

The empirical correlations in these tables illustrate that the approximate magnitude of the chemical shift of a specific nucleus or specific group of atoms is consistent, predictable, and additive. This additivity is extremely useful because it allows estimation of the chemical shift of a proton based on the structure of the compound. For compounds with an isolated functional group, Table 19.3 provides reasonable estimates of chemical shifts for protons attached to adjacent carbons. The aggregate effect of two functional groups on the chemical shift of a methylene group can be determined from Table 19.5 (page 239). The effect of the group alpha to it can be estimated using values in the first column and the effect of the group beta to it by

| TABLE 19.4 | Chemical shifts (δ, ppm) of protons on monosubstituted benzenes | | |
|---|---|---|---|
| Group | ortho | meta | para |
| —OH | 6.83 | 7.22 | 6.93 |
| —OCH$_3$ | 6.91 | 7.29 | 6.95 |
| —OC$_6$H$_5$ | 7.00 | 7.32 | 7.08 |
| —O(C=O)CH$_3$ | 7.09 | 7.38 | 7.23 |
| —O(C=O)C$_6$H$_5$ | 7.22 | 7.43 | 7.27 |
| —CH$_3$ | 7.18 | 7.25 | 7.15 |
| —CH(CH$_3$)$_2$ | 7.22 | 7.28 | 7.16 |
| —CH$_2$Cl | 7.38 | 7.35 | 7.32 |
| —CH=CH$_2$ | 7.40 | 7.32 | 7.24 |
| —CH=CHC$_6$H$_5$ | 7.50 | 7.34 | 7.25 |
| —C≡C—C$_6$H$_5$ | 7.53 | 7.34 | 7.33 |
| —CH=CH(C=O)OH | 7.55 | 7.40 | 7.41 |
| —C$_6$H$_5$ | 7.59 | 7.43 | 7.34 |
| —Cl | 7.34 | 7.29 | 7.23 |
| —Br | 7.49 | 7.23 | 7.28 |
| —NH$_2$ | 6.65 | 7.14 | 6.74 |
| —N(CH$_3$)$_2$ | 6.68 | 7.21 | 6.63 |
| —NH(C=O)CH$_3$ | 7.50 | 7.29 | 7.09 |
| —C≡N | 7.65 | 7.48 | 7.61 |
| —(C=O)C$_6$H$_5$ | 7.81 | 7.48 | 7.59 |
| —(C=O)NH$_2$ | 7.82 | 7.45 | 7.53 |
| —(C=O)H | 7.89 | 7.54 | 7.64 |
| —(C=O)CH$_3$ | 7.96 | 7.46 | 7.56 |
| —(C=O)(C=O)C$_6$H$_5$ | 7.98 | 7.51 | 7.66 |
| —(CO)CH$_2$(CO)C$_6$H$_5$ | 7.99 | 7.48 | 7.55 |
| —(C=O)OCH$_3$ | 8.04 | 7.44 | 7.55 |
| —(C=O)OCH$_2$CH$_3$ | 8.05 | 7.42 | 7.53 |
| —(C=O)Cl | 8.12 | 7.52 | 7.69 |
| —(C=O)OH | 8.13 | 7.47 | 7.61 |
| —NO$_2$ | 8.23 | 7.56 | 7.71 |

using values in the second column. Notice that when the carbon atom bearing the proton is farther away from the functional group, the effect is smaller.

For an example, consider the methylene group of cinnamyl alcohol. It has two alpha groups, a vinyl group and a hydroxyl group. The estimated chemical shift is found by adding the contribution for the vinyl group from the first column in Table 19.5, 0.8 ppm, and the contribution for the hydroxyl group also from the first column, 2.1 ppm, to the base value of the methylene group, 1.2 ppm. The estimated chemical shift is 1.2 + 2.1 + 0.8 = 4.1 ppm, which compares well with the measured value of 4.2 ppm. Deviations from measured values can often be rationalized by considering more subtle and/or long-range contributions. In this case, it is likely that the phenyl group in conjugation with the vinyl group is making a small contribution to the chemical shift. In general, by using Table 19.5 it is possible to estimate chemical shifts to within 0.1–0.4 ppm.

| TABLE 19.5 | **Substituent-shift parameters for alkyl protons compiled from a variety of sources*** | |
|---|---|---|
| **Base values** | | |
| Methyl (R = R′ = H) | 0.9 ppm | |
| Methylene (R′ = H) | 1.2 ppm | |
| Methine | 1.5 ppm | |

| Group (Y) | Alpha substituent RR′C**H**Y | Beta substituent RR′C**H**CH₂Y |
|---|---|---|
| —C=C | 0.8 | 0.2 |
| —C≡C—H, —C≡C—R | 1.0 | 0.2 |
| —C≡C—C₆H₅ | 1.2 | 0.2 |
| —C₆H₅ | 1.3 | 0.4 |
| —(C=O)OH, —(C=O)OR | 1.0 | 0.3 |
| —(C=O)R, —(C=O)H | 1.1 | 0.4 |
| —(C=O)C₆H₅ | 1.6 | 0.6 |
| —I | 2.0 | 0.6 |
| —Br | 2.2 | 0.5 |
| —Cl | 2.3 | 0.4 |
| —F | 3.2 | 0.2 |
| —NH₂ | 1.3 | 0.3 |
| —NH(C=O)R | 1.8 | 0.3 |
| —NH(C=O)C₆H₅ | 2.3 | 0.3 |
| —OR | 1.9 | 0.3 |
| —OH | 2.1 | 0.2 |
| —OC₆H₅ | 2.5 | 0.6 |
| —O(C=O)R | 2.6 | 0.4 |

\* There may be small differences in the chemical-shift values calculated from this table and those measured from individual spectra.

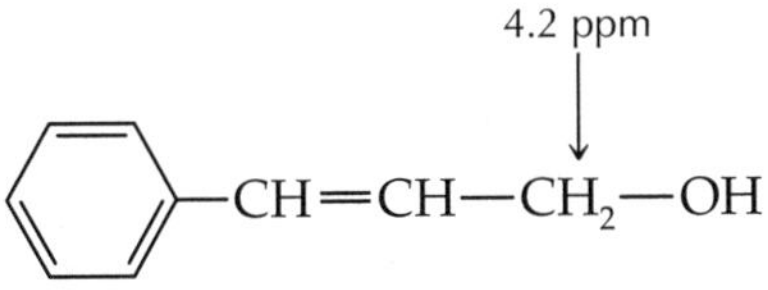

4.2 ppm

⟶

〈benzene ring〉—CH=CH—CH₂—OH

Base value:  1.2
Alpha shift:
—C=C—  0.8
Alpha shift:
—OH  2.1
Estimated:  4.1 ppm
Measured:  4.2 ppm

The effect on the chemical shift by groups beta to the methylene group is smaller than the effect due to alpha substituents. The beta shifts of various substituents are collected in the second column of Table 19.5. The estimations of the chemical shifts of the methylene groups in 3-chloropropiophenone are

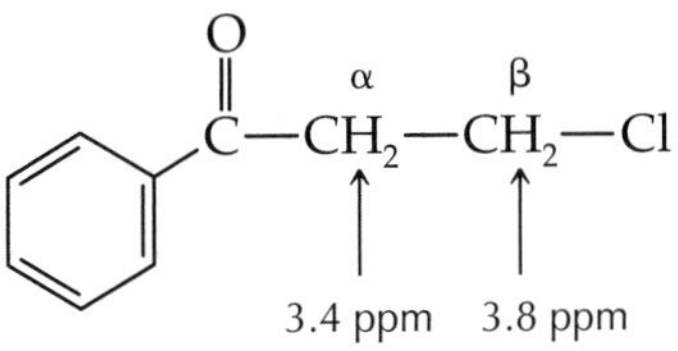

3.4 ppm   3.8 ppm

**Methylene α**

Base value:  1.2
Alpha shift:
—(C=O)C₆H₅  1.6
Beta shift:
—Cl  0.4
Estimated:  3.2 ppm
Measured:  3.4 ppm

**Methylene β**

Base value:  1.2
Alpha shift:
—Cl  2.3
Beta shift:
—(C=O)C₆H₅  0.6
Estimated:  4.1 ppm
Measured:  3.8 ppm

In a similar fashion, it is possible to estimate the chemical shift positions of protons on substituted aromatic rings. To estimate the chemical shift, the contributions of substituents, collected in Table 19.6, are added to a base value of 7.36 ppm. With methyl 3-nitrobenzoate, the contribution to the chemical shift of $H^a$ for an *ortho*-(C=O)OCH₃ group is 0.68 ppm, and the contribution for an *ortho*-nitro group is 0.87 ppm. Adding these values to the base value of 7.36 ppm gives an estimated chemical shift of 8.91 ppm, compared to a measured value of 8.87 ppm.

|  | δHª | δHᵇ | δHᶜ | δHᵈ |
|---|---|---|---|---|
| Base value: | 7.36 | 7.36 | 7.36 | 7.36 |
| —(C=O)OCH₃ | 0.68 | 0.19 | 0.08 | 0.68 |
| —NO₂ | 0.87 | 0.87 | 0.20 | 0.35 |
| Estimated (ppm): | 8.91 | 8.42 | 7.64 | 8.39 |
| Measured (ppm): | 8.87 | 8.42 | 7.67 | 8.38 |

**TABLE 19.6 Substituent-shift parameters for aromatic protons**

Base value for benzene: 7.36 ppm

| Group | ortho | meta | para |
|---|---|---|---|
| —OH | −0.53 | −0.14 | −0.43 |
| —OCH₃ | −0.45 | −0.07 | −0.41 |
| —OC₆H₅ | −0.36 | −0.04 | −0.28 |
| —O(C=O)CH₃ | −0.27 | 0.02 | −0.13 |
| —O(C=O)C₆H₅ | −0.14 | 0.07 | −0.09 |
| —CH₃ | −0.18 | −0.11 | −0.21 |
| —CH(CH₃)₂ | −0.14 | −0.08 | −0.20 |
| —CH₂Cl | 0.02 | −0.01 | −0.04 |
| —CH=CH₂ | 0.04 | −0.04 | −0.12 |
| —CH=CHC₆H₅ | 0.14 | −0.02 | −0.11 |
| —C≡C—C₆H₅ | 0.17 | −0.02 | −0.03 |
| —CH=CH(C=O)OH | 0.19 | 0.04 | 0.05 |
| —C₆H₅ | 0.23 | 0.07 | −0.02 |
| —Cl | −0.02 | −0.07 | −0.13 |
| —Br | 0.13 | −0.13 | −0.08 |
| —NH₂ | −0.71 | −0.22 | −0.62 |
| —N(CH₃)₂ | −0.68 | −0.15 | −0.73 |
| —NH(C=O)CH₃ | 0.14 | −0.07 | −0.27 |
| —C≡N | 0.29 | 0.12 | 0.25 |
| —(C=O)C₆H₅ | 0.45 | 0.12 | 0.23 |
| —(C=O)NH₂ | 0.46 | 0.09 | 0.17 |
| —(C=O)H | 0.53 | 0.18 | 0.28 |
| —(C=O)CH₃ | 0.60 | 0.10 | 0.20 |
| —(C=O)(C=O)C₆H₅ | 0.62 | 0.15 | 0.30 |
| —(CO)CH₂(CO)C₆H₅ | 0.63 | 0.12 | 0.19 |
| —(C=O)OCH₃ | 0.68 | 0.08 | 0.19 |
| —(C=O)OCH₂CH₃ | 0.69 | 0.06 | 0.17 |
| —(C=O)Cl | 0.76 | 0.16 | 0.33 |
| —(C=O)OH | 0.77 | 0.11 | 0.25 |
| —NO₂ | 0.87 | 0.20 | 0.35 |

*Diamagnetic
Shielding*

The chemical shift of a proton is strongly influenced by its environment because the chemical shift depends on the electron density surrounding the nucleus. Circulating electrons in the electron cloud of an atom close to the proton induce a magnetic field in a direction opposite to that of the applied field. Thus, the actual magnetic field that the proton feels is less than the applied field, and the electron cloud is said to shield the nucleus from the magnetic field (Figure 19.7). This effect is called *local diamagnetic shielding.*

Decreasing the electron density around a particular nucleus has the effect of deshielding that nucleus from the applied magnetic field. The chemical shift of a signal increases as the amount of deshielding increases. If the electron density around a particular nucleus is decreased, the opposing magnetic field induced by the circulating electrons is smaller. Because the opposing magnetic field is smaller, the applied field necessary for the NMR signal is less. This effect accounts for the greater chemical shift of protons attached to a carbon atom that has electronegative substituents attached. The effect is also additive. For instance, the chemical shift increases with an increase in the number of chlorine atoms in substituted methane derivatives: methane 0.23 ppm, chloromethane 3.05 ppm, dichloromethane 5.30 ppm, and chloroform 7.24 ppm.

This is a good place to briefly summarize the effect of shielding and deshielding on the chemical shift of protons and introduce some commonly used terms. Increasing the electron density about a nucleus shields it from the applied field, thus requiring an increased applied magnetic field for resonance to occur. This is an *upfield* shift. The observed chemical shift of the signal is smaller

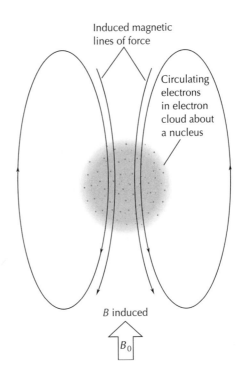

Induced magnetic
lines of force

Circulating
electrons
in electron
cloud about
a nucleus

$B$ induced

$B_0$

**FIGURE 19.7**
Magnetic lines induced in opposition to the applied field due to shielding of the nucleus by circulating electrons.

and, on typical NMR spectra, the signal moves to the *right*. Decreasing the electron density about a nucleus deshields it, causing the chemical shift to increase and moves the signal to the *left*. This is a *downfield* shift. On a typical NMR spectrum, the left side (higher chemical shift) is referred to as the *low-field* region, and the right side (lower chemical shift) is the *high-field* region.

*Anisotropy*

You may have noticed in Table 19.2 that the protons attached to benzene rings absorb at an especially low-field position. This effect is not accounted for simply by the localized deshielding that has been discussed but results from interactions of the π-electrons with the applied field. The protons are substantially deshielded because of the electron circulation set up in the aromatic ring by the applied field. This ring current induces a magnetic field, which deshields the aromatic protons (Figure 19.8). The effect is called *anisotropy*, a term that means having a different effect along a different axis. A similar effect takes place with alkenes (olefins) and aldehydes. In fact, the NMR signal for a proton attached to the carbonyl group of an aldehyde appears even farther downfield than do aromatic protons (see Table 19.2). Protons attached to carbon-carbon double bonds absorb downfield from alkyl hydrogens but not usually as far downfield as the corresponding aromatic protons. One can often determine at a glance whether a compound is aromatic, whether it is an alkene, or whether it contains a CHO group from the positions of the signals in the NMR spectrum.

*Hydrogen Bonding*

The chemical shift for protons attached to oxygen and nitrogen atoms in alcohols and amines can appear over a wide range. The concentration and temperature of the sample usually affect their chemical shift. Therefore, O—H and N—H signals often vary in two NMR spectra of the same compound in the same solvent. This variability depends on the degree of hydrogen bonding within the sample. In dilute samples, there is little or no intermolecular hydrogen bonding and the signals have small chemical shift values. As intermolecular hydrogen bonding increases in concentrated samples, these protons are

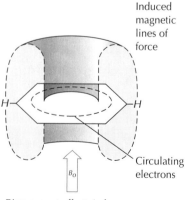

Ring current effects in benzene

**(a)** Induced magnetic fields

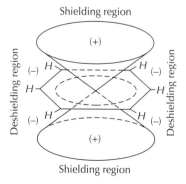

**(b)** Shielding/deshielding areas

**FIGURE 19.8**
Ring current effects in aromatic compounds.

deshielded and shifted to the left (downfield). Hydrogen bonding to the oxygen in an alcohol draws electron density away, deshielding the protons and shifting the signals downfield.

$$H\diagdown O \cdots \diagup^{H}\diagdown O \cdots \diagup^{H}\diagdown O \cdots \diagup^{H}\diagdown O$$

$$\underset{R}{|}\quad\underset{R}{|}\quad\underset{R}{|}\quad\underset{R}{|}$$

The extreme case of hydrogen bonding causing deshielding occurs with carboxylic acids, which have a chemical shift of 10–13 ppm for the O—H proton. In addition, protons of alcohols, amines, and carboxylic acids are capable of chemical exchange with their environment and this may also affect their chemical shift.

To summarize what chemical shifts and integration values of $^1$H NMR signals tell us, let us briefly return to the example of cinnamyl alcohol in Figure 19.3b. The five-proton signal at 7.3 ppm is due to the five protons of a monosubstituted phenyl group. The one-proton signals at 6.6 ppm and 6.3 ppm are consistent with the presence of two different vinyl protons. The signal at 4.2 ppm is assigned to a methylene group flanked by a vinyl group and a hydroxyl group. Finally, the remaining proton signal at 2.8 ppm must be due to the hydroxyl proton itself.

| 19.8 | Spin-Spin Splitting (Coupling) |
|---|---|

The chemical shifts and integrals of NMR signals provide a great deal of information about the structure of a molecule. However, this is often not enough information to determine its structure. Closer examination of NMR signals reveals that they are generally not shapeless blobs, but are highly structured patterns with a multiplicity of lines. The fine structure of these patterns is caused by interactions between the affected proton and its neighboring protons. Examination of these patterns provides highly valuable information about the environments of protons in a molecule.

$$\overset{\displaystyle X\quad Y}{\underset{\displaystyle X\quad Y}{H^a-\overset{|}{\underset{|}{C}}-\overset{|}{\underset{|}{C}}-H^b}}$$

In this hypothetical molecule, the NMR signal of proton $H^a$ is split into two signals by interaction with neighboring proton $H^b$. This split occurs because proton $H^b$ exists in two possible magnetic spin states, one where its spin is aligned with the applied magnetic field and one where its spin is opposed to the applied magnetic field. The aligned orientation leads to a slight deshielding effect and the position of this part of the $H^a$ signal is slightly lower in energy. The opposed orientation leads to a slight shielding

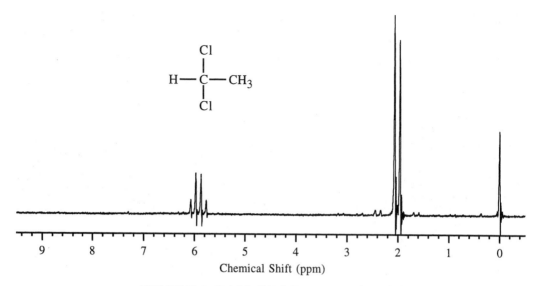

**FIGURE 19.9** 60-MHz ¹H NMR spectrum of 1,1-dichloroethane.

effect and the position of this part of the Hᵃ signal is slightly higher in energy. Thus, the signal for Hᵃ appears as a pair of signals of equal intensity, termed a *doublet.* The distance between the signals of a doublet is called the *coupling constant (J).* Coupling constants are reported in Hz (cycles per second), and their values are independent of the spectrometer operating frequency. In an analogous manner, proton Hᵃ interacts with proton Hᵇ, and Hᵇ also appears as a doublet with the same coupling constant.

The 60-MHz NMR spectrum of 1,1-dichloroethane is shown in Figure 19.9. The NMR signal due to the methyl group, beta to two chlorine atoms, appears at 2.0 ppm and has an integral of three. The signal due to the methine proton, attached to a carbon bearing two chlorine atoms, appears at 5.9 ppm and has an integral of one. The fine structure of the signal at 2.0 ppm reveals two signals of nearly equal intensity, one at 1.96 ppm (117.6 Hz) and one at 2.06 ppm (123.6 Hz). The frequency in Hz is calculated by multiplying the ppm value by the frequency of the NMR spectrometer and dividing by one million ($10^6$).

$$\frac{1.96 \text{ ppm} \times 60 \times 10^6 \text{ Hz (MHz)}}{10^6} = 117.6 \text{ Hz}$$

**Splitting Trees**

A common device for predicting and analyzing the fine structure is a splitting tree. A splitting tree is constructed by mapping the effect of each spin-spin coupling on a signal. The splitting tree for the methyl group is shown in Figure 19.10. The coupling constant is 6.0 Hz (123.6 Hz − 117.6 Hz = 6.0 Hz). Without any splitting, the methyl signal would appear at the center of the doublet, 2.01 ppm [(2.06 ppm + 1.96 ppm)/2]. The signals of the doublet should be of equal intensity. However, careful examination reveals that the

**FIGURE 19.10**
Splitting tree for
methyl group of
1,1-dichloroethane.

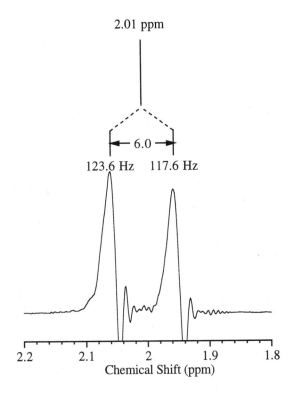

intensity of the upfield signal is slightly less than that of the down-field signal. This difference in intensity is called a *second-order effect* and is discussed further in Technique 19.9. The signals of the doublet would be of equal intensity if the chemical shifts of the interacting nuclei were farther apart. As the chemical shifts of interacting nuclei approach the same value, second-order effects become more apparent. Unless the signals are very close, these effects are small but still informative. In this case, the larger downfield signal of the doublet indicates that the proton coupling with the methyl group is downfield from this signal.

The methine proton at 5.9 ppm is split by coupling with three protons on the adjacent carbon. Because these adjacent protons are equivalent to one another, the signal is split into doublets three times. Since the coupling constant does not depend on direction, the methyl-methine interaction is the same as the methine-methyl interaction, 6.0 Hz. The splitting tree for the methine proton is shown in Figure 19.11. The signal is split into a doublet by the first methyl proton. The coupling with a second methyl proton splits each signal of the doublet into two signals. Because the coupling constants of these two interactions are the same, the position of the high-field signal of one doublet coincides with the position of the low-field signal of the adjacent doublet. If there were no more splitting, the pattern would consist of three equally spaced signals, with the center signal having twice the intensity of the outer signals. This characteristic pattern is called a *triplet.* Coupling with the third methyl proton again splits each of the signals of the triplet into two signals.

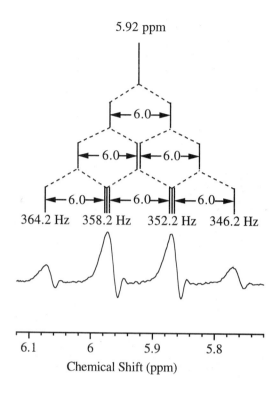

Again, because the coupling constant is the same, the high-field signal of one doublet coincides with the low-field signal of the adjacent doublet. As seen in the splitting tree in Figure 19.11, the resulting pattern is a group of four equally spaced signals. The ratio of the intensities of the signals is $1:3:3:1$. This pattern is called a *quartet.* The center of the quartet is 5.92 ppm [(5.77 ppm + 6.07 ppm)/2].

**N + 1 Rule**

Observations such as these have given rise to the $N + 1$ rule for multiplicity. This rule states that a proton which has $N$ equivalent protons on adjacent carbon atoms will be split into $N + 1$ signals. The ratio of the intensities of the multiplet signals can be obtained from Pascal's triangle (Figure 19.12). The $N + 1$ rule assumes that the coupling constant of each equivalent proton to a different type of proton is the same. On the other hand, if all of the coupling constants to a particular proton are different, the number of peaks is $2^n$, where $n$ equals the number of other protons splitting a particular proton. The ratio of the intensities of the resulting signals is $1:1:1: \ldots . n$.

In the protons of the more typical organic compound—cinnamyl alcohol, for example—some coupling constants are the same and others are different. An expansion of the 200-MHz NMR spectrum of cinnamyl alcohol is shown in Figure 19.13.

The signal at 6.55 ppm, which is due to one proton, appears as a doublet with a coupling constant of 15.9 Hz. The chemical shift indicates that the proton responsible for this signal is a vinyl proton. Because the signal is a doublet, the proton has only one other proton close by on adjacent carbons. Therefore, the signal at 6.55 ppm is assigned to the vinyl proton next to the benzene ring. Note the asym-

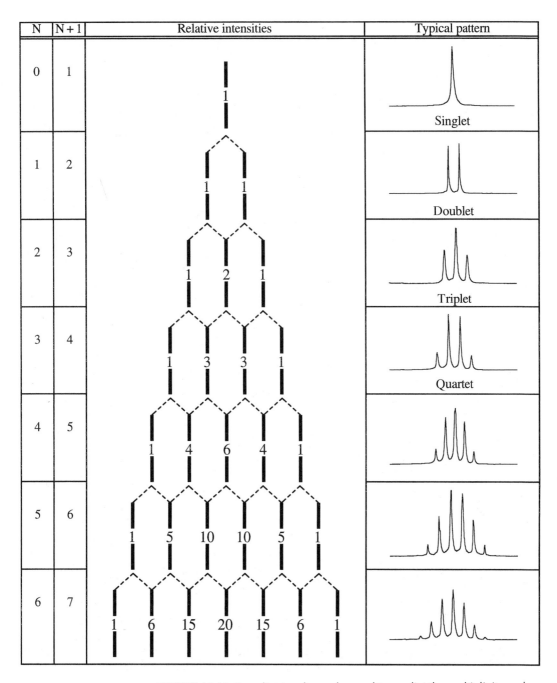

| N | N+1 | Relative intensities | Typical pattern |
|---|-----|---------------------|-----------------|
| 0 | 1 | 1 | Singlet |
| 1 | 2 | 1  1 | Doublet |
| 2 | 3 | 1  2  1 | Triplet |
| 3 | 4 | 1  3  3  1 | Quartet |
| 4 | 5 | 1  4  6  4  1 | |
| 5 | 6 | 1  5  10  10  5  1 | |
| 6 | 7 | 1  6  15  20  15  6  1 | |

**FIGURE 19.12** Pascal's triangle can be used to predict the multiplicity and relative intensities of the signal of any magnetically active nucleus attached to $N$ equivalent nuclei of spin $1/2$ (applies to both $^1H$ and $^{13}C$).

metry of the doublet, indicating that its coupling partner is upfield. From this assignment, it follows that the signal at 6.29 ppm is probably due to the vinyl proton on the carbon beta to the benzene ring. The alpha vinyl proton flanks it on one side and the methylene group flanks it on the other side. The large coupling constant between the

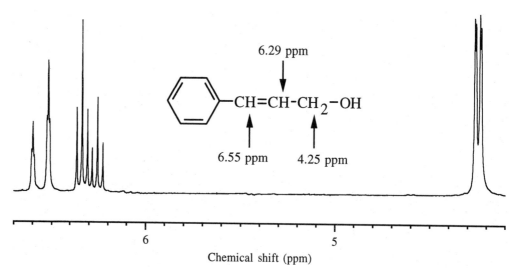

FIGURE 19.13 Expanded section of 200-MHz $^1$H NMR spectrum of cinnamyl alcohol.

two vinyl protons, 15.9 Hz, indicates that the protons are trans to one another; therefore, the C=C of cinnamyl alcohol must be a trans (E) alkene.

If the coupling constants between the vinyl proton and its three adjacent protons were the same, the signal should appear as a quartet ($N = 3$, $N + 1 = 4$). However, the signal at 6.29 ppm is clearly not a quartet but rather looks like a pair of triplets. From the analysis of the 6.55-ppm signal, we have seen that the coupling constant between the two vinyl protons is 15.9 Hz. If this coupling constant were not known, it could be determined from the difference between corresponding peaks of the two triplets, such as the center peaks (1267.2 Hz − 1251.3 = 15.9 Hz). The splitting tree for the proton appearing at 6.29 ppm is shown in Figure 19.14. The coupling constant between the vinyl proton and the methylene group, 5.6 Hz, is obtained from the difference between two adjacent signals of one of the triplets (1272.8 Hz − 1267.2 Hz = 5.6 Hz). Notice here that the triplets centered at 6.29 ppm are not equal in intensity due to the second-order effects between the two vinyl protons.

Finally, the signal at 4.25 ppm in Figure 19.13, which integrates to two protons, appears as a doublet with a coupling constant of 5.5 Hz. This value is within experimental error of 5.6 Hz, the coupling constant seen in Figure 19.14. The correspondence between these two coupling constants (5.6 Hz and 5.5 Hz) is good evidence that the vinyl proton on the beta carbon and the two protons of the methylene group are coupled with each other.

Notice that in this spectrum there is no apparent coupling between the methylene protons and the proton on the adjacent oxygen atom. There is no splitting here because the hydroxyl proton is exchanging rapidly, faster than the NMR time scale. Hydroxyl protons and other exchangeable protons do or do not show coupling

**FIGURE 19.14**
Splitting tree for vinyl hydrogen of cinnamyl alcohol at 6.29 ppm.

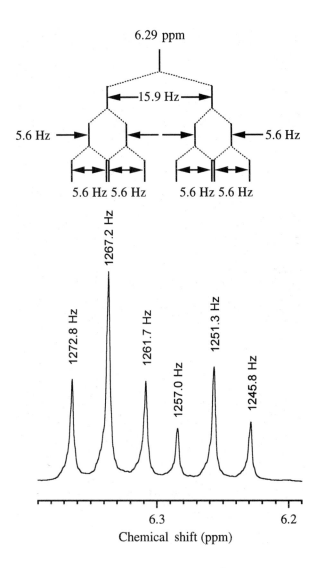

depending on the NMR sample conditions. If even trace amounts of acids or bases are present, no coupling occurs because acid and base catalyze proton exchange faster than the time necessary to set up the NMR coupling between neighboring protons. Neutral, anhydrous samples are likely to show coupling involving a hydroxyl proton.

*Coupling Constants*

Coupling constants can reveal valuable information about the structure of a molecule. Their size is related to the number of bonds between the interacting protons; the more bonds, the smaller the coupling constant. Most of the observed coupling in NMR spectroscopy is due to coupling through three bonds, $H_a$—C—C—$H_b$, called *vicinal coupling.* The magnitude of this coupling ranges from about 2 to 15 Hz. Coupling through two bonds, or *geminal coupling,* occurs between two protons attached to the same carbon atom, $H_a$—C—$H_b$. In many molecules, these two protons are equivalent and coupling is not observed. However, if the protons are different

### TABLE 19.7 Typical proton-proton coupling constants

| Arrangement of protons | $J_{ab}$ (Hz)* | Arrangement of protons | $J_{ab}$ (Hz)* |
|---|---|---|---|
| (geminal) | −8 to −18 | (vinyl geminal) | 0 to 3 |
| Free rotation | 7 | (cis) | 12 to 18 |
| Anti | 8 to 12 | | 6 to 12 |
| Gauche | 2 to 4 | | 0.5 to 2 |
| (axial-axial) | 8 to 13 | | 0 |
| (axial-equatorial) | 2 to 6 | (ortho) | 6 to 9 |
| (equatorial-equatorial) | 2 to 5 | (meta) | 1 to 3 |
| | −11 to −14 | (para) | 0 to 1 |

* The absolute value of a coupling constant is measured from a spectrum. In first-order coupling patterns, the sign of the coupling constant makes no difference in the appearance of the pattern.

and coupling is observed, it is generally on the order of 11–16 Hz. Finally, in certain circumstances, coupling through four or more bonds can be observed, but it is usually very small, 0–2 Hz. One circumstance where coupling through four bonds is often observed is *allylic coupling,* $H_a$—C=C—C—$H_b$. Typical coupling constants for various arrangements of protons are listed in Table 19.7.

Allylic coupling is evident in the NMR spectrum of cinnamyl alcohol. Expansion of the signals at 6.55 ppm and 4.25 ppm reveals further fine structure (Figure 19.15). At 4.25 ppm, the pattern of the NMR signal is a doublet of doublets. The vicinal coupling constant is 5.6 Hz and the allylic coupling constant is 1.3 Hz. The signal at 6.55 ppm is a doublet of triplets with a vicinal coupling constant of 15.9 Hz and allylic coupling constant of 1.3 Hz.

If there is free rotation about the bond connecting the nuclei with the coupled protons, the vicinal or three-bond coupling constants are usually about 7 Hz. If rotation about that bond is restricted, the coupling constant can range from 0 to 15 Hz. The size of a vicinal coupling constant is related to the angle, $\phi$, formed between the C—H bonds of the interacting protons on a Newman

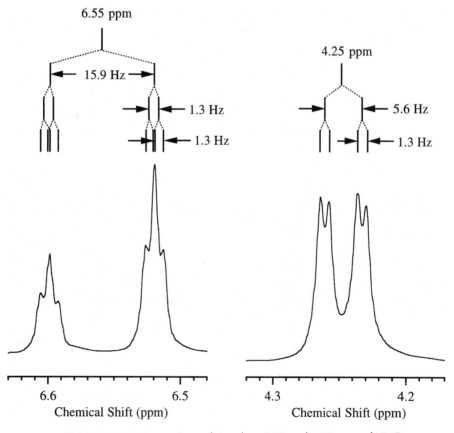

**FIGURE 19.15** Expansions of signals at 6.55 and 4.25 ppm of NMR spectrum of cinnamyl alcohol.

**FIGURE 19.16**
Dihedral angle, $\phi$,
formed by two adjacent
C—H bonds.

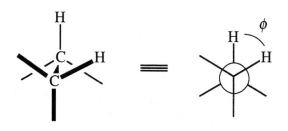

**FIGURE 19.17**
Dependence of
the coupling constant
on dihedral angle
(Karplus relationship).

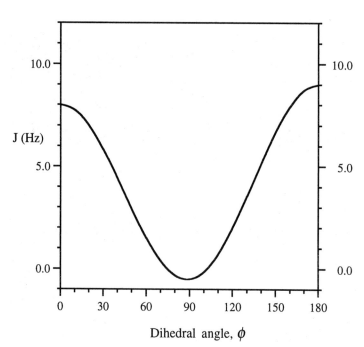

Dihedral angle, $\phi$

projection (Figure 19.16). This angle is called the *dihedral angle.* In the early days of NMR spectroscopy, Martin Karplus (Harvard) studied the relationship between the size of a coupling constant and the dihedral angle. His conclusions are now widely accepted and are often presented as a plot of the coupling constant versus dihedral angle. This plot is called the *Karplus curve* and is shown in Figure 19.17. The important characteristics of the Karplus relationship are the minimum value of the coupling constant at a dihedral angle of 90° and the large values of a vicinal coupling constant at dihedral angles of 0° and 180°.

## 19.9    Sources of Confusion

If you have read through and thought about Techniques 19.4–19.8, you know that using NMR spectroscopy to solve the structure of an organic compound is a perfectly logical process. Sometimes, however, the process of interpreting a spectrum becomes complicated beyond the factors of chemical shift and spin-spin splitting that we have discussed. It is important to be aware of some of the complicating factors so you can make rational choices when confronted with unexpected, confusing, or poorly defined signals in a spectrum.

*History of the NMR Sample: Mixtures of Compounds*

Extra signals in an NMR spectrum are often due to mixtures of the compound with solvents, starting material, reaction side products, and residual proton signals from the deuterated solvent. Much puzzlement and frustration can be avoided by careful consideration of what exactly went into the NMR tube. To determine the source of extra signals, it is important to know the history of the sample. If you have prepared the sample in the laboratory, what solvents and reagents did you use in its preparation? What does the NMR spectrum of the starting material look like? If the sample was purified by recrystallization or chromatography, what solvents were used? What solvent was used to clean the NMR sample tube? Table 19.8 lists some common solvents and the chemical shift positions and multiplicity of their NMR signals.

The NMR spectrum shown in Figure 19.18 is a typical example of a mixture encountered in the laboratory. In this case, the ability of NMR to count the protons in every component of a mixture can be put to good use. The material for the sample was obtained from the base-catalyzed dehydrochlorination of 2-chloro-2-methylpentane.

2-Methyl-1-pentene     2-Methyl-2-pentene

**TABLE 19.8    NMR signals of common solvents**

| Solvent | NMR signals* |
|---|---|
| Acetone | 2.1 (s) |
| Benzene | 7.4 (s) |
| Chloroform | 7.3 (s) |
| Cyclohexane | 1.4 (s) |
| Dichloromethane | 5.3 (s) |
| Diethyl ether | 1.2 (t), 3.5 (q) |
| Dimethyl sulfoxide | 3.6 (s) |
| Ethanol (anhydrous) | 1.2 (t), 3.0 (t) or (s), 3.7 (m) or (q) |
| Ethyl acetate | 1.3 (t), 2.0 (s), 4.1 (q) |
| Hexane | 0.9 (t), 1.3 (m) |
| 2-Propanol | 1.2 (d), 2.6 (s), 4.0 (m) |
| Methanol (anhydrous) | 2.3 (q), 3.4 (d) |
| Tetrahydrofuran | 1.9 (m), 3.8 (m) |
| Toluene | 2.3 (s), 7.2 (s) |
| Water (dissolved) | 1.6 (s) |
| Water (bulk) | 4.5 (br s) |

\* In CDCl$_3$. Multiplicity of signal is shown in parentheses.

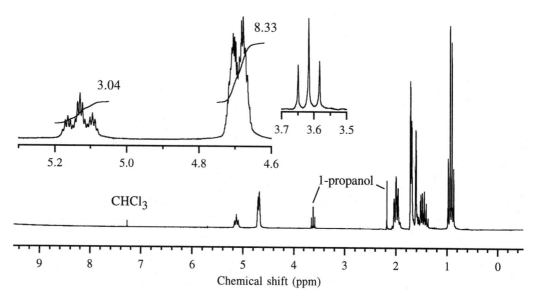

**FIGURE 19.18** 200-MHz $^1$H NMR spectrum of the reaction product from dehydrochlorination of 2-chloro-2-methylpentane, a mixture of 2-methyl-1-pentene and 2-methyl-2-pentene.

The product is a mixture of 2-methyl-1-pentene, 2-methyl-2-pentene, and 1-propanol. The singlet at 2.2 ppm and the triplet at 3.6 ppm are due to the hydroxyl proton of 1-propanol and the methylene group attached to the hydroxyl group. The other signals of 1-propanol are obscured by the signals of the two alkenes. Even though every signal in the spectrum is not distinct, much useful information can be obtained from the spectrum because each component of the mixture exhibits unique features. For example, the ratio of 2-methyl-1-pentene to 2-methyl-2-pentene can be determined from the integrals of the vinyl proton signals at 4.7 ppm and 5.1 ppm. The broad triplet at 5.1 ppm is due to the proton attached to $C_3$ in 2-methyl-2-pentene. The two broad signals at 4.7 ppm are due to the two protons attached to $C_1$ in 2-methyl-1-pentene. To calculate the molar ratio of the two alkenes, the integrals need to be normalized by dividing them by the number of protons causing the signals.

$$\frac{\text{amount (moles) of 2-methyl-1-pentene}}{\text{amount (moles) of 2-methyl-2-pentene}} = \frac{8.33/2}{3.04/1} = \frac{1.37}{1}$$

The calculation shows that the dehydrochlorination of 2-chloro-2-methylpentane produces 58% 2-methyl-1-pentene and 42% 2-methyl-2-pentene.

***Chemical Exchange***   The spectra shown in Figure 19.19 demonstrate another potential source of confusion: chemical exchange. The signal for the hydroxyl proton of 1-pentanol in the 60-MHz NMR spectrum appears as a triplet at 3.19 ppm. In the 200-MHz NMR spectrum, it appears as a

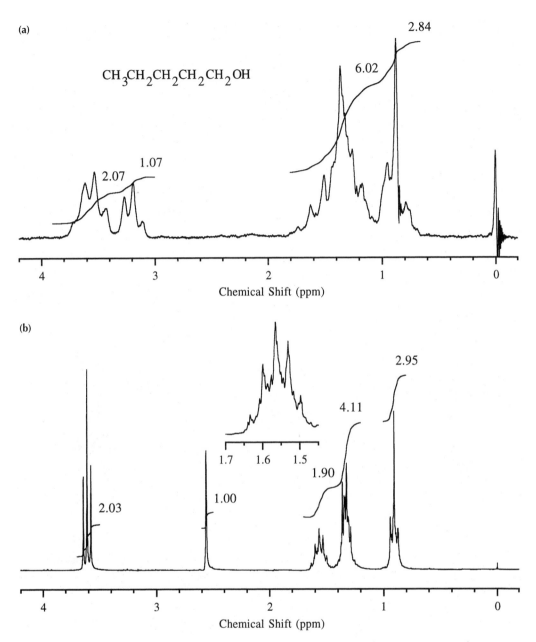

FIGURE 19.19 ¹H NMR spectrum of 1-pentanol at (a) 60 MHz and (b) 200 MHz.

singlet at 2.56 ppm. Protons, which are capable of exchanging from one molecule to another and of hydrogen bonding, show great variability in chemical shift. The position depends on concentration, solvent, temperature, and the presence of water or acid. If, for example, a proton on the oxygen atom of an alcohol exchanges rapidly, and it usually does, no coupling between the hydroxyl proton and protons on the adjacent carbon is observed, as in the case of this 200-MHz NMR spectrum. The hydroxyl proton signal is a singlet and the signal of the adjacent methylene protons appears as a triplet due to splitting by only the $C_2$ methylene group. The sample used to obtain the 60-MHz NMR spectrum was anhydrous and acid-free, and splitting is observed because exchange of the O—H proton is relatively slow. The hydroxyl proton signal is split into a triplet by the adjacent methylene protons. The $C_1$ methylene group appears as an apparent quartet due to splitting by three protons, the hydroxyl proton and the two protons on the adjacent methylene group.

*Overlap of Signals*

Signals from protons with similar chemical shifts may overlap one another, leading to broad and poorly defined patterns. Often patterns are so poorly defined that one is not even tempted to analyze the coupling. An example is demonstrated by the 60-MHz NMR spectrum of 1-pentanol shown in Figure 19.19a. The region 1–2 ppm exhibits a complex signal integrating to six protons due to three similar methylene groups. Sometimes it is possible to unravel these patterns with an NMR spectrum obtained on a higher-field instrument, such as the 200-MHz NMR spectrum of 1-pentanol shown in Figure 19.19b. In this spectrum, the chemical shift of one of the methylene groups is different enough that its signal is separated from the signal for the other two methylene groups and it appears to be a quintet. On even higher-field instruments the other two methylene groups could also exhibit separate signals.

There are also cases where two or more well-defined patterns overlap, producing what at first glance may look like one well-defined pattern but which, on closer examination, has coupling constants and/or signal intensities that do not make sense. A good example of a deceptive pattern is the apparent quartet exhibited at 1.1 ppm in the 60-MHz NMR spectrum of ethyl propanoate shown in Figure 19.20a. The apparent quartet is actually two overlapping triplets, one centered at 1.09 ppm and the other centered at 1.20 ppm. Here again, an NMR spectrum obtained on a higher-field NMR instrument would have clearly shown two separate triplets. In some cases, the chemical shifts may be virtually identical and not capable of being separated using a higher-field instrument. An example of this is found in the 200-MHz NMR spectrum of 2-methyl-1-butanol shown in Figure 19.21 on page 258. The signals in the region of 0.8–1.0 ppm can be assigned to the two methyl groups. They are made up of a triplet at 0.905 ppm and a doublet at 0.907 ppm.

*Second-Order Effects*

As mentioned in the discussion of spin-spin splitting in Technique 19.8, observed splitting patterns may differ somewhat from patterns predicted by simple coupling rules, such as the $N + 1$ rule and Pascal's triangle. These distortions are the result of *second-order effects.* When the effects are small, they can be useful, such as differences in signal intensities that produce "leaning," which indicates the relative position of a coupling partner. As the chemical shifts of the coupled protons become closer to one another, second-order effects become more pronounced. The usual rule of thumb is that second-order effects become apparent in a spectrum when the difference in chemical shifts ($\Delta\nu$, measured in Hz) is less than five times the

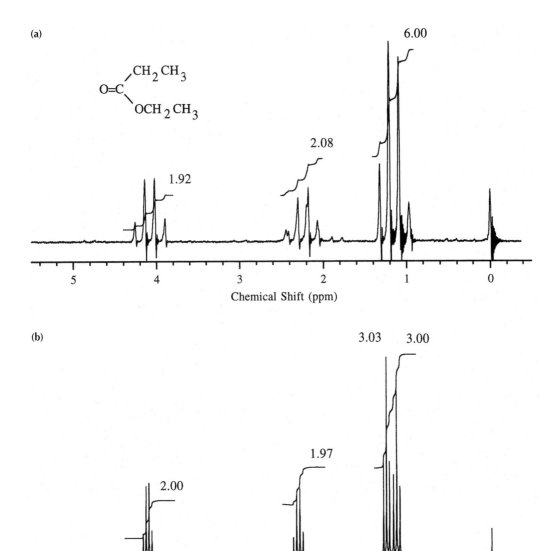

FIGURE 19.20 ¹H NMR spectrum of ethyl propanoate at (a) 60 MHz and (b) 200 MHz.

coupling constant ($\Delta \nu < 5J$). Large second-order effects produce (1) signal intensities that are different from predicted values, (2) additional signals beyond those predicted by simple coupling rules, and (3) coupling constants that cannot be directly measured from differences in signal positions.

Examine the expanded sections of the 60-MHz and 200-MHz spectra of cinnamyl alcohol shown in Figure 19.22 on page 258. In the region from 5.9 to 6.9 ppm the two spectra differ dramatically. On a 200-MHz instrument the individual vinyl protons appear as

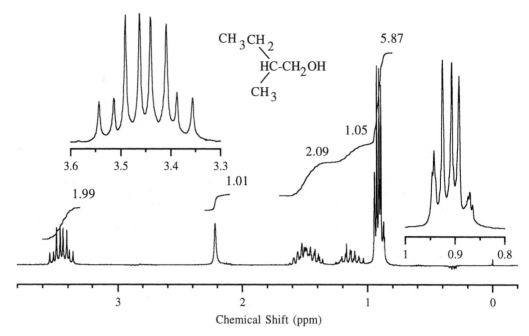

**FIGURE 19.21** 200-MHz $^1$H NMR spectrum of 2-methyl-1-butanol.

well-defined signals at 6.3 ppm and 6.6 ppm, separated by 60 Hz (0.3 ppm × 200 MHz); on a 60-MHz instrument these signals are separated by only 18 Hz (0.3 ppm × 60 MHz). In this instance, even the 200-MHz spectrum exhibits second-order effects: the patterns at 6.3 ppm and 6.6 ppm lean toward each other. In other words, the upfield portion of the 6.6-ppm signal and the down-field portion of the 6.3-ppm signal are larger than the other parts

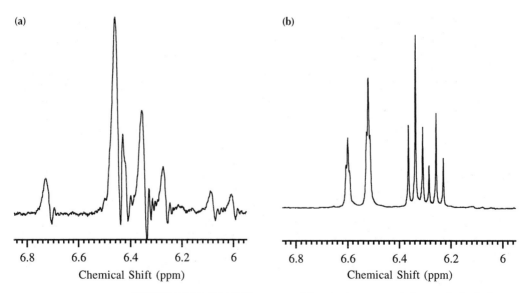

**FIGURE 19.22** $^1$H NMR spectra of the vinyl protons of cinnamyl alcohol at (a) 60 MHz and (b) 200 MHz.

of the two patterns. However, the second-order effects are much larger in the 60-MHz spectrum than they are at 200 MHz. On higher-field instruments, the frequency difference between signals is even larger, so second-order effects are smaller.

**Diasteneotopic Protons**

Structural differences between protons in a molecule may not be obvious at first glance, and this can be a source of confusion. For example, it is very easy to assume that the two protons of a methylene group are equivalent, and in most cases they are. However, if the methylene group is next to a stereocenter, such as an asymmetric carbon atom, the two protons of the methylene group are no longer equivalent. Consider the compound 2-methyl-1-butanol. If there is no coupling to the hydroxyl proton, you might expect the signal for the methylene protons on the carbon to appear as a doublet, due to coupling to the one proton on the adjacent carbon atom. However, the protons of the methylene group are not identical. They cannot be interchanged with one another by any bond rotation or symmetry operation. They are said to be *diastereotopic.* They have different chemical shifts, and they also couple with each other. The appearance of diastereotopic protons is common in the NMR spectra of chiral molecules, those with stereocenters.

Stereocenter

2-Methyl-1-butanol

The NMR spectrum of 2-methyl-1-butanol shown in Figure 19.21 reveals an eight-line pattern for the methylene group. Because the diastereotopic protons are not identical, they also couple with each other. The chemical shifts of the protons of the $C_1$ methylene group are 3.4 ppm and 3.5 ppm. The eight-line pattern between 3.3 and 3.6 ppm is actually a pair of double doublets: the coupling for one set is 6.4 Hz and 10.5 Hz, and the coupling for the other set is 5.6 Hz and 10.5 Hz. The methylene protons of the ethyl group attached to the stereocenter are also diastereotopic, but the patterns are not distinct enough to analyze accurately. If the chemical shifts of diastereotopic protons in a molecule are very close, the splitting pattern degenerates into a pattern closely resembling the one expected if the protons of the methylene group are equivalent. Notice in Figure 19.21 that the two halves of the eight-line pattern at 3.3–3.6 ppm lean into each other. This makes the central lines more intense than the outside lines, even though the splitting tree (Figure 19.23) for this pattern shows equal intensities for all the lines. Use of a higher-field NMR instrument for the spectrum would make all eight lines closer to equal intensity.

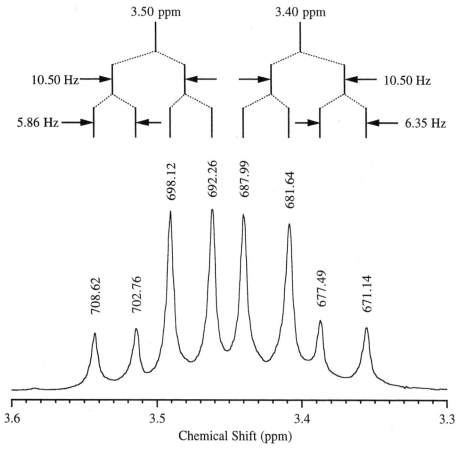

**FIGURE 19.23** Splitting tree for the diastereotopic protons of 2-methyl-1-butanol.

***Sample Preparation and NMR Data Acquisition***

Broad and distorted signals are very often the result of either poor sample preparation or improper operation of the NMR instrument. If all the signals in the spectrum are broad, it is highly likely that one of the following is the problem:

- The sample is not completely dissolved, or there are insoluble impurities in the NMR tube.
- The sample is too concentrated.
- The sample is not spinning or is not spinning evenly during data acquisition.
- The spectrometer is not tuned properly.
- The spectrum has not been phased properly.

Another symptom of poor sample preparation or improper data acquisition is the presence of ***spinning sidebands,*** reflections of a signal that appear symmetrically around the signal. If spinning sidebands are present, every signal in the spectrum has them. An extreme example of spinning sidebands is seen in the NMR spectrum of methyl acetate (Figure 19.24). Spinning sidebands arise

(a)

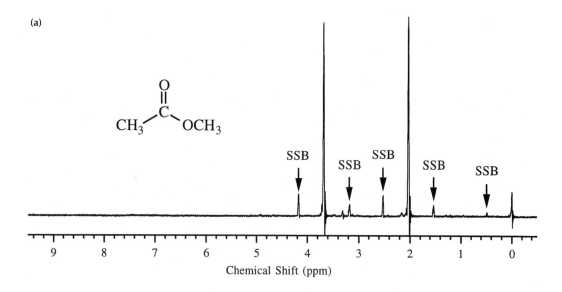

(b)

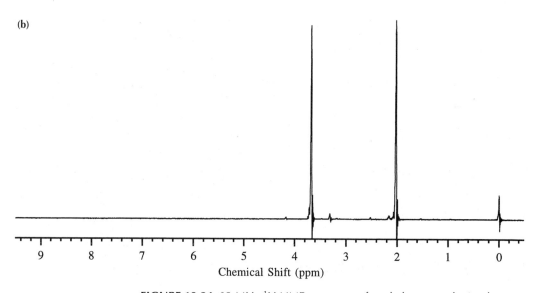

**FIGURE 19.24** 60-MHz ¹H NMR spectrum of methyl acetate obtained on (a) a poorly tuned instrument showing spinning sidebands (SSB) and (b) a properly tuned instrument.

when the magnetic field the sample experiences is not homogeneous. There are several possible causes:

- The height of the sample in the NMR tube is not sufficient.
- The sample tube is not positioned in the spin collar properly.
- The sample is spinning unevenly.
- The NMR sample tube is spinning too rapidly, leading to vortexing of the sample during data acquisition.
- The NMR instrument is not tuned properly.

Some sources of confusion cannot be avoided, but poor lineshape and spinning sidebands can be minimized with careful sample preparation and conscientious data acquisition.

## 19.10     Interpreting $^1$H NMR Spectra

The NMR spectra of organic compounds are important because they give a great deal of information about molecular structure. In this section we look at the spectra of some simple organic molecules and show how the information derived from the spectra can help determine their molecular structures. To start, it may be useful to recap the four major pieces of information that are used in the interpretation of a $^1$H NMR spectrum of a pure compound:

1.  The *number of signals* tells us how many kinds of nonequivalent protons are in the molecule.
2.  The *integration* determines the relative number of nonequivalent protons in a $^1$H NMR spectrum.
3.  The *chemical shift* suggests the environment of a proton. Downfield signals (larger ppm values) suggest nearby deshielding by oxygen atoms, halogen atoms, or π-systems. Tables 19.2–19.6 and Figure 19.6 are useful aids for correlating chemical shift with chemical structure.
4.  The *splitting* of signals, caused by spin-spin coupling of protons to other protons, reveals the presence of nearby protons that produce the coupling. Values of coupling constants can establish coupling connections between protons and can reveal stereochemical relationships (see Table 19.7).

It is important to remember that NMR spectroscopy is most useful in interpreting molecular structure in conjunction with other analytical data, such as infrared spectroscopy (Technique 18) and mass spectroscopy (Technique 20).

Proficiency at interpreting spectroscopic data and deducing chemical structures from the data is a skill that develops with practice. A structured, logical approach to the data assists in learning this skill:

1.  Make inferences and deductions based on the spectral information.
2.  Build up a collection of structure fragments.
3.  Put the pieces together into a chemical structure that is consistent with the data and with the rules of chemical bonding.

*Double-Bond Equivalents*

If you are analyzing a compound that you have synthesized in the lab, you already have a proposed structure. If the history of the sample is not available, a molecular weight may be available from mass spectroscopy. If an infrared spectrum is available, it will give you a good idea of what functional groups are present in the molecule. In problem sets, the molecular formula is often provided. If it is available, the molecular formula can be used to determine the *double-bond equivalents (DBE),* which provide the number of double bonds and/or rings present in the molecule.

$$\text{double-bond equivalents (DBE)} = \frac{2C - H - X + N + 2}{2}$$

C is the number of tetravalent carbon atoms, H is the number of hydrogen atoms, X is the number of halogen atoms, and $N$ is the number of trivalent nitrogen atoms. Other names for double bond equivalents are *degree of unsaturation* and *index of unsaturation*.

**Organizing the Spectral Information**

One way of organizing information from an NMR spectrum is to prepare a table with the following headings: Chemical Shift, $^1$H Type, Integration, Splitting Pattern, Coupling Constant, and Possible Structure Fragment. In the Chemical Shift column, list the positions (or ranges) of all the signals in the spectrum. Some types of protons can appear over a wide range of chemical shifts, but as the analysis is refined one can usually narrow them down. Using Figure 19.6 and Table 19.3 as guides, enter a likely structure assignment into the $^1$H Type column, i.e., Ar—H, =CH, O—CH, and so on. Enter the values of the integrals, rounded to whole numbers, in the Integration column. Remember that integrals must add up to the number of hydrogen atoms in the molecular formula.

In the Splitting Pattern column, enter a description of the splitting pattern. Be as precise as possible, using standard descriptive terms, such as doublet, triplet, quartet, and combinations of these terms, such as doublet of triplets. If you are processing your NMR data or have access to them, you will be able to calculate coupling constants as described in Technique 19.8. If you begin, however, with a printed NMR spectrum, you may find it difficult to estimate coupling constants with any accuracy. Enter the coupling constants in the proper column if you are in a position to measure them. The coupling constants are used to determine which signals are coupled. Because coupling is mutual ($J_{ab} = J_{ba}$), a signal showing large splitting must be coupled to another signal showing equally large splitting; it cannot be coupled to a signal showing only small coupling.

The Possible Structure Fragment column is where you pull all the information together and enter all chemical fragments that are consistent with the data. Be flexible and consider all reasonable possibilities. For example, in the spectrum you may have a quartet that integrates to two protons. The quartet is probably the result of coupling to three protons with equal coupling constants. Two arrangements are consistent with this pattern, CH—**CH₂**—CH₂ and **CH₂**—CH₃ (the boldface indicates the protons exhibiting the quartet). Once this table is constructed, it is a matter of eliminating information that is inconsistent, then putting the pieces together into a reasonable chemical structure. The final structure must be consistent with all spectral data and other chemical data.

**Problem 1**

Using the molecular formula to calculate double-bond equivalents of the compound with the molecular formula $C_4H_9Br$ shown in Figure 19.25, one finds that it contains no double bonds or rings.

$$\text{DBE} = \frac{(2 \times 4) - 9 - 1 + 2}{2} = 0$$

The NMR spectrum shows four signals, at 1.03 ppm, 1.70 ppm, 1.83 ppm, and 4.10 ppm. The first three signals (1.03, 1.70, 1.83 ppm)

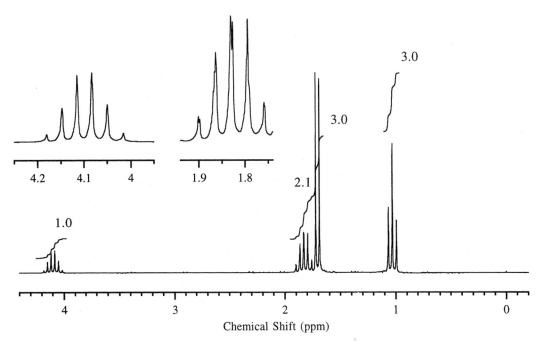

**FIGURE 19.25** 200-MHz $^1$H NMR spectrum of $C_4H_9Br$.

are consistent with protons attached to an $sp^3$ carbon that is singly bonded to carbon, CH—C. The signal at 4.10 ppm is consistent with a proton attached to a carbon atom that is bonded to an element more electronegative than carbon, such as bromine, CH—Br. These data, as well as the integrations, splitting patterns, and coupling constants collected from the NMR spectrum of $C_4H_9Br$, are entered in Table 19.9.

Two possible fragments are consistent with a quintet at 1.83 ppm, but one of them, —$CH_2$—$CH_2$—$CH_2$—, can be eliminated because

**TABLE 19.9**    **Interpreted data from $^1$H NMR spectrum (200 MHz) of $C_4H_9Br$**

| Chemical shift (ppm) | $^1$H type | Integration | Splitting pattern | Coupling constant (Hz) | Possible structure fragment(s) |
|---|---|---|---|---|---|
| 1.03 | CH—C | 3 | Triplet | 7.3 | $CH_3$—$CH_2$— |
| 1.70 | CH—C | 3 | Doublet | 6.6 | $CH_3$—$\overset{\mid}{\underset{\mid}{CH}}$ |
| 1.83 | CH—C | 2 | Quintet | 7.3, 6.6 | —$CH_2$—$CH_2$—$CH_2$— or $CH_3$—$CH_2$—$\overset{\mid}{\underset{\mid}{CH}}$ |
| 4.10 | CH—Br | 1 | Sextet | 6.6 | $CH_3$—$\overset{\overset{\textstyle Br}{\mid}}{\underset{\mid}{CH}}$—$CH_2$— |

integrations show that there is only one methylene group in the molecule. The largest fragment, $CH_3$—$CHBr$—$CH_2$—, is used as the starting point for piecing together the molecule. This fragment accounts for three carbons, six hydrogens, and the one bromine of the molecular formula, leaving only one carbon and three hydrogens to be added. Adding a methyl group to the fragment completes the structure, 2-bromobutane. It is always important to make sure that your answer is consistent with the chemical shift tables given earlier in the chapter (Tables 19.3–19.6). The estimated chemical shifts from Table 19.5 are shown on the right-hand structure.

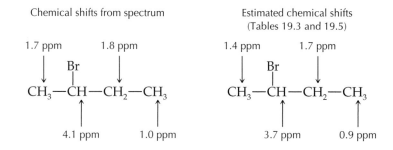

The pattern at 1.83 ppm is not a true quintet because the coupling partners do not all have the same coupling constant. However, the coupling constants are close enough that the pattern approximates a quintet. The signals due to the $C_1$ methyl and the $C_3$ methylene group have been shifted downfield (to the left) by the bromine on the adjacent carbon. Notice that the methine carbon is a stereocenter. Therefore, the methylene protons are diastereotopic. However, as is often the case, the chemical shifts of the diastereotopic protons are so close that their signals appear as if the protons were equivalent.

**Problem 2**

Next, let us analyze the $^1$H NMR spectrum of the compound shown in Figure 19.26. Its molecular formula is $C_5H_{10}O$. Because DBE = 0, the compound contains no rings or double bonds.

$$DBE = \frac{(2 \times 5) - 12 + 2}{2} = 0$$

The molecule has no double bonds and contains an oxygen atom; hence, it must be either an alcohol or an ether. Infrared spectroscopy easily differentiates them. Another spectroscopic method that distinguishes between the two possibilities is to carry out a deuterium exchange experiment with the NMR sample. This experiment is done by obtaining a second NMR spectrum after addition of a drop of $D_2O$ to the sample. Hydroxyl protons in the molecule are replaced by deuterons through chemical exchange. If the compound is an alcohol, the signal due to the hydroxyl proton disappears and a signal due to HOD appears at approximately 4.6 ppm. However, in this case the structure can be solved without resorting to these additional techniques.

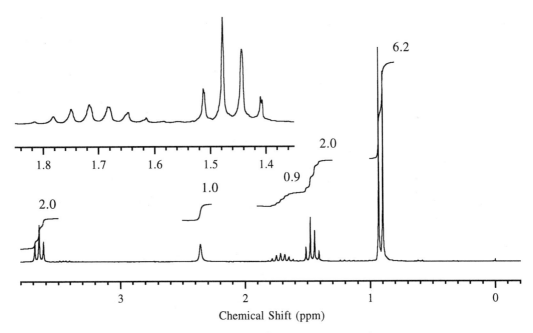

**FIGURE 19.26** 200-MHz $^1$H NMR spectrum of $C_5H_{12}O$.

The data from the spectrum are summarized in Table 19.10. An assignment for the signal at 2.36 ppm is not made because that region is normally where protons on carbons adjacent to carbonyls and alkenes appear, and we know from the DBE calculation that there are no double bonds in the molecule. There are two possible fragments, $CH_3$—$CH_2$— or —$CH_2$—$CH_2$—CH—, to explain the splitting of the signal at 1.46 ppm. The ethyl fragment is eliminated because there is no signal with a pattern consistent with the methyl portion of that fragment, which would have produced a three-proton triplet at approximately 1.0 ppm. The splitting pattern at 1.71 ppm is difficult to know with certainty. Since the outermost signals of highly split patterns are very small relative to the other signals, the multiplet could be either an octet or a nonet. In either case, there must be two methyl groups attached to the methine group. The downfield triplet signal at 3.65 ppm is due to a methylene group attached to an oxygen atom. From the analysis so far, we know that there is an isopropyl group (—$CH(CH_3)_2$), a methylene group flanked by methylene and methine groups (—$CH_2$—$CH_2$—CH—), and a methylene group attached to an oxygen atom and another methylene group (—$CH_2$—$CH_2$—O—). This array of fragments consists of eight carbons, sixteen hydrogens, and one oxygen atom. Obviously, there are some atoms common to more than one fragment. One common feature of two fragments is the methine group. Combining these groups produces a five-carbon fragment (—$CH_2$—$CH_2$—$CH(CH_3)_2$). Adding a hydroxyl group to the fragment completes the structure of the molecule, 3-methyl-1-butanol.

TABLE 19.10 Interpreted data from ¹H NMR spectrum (200 MHz) of $C_5H_{12}O$

| Chemical shift (ppm) | ¹H type | Integration | Splitting pattern | Coupling constant (Hz) | Possible structure fragment(s) |
|---|---|---|---|---|---|
| 0.91 | CH—C | 6 | Doublet | 6.6 | $CH_3$—CH |
| 1.46 | CH—C | 2 | Quartet | 6.9 | $CH_3$—$CH_2$— or —$CH_2$—$CH_2$—CH— |
| 1.71 | CH—C | 1 | Multiplet | 6.6 | $CH_3$ \| —$CH_2$—CH \| $CH_3$ or $CH_3$ \| —CH—CH \| $CH_3$ |
| 2.36 | ? | 1 | Singlet | — | |
| 3.65 | CH—O | 2 | Triplet | 6.9 | —$CH_2$—$CH_2$—O |

Chemical shifts from spectrum

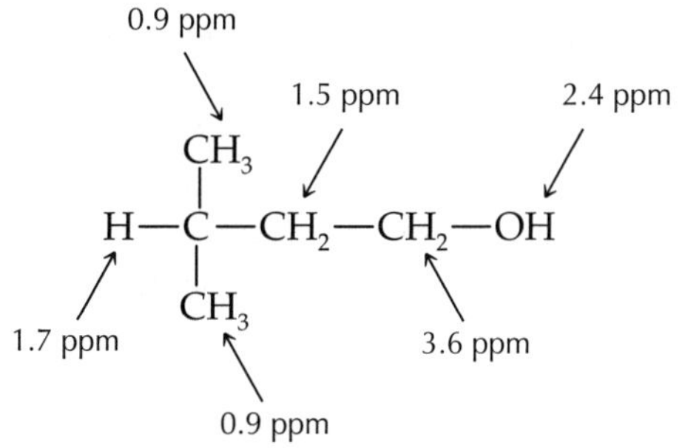

0.9 ppm
1.5 ppm    2.4 ppm

$CH_3$
\|
H—C—$CH_2$—$CH_2$—OH
\|
$CH_3$

1.7 ppm          3.6 ppm

0.9 ppm

Estimated chemical shifts
(Tables 19.3 and 19.5)

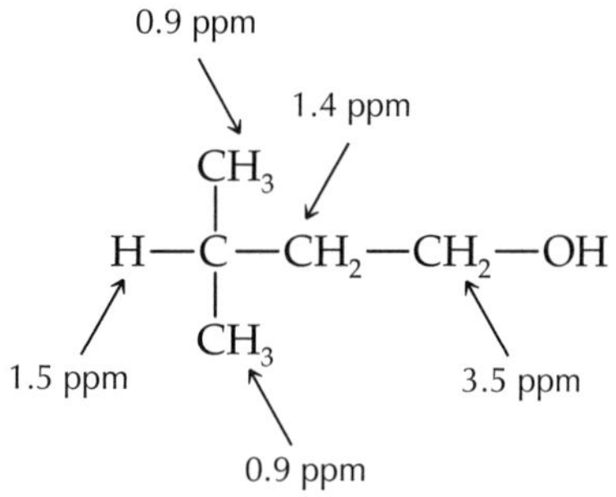

0.9 ppm
1.4 ppm

$CH_3$
\|
H—C—$CH_2$—$CH_2$—OH
\|
$CH_3$

1.5 ppm          3.5 ppm

0.9 ppm

The signal at 2.4 ppm is assigned to the hydroxyl proton of the alcohol. This is consistent with the broad range possible for the chemical shifts of hydroxyl protons. The estimated chemical shifts from Tables 19.3 and 19.5 are shown in the structure on the right; as you can see, the correspondence is very good.

**Problem 3**     The ¹H NMR spectrum of a compound with molecular formula $C_{10}H_{12}O$ is shown in Figure 19.27. The DBE calculation indicates that it contains a combination of five double bonds and rings.

$$DBE = \frac{(2 \times 10) - 12 + 2}{2} = 5$$

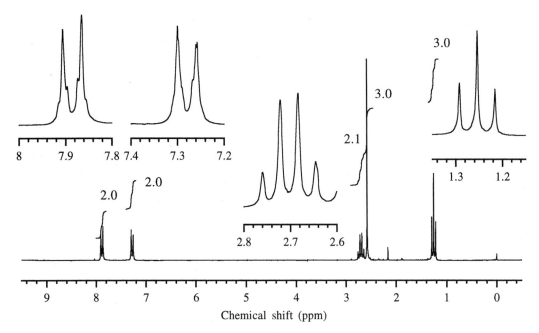

**FIGURE 19.27** 200-MHz $^1$H NMR spectrum of $C_{10}H_{12}O$.

Whenever a large value for DBE is calculated, it is very likely that the structure incorporates one or more benzene rings, because each benzene ring accounts for four DBE, three double bonds and one ring. The NMR spectrum confirms this assumption by the presence of signals in the aromatic proton region, 6.5–8.5 ppm. The integration of the aromatic protons is four, implying that the benzene ring is disubstituted. Moreover, the symmetry of the patterns in the aromatic region, a pair of doublets, indicates two groups of equivalent protons. This pattern is possible only if different substituents are attached to the 1 and 4 positions of the benzene ring (*para* substitution). Several **signature patterns** are observed frequently in NMR spectra, and this is one of the more important ones.

1,4 disubstitution;
two different types
of hydrogens

The data from Figure 19.27 are summarized in Table 19.11. The three-proton triplet at 1.25 ppm and two-proton quartet at 2.70 ppm indicate an ethyl group. This combination is another signature splitting pattern that is important in NMR spectra. The chemical shift of the two-proton quartet indicates the character of the adjacent structure or element. In this case, the position, 2.70 ppm, suggests that the ethyl group could be attached to either a carbonyl

**TABLE 19.11   Interpreted data from $^1$H NMR spectrum (200 MHz) of $C_{10}H_{12}O$**

| Chemical shift (ppm) | $^1$H type | Integration | Splitting pattern | Coupling constant (Hz) | Possible structure fragment(s) |
|---|---|---|---|---|---|
| 1.25 | CH—C | 3 | Triplet | 7.6 | $CH_3—CH_2—$ |
| 2.58 | CH—C=O or CH—Ar | 3 | Singlet | — | $CH_3—C=O$ or $CH_3—Ar$ |
| 2.70 | CH—C=O or CH—Ar | 2 | Quartet | 7.6 | $CH_3—CH_2—C=O$ or $CH_3—CH_2—Ar$ |
| 7.28 | Ar—H | 2 | Doublet | 8.2 | |
| 7.88 | Ar—H | 2 | Doublet | 8.2 | |

group or an aromatic ring. The fragments inferred from the NMR spectrum include a methyl group ($CH_3—$), an ethyl group ($CH_3CH_2—$), and a *para*-disubstituted benzene ring ($C_6H_4$). The atom count in the fragments is nine carbons and twelve hydrogens, leaving only one carbon and one oxygen atom. Because one more DBE is required by the molecular formula, the last fragment for the molecule is a carbonyl group. From these fragments, two possible structures that are consistent with the data can be constructed, 4'-methylphenyl-1-propanone and 4'-ethylphenyl-1-ethanone.

Estimated chemical shifts from Table 19.3, 19.5, and 19.6

4'-Methylphenyl-1-propanone                4'-Ethylphenyl-1-ethanone

The estimated chemical shifts for 4'-ethylphenyl-1-ethanone are closest to the measured values. However, further chemical and spectroscopic measurements would be required to unequivocally establish the structure of this compound.

## 19.11    $^{13}$C NMR

Because $^{13}$C NMR gives direct evidence about the carbon skeleton of an organic molecule, you might think that it would be the favored NMR technique. Why rely on the indirect evidence of $^1$H NMR? You will see in this section that, unless you use sophisticated NMR pulse sequences, $^1$H NMR has some advantages, particularly in the areas of signal integration and spin-spin splitting. However, $^{13}$C NMR is also a valuable instrumental method.

Many of the concepts regarding chemical shifts and spin-spin splitting learned with $^1$H NMR can also be applied to $^{13}$C NMR. The spin multiplicity of carbon ($^{13}$C) is the same as hydrogen ($^1$H); thus we can use the same splitting rules. For example, the carbon of a methylene group ($CH_2$) can be split into a triplet by the attached protons. The chemical shifts for carbon atoms in Table 19.12 are affected by electronegativity and anisotropy in ways similar to those for protons. For example, alkanes appear upfield (lower

| **T A B L E   1 9 . 1 2   Characteristic $^{13}$C NMR chemical shifts** | |
|---|---|
| Compound | Chemical shift (ppm) |
| TMS | 0.0 |
| $CDCl_3$ (t) | 77 |
| Alkane (**C**—$CH_3$) | 7–30 |
| Alkane (**C**—$CH_2$) | 15–40 |
| Alkane (**C**—CH) and (**C**—C) | 15–40 |
| Carboxylic acids, esters, and amides (**C**—C=O) | 27–35 |
| Ketones, aldehydes (**C**—C=O) | 30–45 |
| Amines (**C**—N) | 30–65 |
| Iodides (**C**—I) | 0–45 |
| Bromides (**C**—Br) | 25–65 |
| Chlorides (**C**—Cl) | 35–70 |
| Fluorides (**C**—F) | 80–95 |
| Alcohols (**C**—OH), ethers (**C**—OR) | 50–80 |
| Alkyne (**C**≡C) | 65–90 |
| Alkene (**C**=C) | 80–160 |
| Aromatic | 110–175 |
| Nitriles (**C**≡N) | 110–125 |
| Carboxylic acids, esters, and amides (**C**=O) | 160–180 |
| Ketones, aldehydes (**C**=O) | 185–220 |

chemical shift), nearby electronegative atoms deshield both carbons and protons (moving signals downfield), and aromatic carbons and protons are both strongly deshielded. Carbon NMR, however, opens up new areas for analysis. The shifts of carbonyl carbon atoms are especially useful. Carbon shifts occur over a 200-ppm range compared with the usual 10-ppm range for protons. Because the range is so broad, overlapping signals are not as great a problem in $^{13}$C NMR spectroscopy. Thus, the number of signals in a $^{13}$C NMR spectrum directly determines the number of different carbon atoms in the molecule.

Carbon NMR did not become a useful and routine tool until pulsed Fourier transform instrumentation was developed. Because the magnetically active isotope of carbon, $^{13}$C , is only 1.1% as abundant as $^{12}$C in nature, the signal from carbon is extremely weak. Very few atoms per molecule provide a signal, so even very concentrated samples have weak signals. In addition, $^{13}$C is inherently less sensitive because it has a smaller gyromagnetic ratio than $^{1}$H (about one-fourth that of $^{1}$H). This 1:4 ratio describes another aspect of the proton-carbon comparison; an instrument built to analyze protons at 300 MHz operates at one-fourth the frequency for $^{13}$C nuclei (75 MHz). As instruments with large superconducting electromagnets were developed, their sensitivity increased. With a modern high-field FT NMR instrument and a suitable number of pulses to obtain good NMR signals with the required signal-to-noise ratio, very good $^{13}$C NMR spectra can be obtained routinely with 25 to 50 mg of compound.

Samples are prepared for $^{13}$C NMR much as they are for $^{1}$H NMR. Table 19.3 lists solvents suitable for $^{13}$C NMR. Deuterated chloroform ($CDCl_3$) is again the solvent of choice because it dissolves most organic compounds. Of course, each molecule of $CDCl_3$ has a carbon atom, so a solvent signal always appears at 77.0 ppm in the $^{13}$C NMR spectrum. Unlike $^{1}$H, deuterium nuclei ($^{2}$H) have a spin of 1. Therefore, the $^{13}$C signal of $CDCl_3$ is always a characteristic triplet in which the three lines have equal height.

The splitting of signals in a $^{13}$C NMR spectrum can be used to identify the number of protons attached to a carbon atom. The carbon in a methyl group appears as a quartet, a methylene-carbon signal as a triplet, a methine-carbon signal as a doublet, and a quaternary-carbon signal as a singlet, as clearly demonstrated in the spectrum of bromoethane shown in Figure 19.28. The upfield methyl carbon is a quartet because it has three attached protons, and the downfield methylene group is a triplet because it has two attached protons. Note also that the lower field position for the methylene carbon resonance is consistent with patterns we have learned for protons; that is, the methylene carbon is more deshielded by the directly attached bromine. Because the protons are directly attached to the $^{13}$C atoms, their coupling is very large, on the order of 130 Hz.

In molecules containing many carbon atoms, the spectrum can become extremely complex because of multiple overlapping splitting patterns. The usual practice is to reduce this complexity by a technique called **broadband decoupling.** Irradiation of the sample

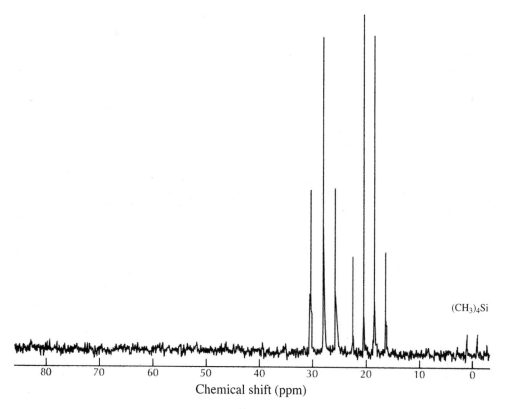

$(CH_3)_4Si$

Chemical shift (ppm)

**FIGURE 19.28** 62.8-MHz $^{13}C$ NMR spectrum of bromoethane. The upfield quartet at 18.3 ppm is due to the methyl group, and the downfield triplet is due to the methylene. The coupling due to direct proton attachment is 118 MHz (methyl) and 151 MHz (methylene). The tetramethylsilane (TMS) signal is a quartet whose outer lines have nearly been lost.

with a broad band of energy during data acquisition collapses the multiplets to singlets. The decoupled spectrum of bromoethane is shown in Figure 19.29. Besides simplifying the spectrum, broad-band decoupling enhances the signal-to-noise ratio. Carbon-carbon coupling is not observed in small molecules because the probability that two attached carbons will both be $^{13}C$ is vanishingly small.

Table 19.12 reveals chemical-shift trends that we have seen before with $^1H$ NMR spectroscopy. Carbons attached to electronegative atoms, such as oxygen, are much more deshielded than the corresponding carbons attached only to $sp^3$ carbon atoms. Moreover, alkene and aromatic carbons are also downfield (recall that the same was true for protons on these $sp^2$ carbons). Finally, carbonyl-carbon atoms show large downfield shifts, the position of which depends on the type of functional group containing the carbon atom (for example, ketone versus ester).

Some shielding effects are unique to $^{13}C$ NMR. First, carbon atoms are quite strongly deshielding. Specifically, it is known that carbons both $\alpha$ and $\beta$ to a given carbon atom (*) deshield that carbon by about 9 ppm. That is, both the $\alpha$ and $\beta$ carbons of the $—C^*—C^\alpha—C^\beta—$ structural unit increase the chemical shift

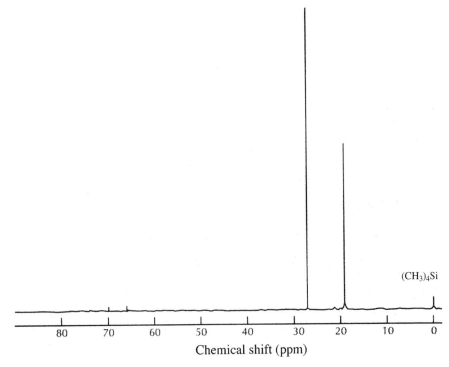

(CH$_3$)$_4$Si

Chemical shift (ppm)

**FIGURE 19.29** Decoupled 62.8-MHz $^{13}$C NMR spectrum of bromoethane. Protons in this sample have been irradiated by a broadband decoupling signal centered at 250 MHz to remove all $^1$H–$^{13}$C coupling. All three types of carbon atoms appear as singlets.

of the C* carbon atom by 9 ppm compared with the shift of methane. Also, although strongly electronegative halogens deshield carbon (the carbon of fluoromethane occurs at 75.4 ppm), halogens with high atomic numbers sometimes shield nearby carbon atoms; iodomethane has a $^{13}$C chemical shift of −20.7 ppm, which is over 20 ppm upfield of TMS. This shielding effect has been assigned to "steric compression." In such cases, steric factors apparently cause the electrons in the orbitals of the carbon atom to be compacted into a smaller volume, closer to the nucleus, thus making it more highly shielding of the carbon nucleus.

Consider the $^{13}$C NMR spectra of 2-butanone and ethyl acetate. Decoupled spectra of both compounds show four signals for different carbon atoms. In comparing the chemical shifts of the two methyl groups in 2-butanone, you can see that the methyl group attached to the carbonyl carbon is shifted approximately 20 ppm downfield relative to the methyl group attached to the alkyl methylene group. Now compare the positions of the methylene groups in the two different compounds. The methylene group attached to the oxygen in ethyl acetate is approximately 25 ppm further downfield relative to the methylene group attached to the carbonyl carbon in 2-butanone. Finally, the different shifts of the downfield signals of 2-butanone (209 ppm) and ethyl acetate (171 ppm) clearly indicated different types of carbonyl-carbon atoms.

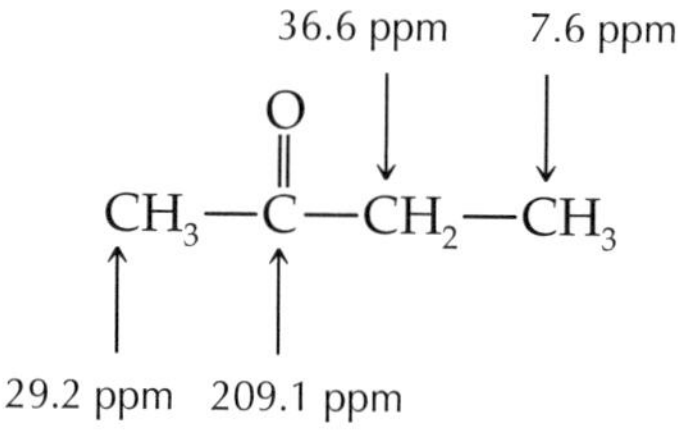

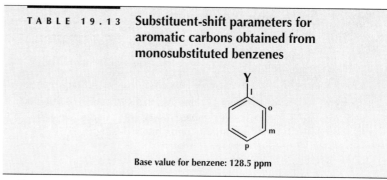

36.6 ppm    7.6 ppm

$$CH_3 - \overset{\overset{O}{\|}}{C} - CH_2 - CH_3$$

29.2 ppm   209.1 ppm

2-Butanone

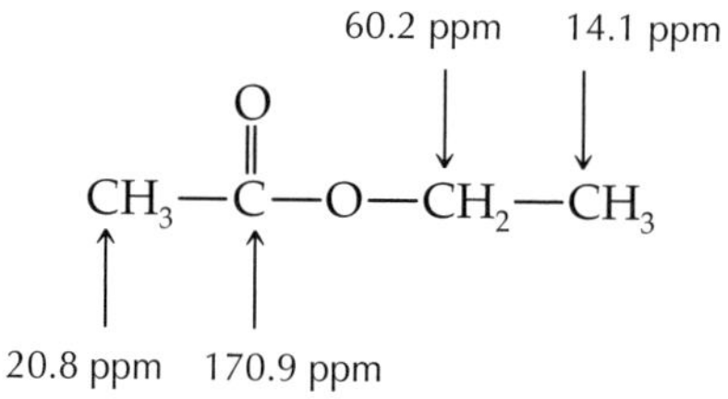

60.2 ppm    14.1 ppm

$$CH_3 - \overset{\overset{O}{\|}}{C} - O - CH_2 - CH_3$$

20.8 ppm   170.9 ppm

Ethyl acetate

The chemical shifts of carbon atoms in substituted benzene rings can be estimated using the parameters listed in Table 19.13; this table can be used just as Table 19.6 can be used for $^1$H NMR chemical shifts. Using methyl 3-nitrobenzoate as an example, the

| TABLE 19.13 | Substituent-shift parameters for aromatic carbons obtained from monosubstituted benzenes |

Base value for benzene: 128.5 ppm

| Group | C1 | ortho | meta | para |
|---|---|---|---|---|
| —I | −34.1 | 8.9 | 1.6 | −1.1 |
| —Br | −5.8 | 3.2 | 1.6 | −1.6 |
| —Cl | 6.3 | 0.4 | 1.4 | −1.9 |
| —F | 34.8 | −13.0 | 1.6 | −1.1 |
| —H | 0.0 | 0.0 | 0.0 | 0.0 |
| —(C=O)OCH$_3$ | 2.0 | 1.2 | −0.1 | 4.3 |
| —(C=O)OH | 2.1 | 1.6 | −0.1 | 5.2 |
| —(C=O)H | 8.2 | 1.2 | 0.5 | 5.8 |
| —(C=O)CH$_3$ | 8.9 | 0.1 | −0.1 | 4.4 |
| —CH=CH$_2$ | 8.9 | −2.3 | −0.1 | −0.8 |
| —CH$_3$ | 9.2 | 0.7 | −0.1 | −3.0 |
| —CH$_2$Cl | 9.3 | 0.3 | 0.2 | 0.0 |
| —C$_6$H$_5$ | 13.1 | −1.1 | 0.5 | −1.1 |
| —CH$_2$CH$_3$ | 15.7 | −0.6 | −0.1 | −2.8 |
| —CH(CH$_3$)$_2$ | 20.2 | −2.2 | −0.3 | −2.8 |
| —C(CH$_3$)$_3$ | 22.4 | −3.3 | −0.4 | −3.1 |
| —NH(C=O)CH$_3$ | 9.7 | −8.1 | 0.2 | −4.4 |
| —NH$_2$ | 18.2 | −13.4 | 0.8 | −10.0 |
| —NO$_2$ | 19.9 | −4.9 | 0.9 | 6.1 |
| —O(C=O)CH$_3$ | 22.4 | −7.1 | 0.4 | −3.2 |
| —OH | 26.9 | −12.8 | 1.4 | −7.4 |
| —OC$_6$H$_5$ | 27.6 | −11.2 | −0.3 | −6.9 |
| —OCH$_3$ | 31.4 | −14.4 | 1.0 | −7.7 |

estimates of the chemical shifts provide a useful guide for assigning the signals in the spectrum to the appropriate carbons.

|  | C1 | C2 | C3 | C4 | C5 | C6 |
|---|---|---|---|---|---|---|
| Base value: | 128.5 | 128.5 | 128.5 | 128.5 | 128.5 | 128.5 |
| —COOCH$_3$ | 2.0 | 1.2 | −0.1 | 4.3 | −0.1 | 1.2 |
| —NO$_2$ | 0.9 | −4.9 | 19.9 | −4.9 | 0.9 | 6.1 |
| Estimated (ppm): | 131.4 | 124.8 | 148.3 | 127.9 | 129.3 | 135.8 |
| Measured (ppm): | 135.0 | 124.3 | 148.4 | 127.1 | 129.4 | 131.6 |

As mentioned before, the number of signals in a $^{13}$C NMR spectrum indicates the number of different types of carbon atoms present in the molecule. If the number of signals is less than the number of carbon atoms, there must be some element of symmetry in the molecule. This observation can be used advantageously to select a structure from several possibilities. For example, consider an aromatic compound with the molecular formula $C_8H_{10}$. There are four possible structures that are consistent with this formula: ethyl benzene, 1,2-dimethylbenzene, 1,3-dimethylbenzene, and 1,4-dimethylbenzene. Each structure possesses at least one plane of symmetry, and 1,4-dimethylbenzene has two planes of symmetry.

Each compound contains a different number of unique carbon atoms: ethyl benzene has six, 1,2-dimethylbenzene has four, 1,3-dimethylbenzene has five, and 1,4-dimethylbenzene has three. The structure of the compound could be determined by obtaining the $^{13}$C NMR spectrum and simply counting the number of signals. In addition, ethyl benzene has four signals in the aromatic region and two signals in the alkyl region, whereas the $^{13}$C NMR spectrum of Figure 19.30 shows that 1,3-dimethylbenzene has four signals in the aromatic region and only one signal in the alkyl region.

Obtaining the quantitative number of a given type of carbon atom by using integration is, however, far less straightforward than it is for $^1$H NMR. The area of a $^{13}$C signal is the result not only of the number of carbon atoms causing the signal, but also from variations in the so-called spin-lattice relaxation times ($T_1$'s) and the nuclear Overhauser effect (NOE), factors not elaborated in this book. In

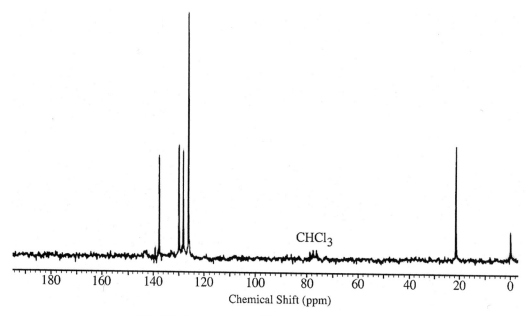

**FIGURE 19.30** 25-MHz $^{13}$C NMR spectrum of 1,3-dimethylbenzene.

general, $^{13}$C integration is not a useful technique for counting the relative numbers of carbon atoms in a compound. In particular, carbon atoms that have no protons attached are troublesome; they often give much smaller signals than carbon atoms with protons attached.

## 19.12    DEPT Experiments

As stated previously, the number of protons attached to carbons in a compound can be determined from the multiplicity of the signals in its $^{13}$C NMR spectrum. However, in compounds with more than just a few different carbon atoms, the splitting due to the protons usually results in spectra that are extremely complex and difficult to interpret. Typically, $^{13}$C NMR spectra are obtained using broadband decoupling because it collapses the signals to singlets and enhances the signal-to-noise ratio. With modern NMR spectrometers, the number of protons attached to a carbon atom is obtained using a method called **DEPT** (distortionless enhancement by polarization transfer). This is a suite of pulse programs routinely available on most FT NMR spectrometers. Detailed description of DEPT experiments is left to higher level textbooks. Here, we are interested only in the results of the experiments since the number of protons attached to a carbon atom is easily and confidently determined with DEPT.

In a typical broadband-decoupled $^{13}$C NMR spectrum, each different carbon atom in the sample appears as a single peak. In a DEPT experiment, signals in the $^{13}$C NMR spectrum may be suppressed or even inverted depending on the number of protons attached to the carbon and the conditions set in the pulse program.

The DEPT45 version of the experiment provides a $^{13}$C NMR spectrum in which only carbon atoms that have protons attached to them appear. In other words, signals due to quaternary carbons are not observed. The spectrum produced by the DEPT90 pulse program exhibits only signals from carbon atoms that have one hydrogen attached (methine carbons). Signals due to all carbon atoms with attached protons are observed in the $^{13}$C NMR spectrum from a DEPT135 experiment; however, the signals due to carbon atoms with two protons attached (methylene carbons) are inverted.

| Type of $^{13}$C spectrum | $-CH_3$ | $-CH_2-$ | $-\overset{\mid}{\underset{\mid}{C}}-H$ | $-\overset{\mid}{\underset{\mid}{C}}-$ |
|---|---|---|---|---|
| Broadband-decoupled $^{13}$C | + | + | + | + |
| DEPT45 | + | + | + | 0 |
| DEPT90 | 0 | 0 | + | 0 |
| DEPT135 | + | − | + | 0 |

Comparing the spectra from a set of DEPT experiments allows you to determine the number of protons attached to every carbon atom in a molecule. Signals that appear in the broadband-decoupled $^{13}$C NMR spectrum but are not present in DEPT45 and DEPT135 spectra are due to quaternary carbons. Signals that appear in the DEPT90 spectrum are due to methine carbons. Signals that are inverted in the DEPT135 spectrum are due to methylene carbons. Any signals in the DEPT135 spectrum not accounted for previously are due to methyl groups.

Many NMR data processing software packages allow the results of DEPT experiments to be displayed in "edited" form, in which each "edited" spectrum contains only signals for one type of carbon.

| Edited spectrum | Combination of DEPT spectra |
|---|---|
| Methine carbon, $-\overset{\mid}{\underset{\mid}{C}}-H$ | DEPT90 |
| Methylene carbon, $-CH_2-$ | (DEPT45) − $a$(DEPT135) |
| Methyl carbons, $-CH_3$ | (DEPT45) + $b$(DEPT135) − $c$(DEPT90) |

The symbols $a$, $b$, and $c$ are coefficients used to correct for differences in signal intensities in individual DEPT spectra.

# References

1. Atta-ur-Rahman. *Nuclear Magnetic Resonance: Basic Principles*; Springer-Verlag: New York, 1986.
2. Duddeck, H.; Dietrich, W. *Structure Elucidation by Modern NMR*; Springer-Verlag: New York, 1989.
3. Pouchert, C. J.; Behnke, J. (Eds.) *The Aldrich Library of $^{13}$C and $^{1}$H FT-NMR Spectra*; Aldrich Chemical Co.: Milwaukee, WI, 1993; 3 volumes.
4. Pretsch, E; Seibl, J.; Clerc, T; Simon, W., Biemann, K. (Trans.) *Tables of Spectral Data for*

*Structure Determination of Organic Compounds;* 2nd English ed.; Springer-Verlag: New York, 1989.

5. Sanders, J. K.; Hunter, B. K. *Modern NMR Spectroscopy*; 2nd ed.; Oxford University Press: Oxford, 1993.

6. Silverstein, R. M.; Webster, F. X. *Spectrometric Identification of Organic Compounds*; 6th ed.; Wiley: New York, 1998.

## Questions

1. The $^1$H NMR spectrum of a compound of molecular formula $C_4H_9Br$ is recorded in Figure 19.31. Deduce the structure of the compound and assign all NMR signals.

2. The $^1$H NMR spectrum of a compound of molecular formula $C_5H_8O_2$ is recorded in Figure 19.32. Deduce the structure of the compound and assign all NMR signals.

3. A compound of molecular formula $C_3H_8O$ produces the $^1$H NMR spectrum shown in Figure 19.33. In addition, when this compound is treated with $D_2O$, the $^1$H NMR signal at 2.0 ppm disappears and another signal at 4.6 ppm appears. Moreover, when the $C_3H_8O$ compound is highly purified and care is taken to remove all traces of acid in the

NMR solvent, the singlet at 2.0 ppm is replaced by a doublet. Finally, the chemical shift of the 2.0 ppm signal is highly concentration dependent; an increase in the concentration of $C_3H_8O$ in the NMR sample results in a downfield shift of this signal. Deduce the structure of $C_3H_8O$, assign all NMR signals, and explain the changes observed for the 2.0 ppm signal.

4. Evaluate the $^1$H NMR spectra of 1,3-dimethylbenzene recorded in Figure 19.34 on page 280 at two substantially different field strengths (60 and 200 MHz). Assign the signals and explain the differences in resolution in the signals centered near 7.0 ppm.

5. The $^1$H NMR spectrum of oil of wintergreen, molecular formula $C_8H_8O_3$, is

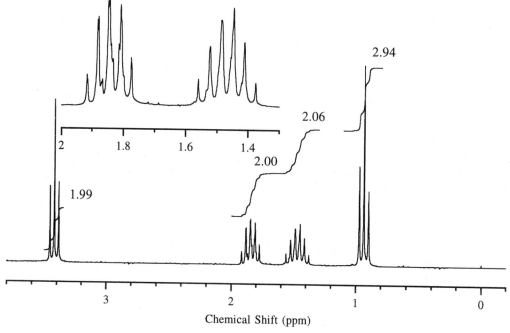

**FIGURE 19.31** 200-MHz $^1$H NMR spectrum of a compound of molecular formula $C_4H_9Br$.

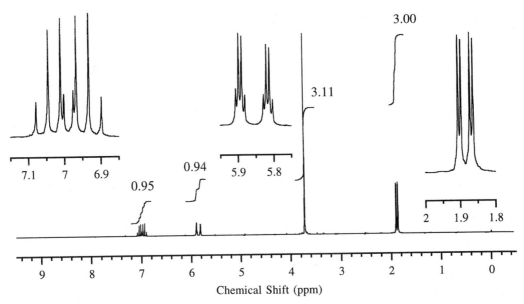

**FIGURE 19.32** 200-MHz $^1$H NMR spectrum of a compound of molecular formula $C_5H_8O_2$.

shown in Figure 19.35 on page 281. **Note:** This compound is a phenol with the O—H signal appearing at 10.8 ppm; in this spectrum intramolecular hydrogen bonding shifts it farther downfield than is usual for phenols. Deduce the structure of the compound and assign all NMR signals. Compare chemical

shifts of the four protons in the aromatic region. Estimate the chemical shifts of the aromatic protons using the parameters in Table 19.6 and compare them with those measured from the spectrum.

6. A compound of molecular formula $C_{10}H_{14}$ produces the $^1$H NMR spectrum shown in Figure 19.36 on page 281. The

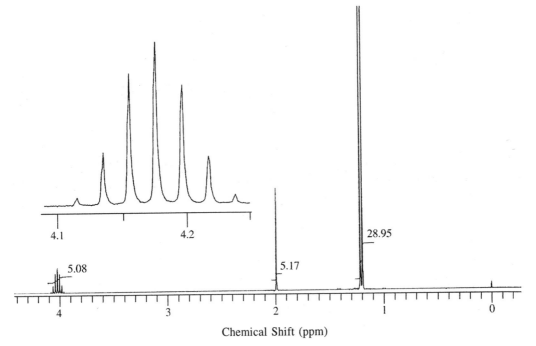

**FIGURE 19.33** 300-MHz $^1$H NMR spectrum of a compound of molecular formula $C_3H_8O$.

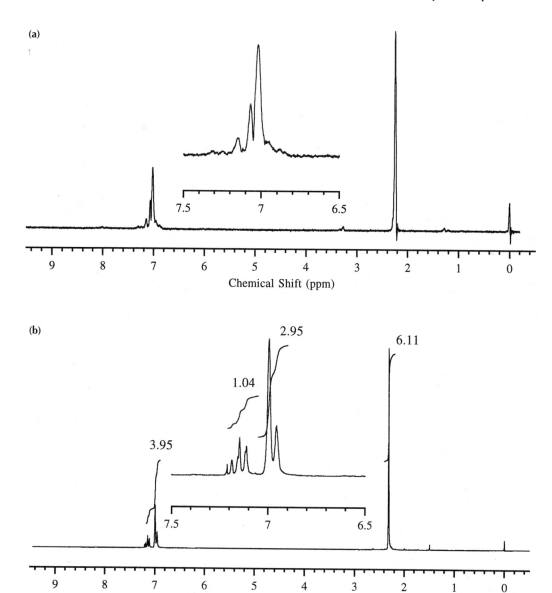

**FIGURE 19.34** $^1$H NMR spectra of 1,3-dimethylbenzene at (a) 60 MHz and (b) 200 MHz.

proton-decoupled $^{13}$C NMR spectrum reveals seven singlets (21.3 ppm, 24.5 ppm, 34.2 ppm, 126.6 ppm, 129.5 ppm, 135.4 ppm, and 145.2 ppm). Two of the signals (145.2 ppm and 135.4 ppm) are very weak. Use these data to deduce the structure of $C_{10}H_{14}$ and assign all the $^1$H NMR signals.

7. Dimedone (5,5-dimethylcyclohexan-1,3-dione) in solution is a mixture of keto and enol structures. The $^1$H NMR

spectrum of a solution of dimedone in $CDCl_3$ is shown in Figure 19.37 on page 282. In the sample the two enols are equilibrating very fast compared with the NMR time scale. Assign the NMR signals and use NMR integrations to determine the composition of the mixture.

8. A compound of molecular formula $C_8H_{14}O$ produces the $^1$H NMR spectrum shown in Figure 19.38. Deduce

*(continued on page 283)*

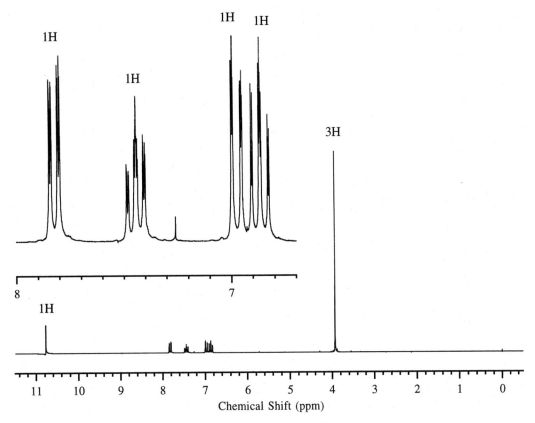

**FIGURE 19.35** 200-MHz $^1$H NMR spectrum of wintergreen oil.

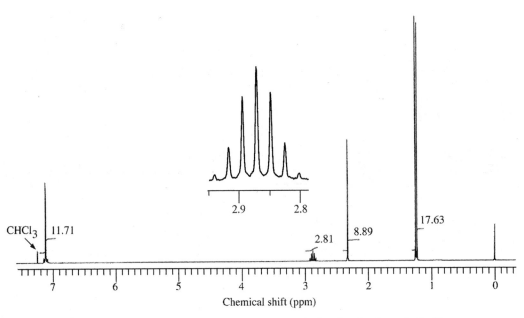

**FIGURE 19.36** 300-MHz $^1$H NMR spectrum of a compound of molecular formula $C_{10}H_{14}$.

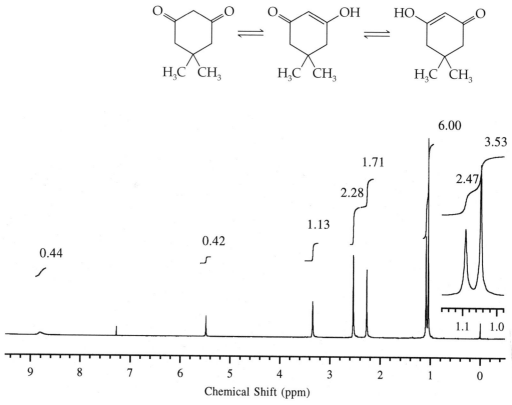

**FIGURE 19.37** 200-MHz ¹H NMR spectrum of 5,5-dimethylcyclohexan-1,3-dione.

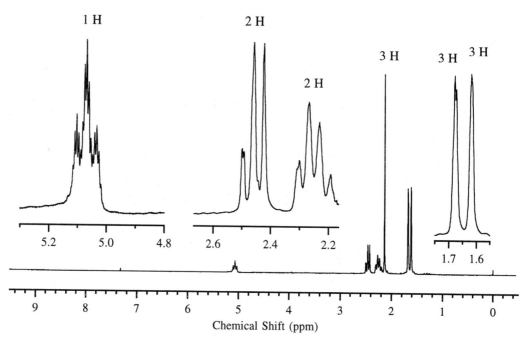

**FIGURE 19.38** 200-MHz ¹H NMR spectrum of compound of molecular formula $C_8H_{14}O$.

the structure of the compound and assign all NMR signals.

9. Ibuprofen is the active ingredient in several nonsteroid anti-inflammatory drugs (NSAIDs). The $^1H$ NMR spectrum of the methyl ester of ibuprofen is shown in Figure 19.39. The molecular formula of the methyl ester of ibuprofen is $C_{14}H_{20}O_2$. Deduce the structure of this compound and assign all NMR signals. Compare chemical shifts estimated using the parameters in Tables 19.5 and 19.6 with those measured from the spectrum.

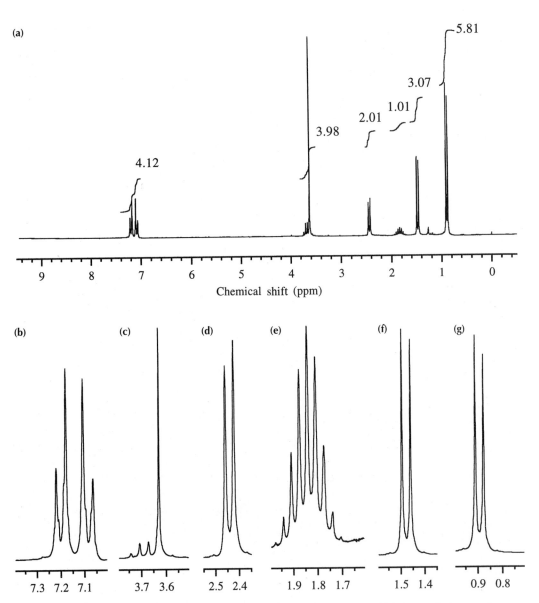

**FIGURE 19.39** 200-MHz $^1H$ NMR spectrum of (a) the methyl ester of ibuprofen with (b–g) expanded sections.

# 20

# MASS SPECTROMETRY

Most spectrometric techniques used by organic chemists are based on the ability of a carbon compound to absorb light of various energies. This absorption yields structural information. Mass spectrometry (MS), however, is different in that it does not involve absorption of light. In electron impact mass spectrometry, the most common kind of MS, a stream of electrons is used to bombard the compound, converting it to a radical cation. The cation formed in this way from an intact molecule is called a *molecular ion (M⁺·)* or *parent ion.*

## 20.1    Mass Spectrometer

In electron impact MS, the molecule loses one of its own electrons when struck by an external electron and becomes a molecular ion bearing a single positive charge. Once formed, the highly energetic molecular ion often divides, forming both uncharged and charged fragments. Only the positively charged ions are analyzed.

$$e^- \searrow$$

$$M \longrightarrow M^{+\bullet} \xrightarrow{\text{fragmentation}} f^+ + f'^{\bullet}$$

Radical cation
form of molecule

$$e^- \quad e^-$$

Normally, a molecule is converted to its gaseous form for simple mass spectrometry. Thus the electron impact technique is limited to compounds with significant vapor pressures at room temperature. However, special ionization techniques can be used to ionize samples directly in the solid state. These special techniques make it possible to study samples with high molecular weights and very low vapor pressures.

As shown in Figure 20.1, we can take advantage of the radical cation's charge by applying a magnetic field perpendicular to the ion's flight path (perpendicular to the page). The magnetic field causes the pathway of each cation to be curved. The amount of curvature is a function of the mass of the ion and the strength of the magnetic field. For an ion to strike the ion collector, it must follow a path consistent with the radius of the mass analyzer portion of the mass spectrometer. The larger the mass of the ion, the larger the magnetic field strength required to direct the ion to the collector. The signal at the collector is recorded as a function of magnetic field

**FIGURE 20.1**
Schematic
representation of a
mass spectrometer that
uses electron impact
ionization. The flow of
molecular ion species
is from the upper right
to the lower left. From
Vollhardt, K. P. C.;
Schore, N. E. *Organic
Chemistry,* 3rd ed.;
W. H. Freeman and
Company: New York,
1999; p. 908.

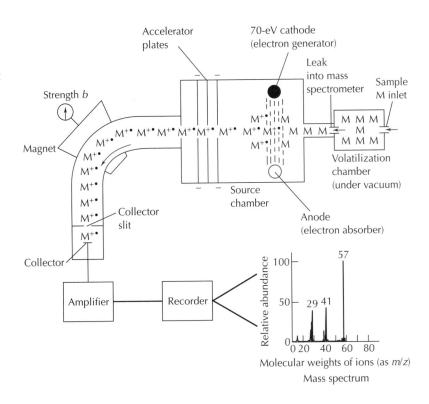

strength, which is converted to *m/z*. Because each ion usually bears a single positive charge, *m/z* simplifies to *m*, the mass of the ion.

In the past twenty years, hybrid instruments combining a gas chromatograph with a mass spectrometer (**GC-MS**) have become available. In these instruments, small samples of the effluent stream from a gas chromatograph are directed into a mass spectrometer. The molecules in the sample are ionized by electron bombardment, and the resulting ions are accelerated and passed into the mass analyzer of a mass spectrometer. The result is a mass spectrum for every compound eluting from the gas chromatograph. This technique is very efficient for analyzing mixtures of compounds since it provides the number of components in the mixture, a rough measure of their relative amounts, and the possible identities of the components. Although the mass analyzer of the GC-MS instrument uses a quadrupole system to sort the ions according to mass, the resulting spectra are comparable to other mass spectra. These systems have excellent resolution and are ideally suited for compounds with molecular weights up to 300–400 amu. The GC-MS has quickly become a major workhorse of modern organic and analytical chemistry laboratories.

Mass spectral data can be organized into tables of signal intensities at *m/z* values. Often, the data is presented in graphical form as a histogram of intensity along the ordinate versus *m/z* along the abscissa. The intensities are represented as percentages of the most intense peak of the spectrum, called the **base peak.** An example of

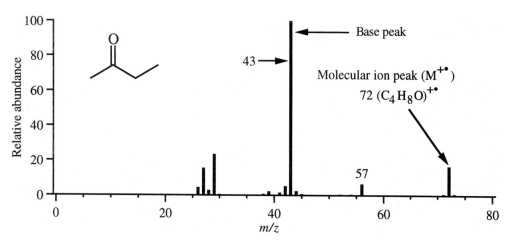

**FIGURE 20.2** Mass spectrum of 2-butanone.

the mass spectrum of 2-butanone is shown in Figure 20.2. In this spectrum, the base peak is at $m/z$ 43.

## 20.2    The Molecular Ion

The first area of interest in a mass spectrum is the molecular ion. If the molecular ion does not fragment before being detected, its $m/z$ provides the molecular weight of the compound, which is a highly valuable piece of information about any unknown organic compound. In the mass spectrum of 2-butanone, the peak at $m/z$ 72 corresponds to the molecular ion.

Most elements occur in nature as mixtures of isotopes. Table 20.1 provides a list of isotopes of elements commonly found in organic compounds. If the molecular ion is reasonably intense, signals for M+1 and M+2 ions can be observed. The ratios of the intensities of the M+1 and M+2 peaks to that of the M peak depend on the

**TABLE 20.1    Relative isotope abundances of common elements (%)**

| Element | Isotope | Relative abundance | Isotope | Relative abundance | Isotope | Relative abundance |
|---------|---------|--------------------|---------|--------------------|---------|--------------------|
| Hydrogen | $^1H$ | 100 | $^2H$ | 0.016 | | |
| Carbon | $^{12}C$ | 100 | $^{13}C$ | 1.08 | | |
| Nitrogen | $^{14}N$ | 100 | $^{15}N$ | 0.38 | | |
| Oxygen | $^{16}O$ | 100 | $^{17}O$ | 0.04 | $^{18}O$ | 0.20 |
| Fluorine | $^{19}F$ | 100 | | | | |
| Silicon | $^{28}Si$ | 100 | $^{29}Si$ | 5.10 | $^{30}Si$ | 3.35 |
| Phosphorus | $^{31}P$ | 100 | | | | |
| Sulfur | $^{32}S$ | 100 | $^{33}S$ | 0.78 | $^{34}S$ | 4.40 |
| Chlorine | $^{35}Cl$ | 100 | | | $^{37}Cl$ | 32.5 |
| Bromine | $^{79}Br$ | 100 | | | $^{81}Br$ | 98.0 |
| Iodine | $^{127}I$ | 100 | | | | |

isotopic abundances of the atoms in a molecule and the number of each kind of atom. Isotopes of carbon, hydrogen, oxygen, and nitrogen, the elements that make up most organic compounds, make small contributions to the M+1 and M+2 peaks, and the resulting intensities have been used to reveal the molecular formulas of organic compounds. For example, the intensity of the M+1 peak relative to the intensity of the M peak in 2-butanone (molecular formula $C_4H_8O$) should be $(4 \times 1.08\%) + (8 \times 0.016\%) + (1 \times 0.04\%) = 4.49\%$.

Tables listing molecular formulas and the ratios expected for these formulas have been constructed (Refs. 1 and 2). Using the spectrum in Figure 20.2 as an example, a molecular ion of 72 can arise from one of several molecular formulas, $C_2H_4N_2O$, $C_3H_4O_2$, $C_3H_8N_2$, $C_4H_8O$, and $C_5H_{12}$. The relative abundances of the M+1 and M+2 peaks (abundance of M peak = 100) corresponding to these molecular formulas are as follows:

| Formula | M+1 | M+2 | (M+1)/(M+2) |
|---------|-----|-----|-------------|
| $C_2H_4N_2O$ | 3.03 | 0.23 | 13.2 |
| $C_3H_4O_2$ | 3.38 | 0.44 | 7.7 |
| $C_3H_8N_2$ | 4.13 | 0.07 | 58.9 |
| $C_4H_8O$ | 4.49 | 0.28 | 16.0 |
| $C_5H_{12}$ | 5.60 | 0.13 | 43.1 |

The relative abundances of the M, M+1, and M+2 peaks in the mass spectrum shown in Figure 20.2 are as follows:

| Ion | m/z | Relative abundance (base peak = 100) | Relative abundance (molecular ion = 100) |
|-----|-----|--------------------------------------|------------------------------------------|
| $M^{+\bullet}$ | 72 | 16.5 | 100 |
| $(M+1)^{+\bullet}$ | 73 | 0.77 | 4.67 |
| $(M+2)^{+\bullet}$ | 74 | 0.04 | 0.24 |

For this spectrum, the ratio of M+1 to M+2 is $4.67/0.24 = 19.5$. The relative abundances of M+1 and M+2 and the ratio are most consistent with a molecular formula of $C_4H_8O$.

Unfortunately, practical experience has shown that for many C, H, N, and O compounds, the expected ratios are often in error. Errors can occur for many reasons. For example, the sample may undergo ion-molecule collisions that provide additional intensity to the M+1 peak.

$$M^{+\bullet} + RH \longrightarrow MH^{+\bullet} + R^{\bullet}$$

Even when the sample is pure, it can often be necessary to carry out the time-consuming task of obtaining the entire spectrum several times and averaging the abundance values. Furthermore, if the molecular ion is weak, the precision of the data is insufficient to give reliable ratios.

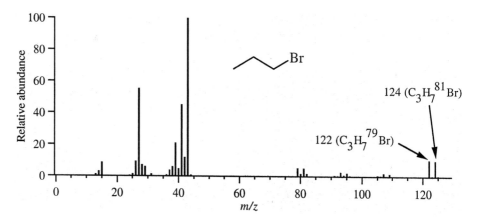

**FIGURE 20.3** Mass spectrum of 1-bromopropane.

Although it can be difficult to use M+1/M+2 data to determine accurate molecular formulas, MS is highly valuable for qualitative elemental analysis. In particular, it is quite easy to use the M+2 peak to identify the presence of bromine and chlorine in organic compounds. The appearance of a large M+2 peak in a mass spectrum is evidence for the presence of one of these elements. The relative intensities indicate which element, as illustrated in the mass spectrum of 1-bromopropane (Figure 20.3). An important aspect of this spectrum is the existence of the two major peaks in the molecular ion region at $m/z$ 122 and 124 with an approximately 1:1 ratio. You can see from Table 20.1 that natural bromine exists as a mixture of $^{79}$Br and $^{81}$Br in a ratio very close to 1:1. The peak at $m/z$ 122 is by convention defined as the molecular ion peak (M); in this case, the bromine atom has a mass of 79. The $m/z$ 124 peak is also a molecular ion peak because it corresponds to an intact molecule of +1 charge. Even though the mass 124 species has an atomic composition corresponding to the molecular formula $C_3H_7Br$, it is referred to as the M+2 peak because the bromine atom is $^{81}$Br. Isotopic contributions of carbon, hydrogen, and oxygen to the M+1 and M+2 peaks are comparatively small. Thus a ratio very close to 1:1 clearly shows

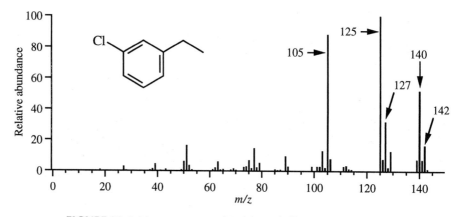

**FIGURE 20.4** Mass spectrum of 3-chloroethylbenzene.

that the molecule contains one bromine atom. Tiny contributions of the isotopes of carbon and hydrogen can be observed in Figure 20.3; they are revealed as a small $m/z$ 123 peak (M+1 peak).

A monochloro compound is expected to have an M+2 peak that is 32.5% as intense as the M peak, as shown in Figure 20.4, the mass spectrum of 3-chloroethylbenzene. The peak at $m/z$ 142 is approximately one-third the intensity of the molecular ion peak at $m/z$ 140.

## 20.3    High-Resolution Mass Spectrometry

In modern research laboratories, molecular formulas are usually determined by *high-resolution mass spectrometry*. The expensive high-resolution instruments used for this purpose have both electric and magnetic fields for focusing the ion pathways. These double-resolution instruments measure masses to four figures beyond the decimal point. Table 20.2 provides the masses that should be used for this approach. The exact mass of an isotope is established using carbon-12 as the standard, and the exact mass of $^1H$ is 1.00783. The exact mass of a molecule is determined by summing the exact masses of all of the isotopes contained in that molecule. For example, the exact mass of the molecular ion of 2-butanone is

**TABLE 20.2  Exact masses of isotopes**

| Element | Atomic weight | Nuclide | Mass |
|---|---|---|---|
| Hydrogen | 1.00794 | $^1H$ | 1.00783 |
|  |  | $D(^2H)$ | 2.01410 |
| Carbon | 12.01115 | $^{12}C$ | 12.00000 (std) |
|  |  | $^{13}C$ | 13.00336 |
| Nitrogen | 14.0067 | $^{14}N$ | 14.0031 |
|  |  | $^{15}N$ | 15.0001 |
| Oxygen | 15.9994 | $^{16}O$ | 15.9949 |
|  |  | $^{17}O$ | 16.9991 |
|  |  | $^{18}O$ | 17.9992 |
| Fluorine | 18.9984 | $^{19}F$ | 18.9984 |
| Silicon | 28.0855 | $^{28}Si$ | 27.9769 |
|  |  | $^{29}Si$ | 28.9765 |
|  |  | $^{30}Si$ | 29.9738 |
| Phosphorus | 30.9738 | $^{31}P$ | 30.9738 |
| Sulfur | 32.066 | $^{32}S$ | 31.9721 |
|  |  | $^{33}S$ | 32.9715 |
|  |  | $^{34}S$ | 33.9679 |
| Chlorine | 35.4527 | $^{35}Cl$ | 34.9689 |
|  |  | $^{37}Cl$ | 36.9659 |
| Bromine | 79.9094 | $^{79}Br$ | 78.9183 |
|  |  | $^{81}Br$ | 80.9163 |
| Iodine | 126.9045 | $^{127}I$ | 126.9045 |

$(4 \times 12.00000) + (8 \times 1.00783) + (1 \times 15.9949) = 72.0575$. Looking at the exact masses of molecules whose nominal molecular weight is 72, it is obvious that the exact molecular formula can be determined from the exact masses measured to four decimal places.

| Formula | Exact mass |
|---------|-----------|
| $C_2H_4N_2O$ | 72.0324 |
| $C_3H_4O_2$ | 72.0211 |
| $C_3H_8N_2$ | 72.0688 |
| $C_4H_8O$ | 72.0575 |
| $C_5H_{12}$ | 72.0939 |

## 20.4  Fragmentation of the Molecule

A molecular ion often fragments, and the resulting spectrum can be complex because any one of a number of covalent bonds might be broken during fragmentation. From Figures 20.2–20.4, it is apparent that a large number of peaks arise even for relatively low molecular weight organic compounds. The array of fragmentation peaks constitutes a fingerprint that can be used for identification. Modern mass spectrometers are routinely equipped with computer libraries containing tens of thousands of such fingerprints for matching purposes. Typically, a computer program compares the experimental spectrum with spectra in the library and produces a ranked "hit list" of compounds with similar mass spectra. The ranking is based on how close the match is in terms of the presence of peaks and the intensities of the peaks.

In cases where you are working with compounds not included in a library, fragmentation pathways can provide important clues to the underlying molecular structure. Rules of fragmentation have been established, but this topic is too wide in scope to be adequately covered in this book. However, a few of the most useful fragmentation patterns for common functional groups are described in the following paragraphs.

Mechanisms depicting fragmentation are enhanced by the efficient use of fishhooks (curved arrows with half-heads) representing the migration of a single electron. This notation is similar to that used in free radical or photochemical processes:

$$M \longrightarrow \underbrace{M^{+\bullet} = [f\!-\!f']^+}_{\substack{\text{Molecular} \\ \text{cation radical}}} \longrightarrow \underbrace{f^+ + f'^\bullet}_{\text{Fragments}}$$

Forces that contribute to the ease with which fragmentation processes occur include the strength of bonds in the molecule (for example, the f'—f bond) and the stability of the free radicals (f'•) and carbocations (f⁺) produced by fragmentation. Although these fragments are formed in the gas phase, we can still apply our "chemical intuition," based on reactions in solution.

**Aromatic Hydrocarbons**

Aromatic hydrocarbons are prone to fragmentation at the bond β to the aromatic ring, yielding a benzylic cation that rearranges to a very stable $C_7H_7$ carbocation called a tropylium ion.

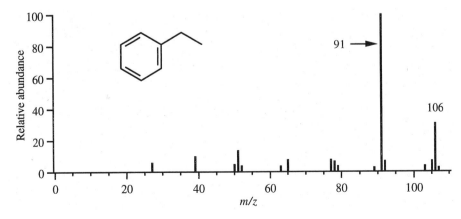

*m/z* 91

For monosubstituted alkylbenzenes, the peak at *m/z* 91 is a very large signal, often the base peak. In the mass spectrum of ethylbenzene shown in Figure 20.5, the base peak of 91 is the result of loss of a methyl group (G• = •CH₃) to produce the tropylium ion.

**Alkenes**

Alkenes are similarly prone to fragmentation at the bond β to the double bond to give a stabilized allylic cation or allylic radical.

In the spectrum of 1-hexene shown in Figure 20.6, the allylic cation fragment ($CH_2=CH—CH_2^+$) is observed at *m/z* 41 and the propyl cation ($CH_3CH_2CH_2^+$) is observed at *m/z* 43.

**FIGURE 20.5** Mass spectrum of ethylbenzene.

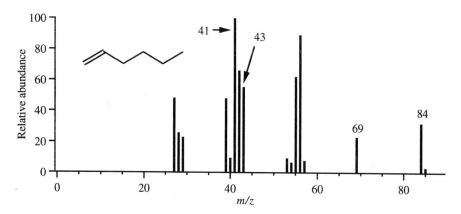

**FIGURE 20.6** Mass spectrum of 1-hexene.

*Alcohols*

Alcohols fragment easily and, as a result, the molecular ion peak is often very weak. In many cases, the molecular ion is not even apparent on the mass spectrum. One pathway is the loss of hydroxyl radical (•OH) to produce a carbocation. However, the most important fragmentation pathway is the loss of an alkyl group from the molecular ion to form resonance-stabilized oxonium ions.

$$\text{R}-\overset{\overset{\displaystyle R'}{|}}{\underset{\underset{\displaystyle R''}{|}}{C}}-\overset{\bullet+}{\underset{\bullet\bullet}{O}}-H \longrightarrow R\bullet + \underset{R''}{\overset{R'}{>}}C=\overset{+}{\underset{\bullet\bullet}{O}}-H$$

The mass spectrum of 2-methyl-2-butanol shown in Figure 20.7 exemplifies the various fragmentation pathways available to alcohols. Notice that the molecular ion (*m/z* 88) is not present in the spectrum.

$$CH_3CH_2-\overset{\overset{\displaystyle CH_3}{|}}{\underset{\underset{\displaystyle CH_3}{|}}{C}}-\overset{\bullet+}{\underset{\bullet\bullet}{O}}-H \longrightarrow CH_3CH_2-\overset{\overset{\displaystyle CH_3}{}}{\underset{\underset{\displaystyle CH_3}{}}{C}}+ \quad + \bullet\overset{\bullet\bullet}{O}H$$

*m/z 71*

$$CH_3CH_2-\overset{\overset{\displaystyle CH_3}{|}}{\underset{\underset{\displaystyle CH_3}{|}}{C}}-\overset{\bullet+}{\underset{\bullet\bullet}{O}}-H \longrightarrow CH_3CH_2\bullet + \underset{CH_3}{\overset{CH_3}{>}}C=\overset{+}{\underset{\bullet\bullet}{O}}-H$$

*m/z 59*

$$CH_3CH_2-\overset{\overset{\displaystyle CH_3}{|}}{\underset{\underset{\displaystyle CH_3}{|}}{C}}-\overset{\bullet+}{\underset{\bullet\bullet}{O}}-H \longrightarrow CH_3\bullet + \underset{CH_3}{\overset{CH_3CH_2}{>}}C=\overset{+}{\underset{\bullet\bullet}{O}}-H$$

*m/z 73*

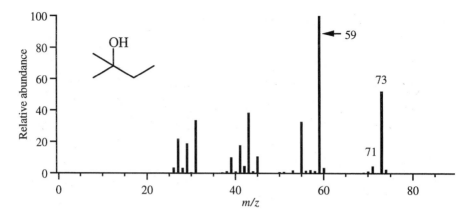

**FIGURE 20.7** Mass spectrum of 2-methyl-2-butanol.

Primary alcohols show an intense $m/z$ 31 peak that arises from a cleavage in which the electron pair on the oxygen atom stabilizes the fragment by resonance:

$$R\overset{\frown}{\phantom{.}}CH_2\overset{\frown}{\phantom{.}}\overset{+}{\overset{..}{O}}H \longrightarrow R\cdot + \left[CH_2 =\overset{+}{\overset{..}{O}}H \longleftrightarrow \overset{+}{C}H_2 -\overset{..}{\overset{..}{O}}H\right]$$
$$m/z\ 31$$

Other heteroatoms also provide the driving force for similar types of cleavage. Appropriately structured amines, ethers, and sulfur compounds undergo fragmentations analogous to those exhibited by alcohols.

$$R\overset{\frown}{\phantom{.}}CH_2\overset{\frown}{\phantom{.}}\overset{+}{\overset{..}{Y}} \longrightarrow R\cdot + \left[CH_2 =\overset{+}{Y} \longleftrightarrow \overset{+}{C}H_2 -\overset{..}{\overset{..}{Y}}\right]$$

$Y = NH_2,\ NHR,\ NR_2,$
OR, SH, or SR

***Carbonyl Compounds*** Ketones and other carbonyl compounds, such as esters, fragment by cleavage of bonds α to the carbonyl group.

$$\begin{matrix}R \\ \diagdown \\ \phantom{.}C=\overset{+}{\overset{..}{O}}\cdot \\ \diagup \\ R'\end{matrix} \longrightarrow R\cdot + \left[\begin{matrix}R'-C\equiv\overset{+}{\overset{..}{O}} \\ \updownarrow \\ R'-\overset{+}{C}=\overset{..}{\overset{..}{O}}\end{matrix}\right]$$

$$\begin{matrix}RO \\ \diagdown \\ \phantom{.}C=\overset{+}{\overset{..}{O}}\cdot \\ \diagup \\ R'\end{matrix} \longrightarrow RO\cdot + \left[\begin{matrix}R'-C\equiv\overset{+}{\overset{..}{O}} \\ \updownarrow \\ R'-\overset{+}{C}=\overset{..}{\overset{..}{O}}\end{matrix}\right]$$

In the spectrum of 2-butanone shown in Figure 20.2, we see fragmentations on both sides of the carbonyl group.

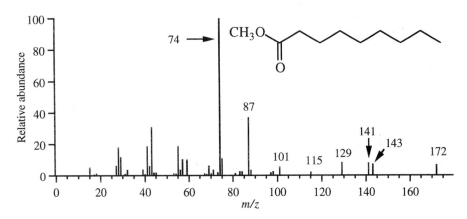

$$CH_3CH_2\text{—} \quad C{=}\ddot{O}{\cdot} \longrightarrow {\cdot}CH_2CH_3 + \left[ \begin{array}{c} CH_3\text{—}C{\equiv}\overset{+}{O} \\ \updownarrow \\ CH_3\text{—}\overset{+}{C}{=}\ddot{O} \end{array} \right] \quad m/z\ 43$$

$$CH_3CH_2 \quad C{=}\ddot{O}{+} \longrightarrow {\cdot}CH_3 + \left[ \begin{array}{c} CH_3CH_2\text{—}C{\equiv}\overset{+}{O} \\ \updownarrow \\ CH_3CH_2\text{—}\overset{+}{C}{=}\ddot{O} \end{array} \right] \quad m/z\ 57$$

In the mass spectrum of the ester methyl nonanoate (MW 172), shown in Figure 20.8, a significant peak at $m/z$ 141 is due to the loss of a fragment with a mass of 31, corresponding to a methoxyl radical.

$$CH_3O\text{—} \quad C{=}\ddot{O}{\cdot} \longrightarrow CH_3O{\cdot} + \left[ \begin{array}{c} C_8H_{17}\text{—}C{\equiv}\overset{+}{O} \\ \updownarrow \\ C_8H_{17}\text{—}\overset{+}{C}{=}\ddot{O} \end{array} \right] \quad m/z\ 141$$
$$C_8H_{17}$$

The base peak at $m/z$ 74 is due to the loss of a fragment with a mass of 98, corresponding to loss of a neutral molecule with a molecular formula $C_7H_{14}$. Carbonyl compounds with alkyl groups containing a chain of three or more carbon atoms can cleave at the $\beta$ bond. This pathway, called the *McLafferty*

FIGURE 20.8 Mass spectrum of methyl nonanoate.

requires the presence of a hydrogen atom on the γ carbon atom.

*m/z 172*                      *m/z 74*

The mass spectrum of methyl nonanoate also demonstrates fragmentations characteristic of organic compounds with straight chain alkyl groups. Carbon-carbon bonds can break at any point along the chain leading to the loss of alkyl radicals.

| *m/z* | | Radical fragment lost from the molecular ion |
|---|---|---|
| 143 | (M − 29) | $CH_3CH_2{}^\bullet$ |
| 129 | (M − 43) | $CH_3CH_2CH_2{}^\bullet$ |
| 115 | (M − 57) | $CH_3(CH_2)_2CH_2{}^\bullet$ |
| 101 | (M − 71) | $CH_3(CH_2)_3CH_2{}^\bullet$ |
| 87 | (M − 85) | $CH_3(CH_2)_4CH_2{}^\bullet$ |

*Fundamental Nitrogen Rule*

The *fundamental nitrogen rule* states that a compound (containing nothing other than C, H, N, or O) whose molecular weight is an even number must contain either no nitrogen atoms or an even number of nitrogen atoms. A compound whose molecular weight is an odd number must contain an odd number of nitrogens. The following compounds support the nitrogen rule:

| Pyridine | Triethylamine | Benzylamine | Ethylenediamine |
|---|---|---|---|
| MW 79 | MW 101 | MW 107 | MW 60 |

A corollary to this rule applies to fragmentation. The corollary states that when a compound with an even molecular weight gives a fragment whose mass is an odd number, simple fragmentation (cleavage of just one covalent bond) must have occurred. Simple cleavage of a molecule whose molecular weight is odd gives a fragment with a mass that is an even number:

$$[H_3C-CH_3]^{+} \longrightarrow CH_3^{+} + CH_3 \cdot$$
MW 30                 m/z 15

$$(CH_3CH_2)_2\overset{+\cdot}{N}\overset{\frown}{CH_2}-CH_3 \longrightarrow (CH_3CH_2)_2\overset{+}{N}=CH_2 + CH_3 \cdot$$
MW 101                                  m/z 86

Although there are exceptions to these rules, in general they are useful in the analysis of mass spectrometric data.

---

**20.5**                    Case Study

In summary, mass spectrometry can be used to identify organic compounds. Molecular weights and molecular formulas can be obtained; both are frequently significant clues to the identity and the structure of an organic compound. Moreover, the basic mechanisms for molecular fragmentation have been extensively studied; the detailed patterns provided by a compound's fragmentation profile can very often establish the identity of the compound. Mass spectrometry is most often used in conjunction with other spectroscopic methods, such as infrared and NMR spectroscopy. Determining a structure from a mass spectrum alone is challenging and in most cases requires supplementary information.

As an example of how to approach MS analysis, consider the mass spectrum shown in Figure 20.9. The molecular ion peak appears at $m/z$ 114. The lack of a significant M+2 peak rules out the presence of any halogen. The IR spectrum of the compound has an intense peak at 1715 cm$^{-1}$, which indicates the presence of a C=O group. Knowing that you are working with a carbonyl compound suggests that the base peak at $m/z$ 57 is a stabilized oxonium-ion fragment, formed by an a cleavage and loss of a $C_4H_9 \cdot$ radical.

$$\overset{C_2H_5}{\underset{C_4H_9}{\diagdown}}C=\overset{..}{\overset{+}{O}} \longrightarrow \cdot C_4H_9 + C_2H_5-C\equiv\overset{+}{\underset{..}{O}}$$
m/z 57

*Rule of Thirteen*            A method known as the *rule of thirteen* can be used to generate the chemical formula of a hydrocarbon, $C_nH_m$, with the requisite molecular weight. The integer obtained by dividing the molecular weight by thirteen (atomic weight of carbon + atomic weight of hydrogen) corresponds to the number of carbon atoms. The remainder from the division is added to the integer to give the number of hydrogen atoms. In our example, the molecular formula for the hydrocarbon is $C_8H_{18}$ ($n = 114/13 = 8$ with a remainder of 10 and $m = 8 + 10 = 18$). Because the compound we are working with contains one or more oxygen atoms, we have to subtract one

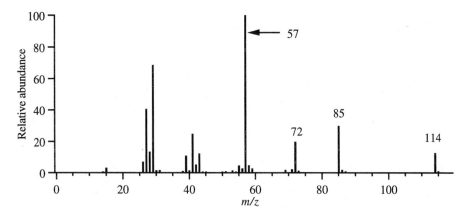

FIGURE 20.9 Mass spectrum of unknown for case study.

or perhaps two carbon atoms and adjust the number of hydrogen atoms to give the appropriate molecular weight. A short list of possible molecular formulas is $C_7H_{14}O$ and $C_6H_{10}O_2$.

Next, an inventory of the significant MS peaks is put together, along with the masses lost on fragmentation.

| m/z | Mass | Possible fragments |
|-----|------|--------------------|
| 114 | M | |
| 85 | M − 29 | $C_2H_5$ |
| 72 | M − 42 | $C_3H_6$ |
| 57 | M − 57 | $C_4H_9$ |

It is likely that this carbonyl compound is a ketone, because the peak at $m/z$ 85 is also consistent with an α cleavage and loss of an ethyl radical to give a stabilized oxonium cation.

$$C_2H_5 + C_4H_9-C\equiv\overset{+}{O}$$
$$m/z\ 85$$

Another important clue in this spectrum is the loss of a neutral molecule, $C_3H_6$, producing the peak at $m/z$ 72. This is the result of a McLafferty rearrangement and requires the presence of a carbonyl group. Moreover, the fragment lost, propene, implies the partial structure, $C_4H_9C{=}O$.

Propene could be lost if a methyl group is attached to the γ carbon atom or to the β carbon atom. Therefore, there are two compounds consistent with all of the evidence, 3-heptanone and 5-methyl-3-hexanone. An NMR spectrum of the compound would establish the structure unambiguously.

3-Heptanone                                        5-Methyl-3-hexanone

## 20.6    Sources of Confusion

As with IR and NMR spectroscopy, it is important to be aware of factors that can lead to unexpected, confusing, or poorly defined peaks in a mass spectrum. Some of the problems can be avoided by proper sample preparation or modifying sampling conditions. Some "problems" are inherent features of the technique.

Small amounts of other compounds can produce MS peaks in regions of the mass spectrum that should be blank, particularly important when you are trying to determine the $m/z$ of the molecular ion. In GC-MS the impurities may be due to residual material from a previous sample injected into the spectrometer or from degradation of the GC column itself. Small peaks at $m/z$ values higher than the molecular weight of any compound in your sample may appear. So it is necessary to be judicious in your assignment of a molecular ion peak. It is also important to allow enough time between samples to clear the previous sample out of the instrument. A background scan can be used to identify peaks due to residual materials in the mass spectrometer.

Many compounds fragment so easily that there is no discernible molecular ion. Examples of these types of compounds are tertiary alcohols, which dehydrate easily, and most alkyl bromides and chlorides, which readily cleave off the bromine or chlorine. Sometimes it is possible to coax a small molecular ion peak from a sample by adjusting instrumental conditions, such as decreasing the ionizing voltage.

As mentioned in Technique 20.2, M+1 and M+2 peaks of compounds containing only carbon, hydrogen, oxygen, and nitrogen can sometimes offer guidance or confirmation regarding a molecular formula. However, the range of error is so broad that molecular formulas cannot be based on these data alone.

Sometimes a mass spectrum of a pure compound exhibits significant peaks that are difficult to rationalize. These fragments may be the result of multiple-step fragmentations or may have been formed in a number of steps involving complex rearrangements. Do not dwell on these peaks.

## References

1. Beynon, J. H.; Saunders, R. A.; Williams, A. E. *The Mass Spectrometry of Organic Molecules*; Elsevier: New York, 1968.
2. Beynon, J. H.; Williams, A. E. *Mass and Abundance Tables for Use in Mass Spectrometry*; Elsevier: New York, 1963.
3. Silverstein, R. M.; Webster, F. X. *Spectrometric Identification of Organic Compounds*; 6th ed.; Wiley: New York, 1998.

## Questions

1. Match the compounds azobenzene, ethanol, and pyridine to their molecular weights: 46, 79, 182. How does the fact that in one case the molecular weight is odd and in the other two cases the molecular weight is even help in the selection process?
2. The mass spectrum of 1-bromopropane is shown in Figure 20.3. Propose a structure for the $m/z$ 43 peak.
3. 1-Pentanol has its base peak at $m/z$ 31, whereas that for 2-pentanol is at $m/z$ 45. Explain briefly.
4. Similar types of cleavages give rise to two peaks for 4-chlorobenzophenone, one at $m/z$ 105 (the base peak), and one at $m/z$ 139 (70% of base). A clue to their identities is the fact that, whereas the 139 peak is accompanied by a peak at 141 that is about one-third the intensity of the 139 peak, the 105 peak has no such partner. What structures correspond to these peaks?

4-Chlorobenzophenone

5. What fragmentations lead to the peaks at $m/z$ 127, 125, and 105 in Figure 20.4?
6. An unknown compound has the mass spectrum shown in Figure 20.10. The molecular ion peak is at $m/z$ 121. The infrared spectrum of the unknown shows a broad band of medium intensity at 3300 $cm^{-1}$. Determine the structure of the unknown. What fragmentations lead to the peaks at $m/z$ 120, 91, and 44?

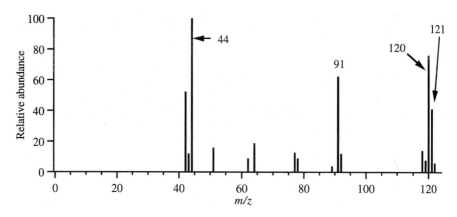

FIGURE 20.10 Mass spectrum of unknown for Question 6.

# INTEGRATED SPECTROSCOPY PROBLEMS

1. A compound of molecular formula $C_4H_8O$ shows a peak of substantial size in its mass spectrum at $m/z$ 43. This compound produces an infrared spectrum that shows, among other absorptions, four peaks in the 2999–2850 $cm^{-1}$ range and a strong peak at 1715 $cm^{-1}$. There are no IR peaks at greater than 3000 $cm^{-1}$. The $^1H$ NMR spectrum contains a triplet (t) at 1.08 ppm (3H), a singlet (s, 3H) at 2.15 ppm, and a quartet (q) at 2.45 ppm (2H). The magnitudes of the splitting of both the quartet and triplet are identical. Deduce the structure of this compound and assign all peaks.

2. A compound shows a molecular ion peak in its mass spectrum at $m/z$ 92 and a satellite peak at $m/z$ 94 that is 36% the intensity of the $m/z$ 92 peak. The $^1H$ NMR spectrum contains only one signal, a singlet at 1.65 ppm. The proton-decoupled $^{13}C$ NMR spectrum reveals a strong peak at 34.6 ppm and a weak peak at 67.2 ppm. Deduce the structure of this compound and assign all peaks.

3. A compound of molecular formula $C_7H_{16}O$ produces an infrared spectrum that shows among other absorptions a strong, very broad peak at 3300 $cm^{-1}$ and a reasonably strong peak at 1080 $cm^{-1}$. The 300-MHz $^1H$ NMR spectrum contains a distorted triplet at 0.9 ppm (3H), a somewhat broadened singlet at 1.22–1.35 ppm (8H), a distorted quintet (2H) at just above 1.4 ppm, a broadened singlet at 2.3 ppm (1H), which has a chemical shift that is strongly concentration dependent, and a 2H triplet at 3.65 ppm. The proton-decoupled $^{13}C$ NMR spectrum reveals singlets at 63, 33, 32, 29, 26, 23, and 14 ppm. Deduce the structure of this compound and assign all peaks.

4. A compound shows a molecular ion peak in its mass spectrum at $m/z$ 112 and strong satellite peaks at $m/z$ 114 (66% the intensity of the $m/z$ 112 peak) and at $m/z$ 116 (11% the intensity of the $m/z$ 112) peak. The $^1H$ NMR spectrum contains a 3H doublet (d) at 1.63 ppm, two 1H doublets of doublets—one at 3.58 ppm ($J = 9$, 7 Hz) and one at 3.78 ppm ($J = 7$, 5 Hz)—and a 1H complex multiplet at 4.16 ppm. The proton-decoupled $^{13}C$ NMR spectrum reveals singlets at 22.6, 49.3 and 56 ppm. DEPT analysis shows these signals to be due respectively to methyl, methylene and methine carbons. Deduce the structure of this compound and assign all peaks.

5. A compound whose mass spectrum reveals its molecular ion peak at $m/z$ 108 produces an infrared spectrum that shows, among other absorptions, a strong, very broad peak at 3300 $cm^{-1}$, a strong peak at 1030 $cm^{-1}$, three weak peaks in the 3100–3000 $cm^{-1}$ range, and two weak peaks in the 2950–2850 $cm^{-1}$ range. The $^1H$ NMR spectrum contains a somewhat broadened singlet at 7.2–7.5 ppm (5H), a singlet (2H) at 4.65 ppm, and a broadened singlet at 2.5 ppm (1H). Deduce the structure of this compound and assign all peaks.

6. A compound shows a molecular ion in its mass spectrum at $m/z$ 108. This compound produces an infrared spectrum that shows, among other absorptions, a strong, broad peak at 3350 $cm^{-1}$, a weak peak at 3020 $cm^{-1}$, a strong peak at 1230 $cm^{-1}$, and weak peaks at 2900 and 2850 $cm^{-1}$. The $^1H$ NMR spectrum contains doublets (d) at 6.7 and 7.05 ppm (2H each), a broadened singlet (s, 1H) at 4.7 ppm, and a singlet (s) at 2.3 ppm (3H). The magnitude of the splitting of the two doublets is identi-

cal. The proton-decoupled $^{13}$C NMR spectrum reveals peaks at 154.6, 130.0, 128.9, 115.2, and 20.4 ppm. DEPT analysis shows the signals at 154.6 and 128.9 ppm to be due to quaternary carbons, the signals at 130.0 and 115.2 ppm to be due to methine carbons, and the signal at 20.4 ppm to be due to a methyl carbon. Deduce the structure of this compound and assign all peaks.

7. A compound shows a molecular ion in its mass spectrum at $m/z$ 101 and a base peak at $m/z$ 30. This compound produces an infrared spectrum that shows two medium-intensity peaks at 3376 cm$^{-1}$. The 300-MHz $^1$H NMR spectrum contains a triplet at 0.9 ppm (6H), a multiplet at 1.2 ppm (1H), a pentet at 1.3 ppm (4H), a doublet at 2.65 ppm (2H), and a broadened singlet at 1.5 ppm (2H) that has a chemical shift that is strongly concentration dependent. The proton-decoupled $^{13}$C NMR spectrum reveals singlets at 11.0, 23.6, 44.1, and 44.5 ppm. DEPT analysis shows the signal at 11.0 ppm to be due to a methyl carbon, the signals at 23.6 and 44.5 ppm to be due to methyl-

ene carbons, and the signal at 44.1 ppm to be due to a methine carbon. Deduce the structure of this compound and assign all peaks.

8. A compound of molecular formula $C_7H_{12}O_2$ shows a peak of substantial size in its mass spectrum at $m/z$ 71. This compound produces an infrared spectrum that shows a strong peak at 1740 cm$^{-1}$ and a medium-intensity peak at 1650 cm$^{-1}$. The 300-MHz $^1$H NMR spectrum contains a triplet at 1.0 ppm (3H), a multiplet at 1.7 ppm (2H), a triplet at 2.3 ppm (2H), a doublet at 4.6 ppm (2H), a doublet at 5.2 ppm (1H), a doublet at 5.3 ppm (1H), and a multiplet at 5.9 ppm (1H). The proton-decoupled $^{13}$C NMR spectrum reveals singlets at 13.7, 18.5, 36.2, 64.9, 117.9, 132.4, and 173.2 ppm. DEPT analysis shows the signal at 13.7 ppm to be due to a methyl carbon, the signals at 18.5, 36.2, 64.9, and 117.9 ppm to be due to methylene carbons, and the signal at 132.4 ppm to be due to a methine carbon. The signal at 173.2 ppm is due to a quaternary carbon. Deduce the structure of this compound and assign all peaks.

# The Literature
# of Organic Chemistry

Before organic chemists carry out reactions in the laboratory, they need to be aware of what is known about the chemicals and reactions they will be using, all of which can be found in the chemistry library. Carrying out any laboratory project requires the use of chemical information sources.

Four types of volumes are found in all chemistry libraries: textbooks, reference books, chemistry journals, and abstracts and indexes of chemical journals. Textbooks and reference books are frequently called secondary sources of information because they compile information from the primary sources, the journals in which the original works were published. Abstracts and indexes aid in locating a journal article about a specific topic or compound.

Now that computer databases are widely used, many of these resources can be accessed by electronic means. Use of databases may allow a search of the chemistry literature to be completed more rapidly and in some cases more comprehensively than by a conventional manual search of abstracts and indexes. We do not provide database information that is current as this book is being written because the databases may well have changed by the time you read the book. The *Journal of Chemical Education* maintains a feature on evaluated Web sites.

## Textbooks

Textbooks contain general information. For example, an undergraduate organic chemistry text will tell you that aldehydes can be oxidized with chromic acid to carboxylic acids or that aromatic amines are weak bases. Specific information about compounds, such as the melting point or the $^1H$ NMR spectrum, is not commonly found in undergraduate textbooks. Advanced textbooks provide a more comprehensive treatment of organic reactions and their applications, and they usually include numerous references to the original literature and to review articles. Some particularly useful advanced texts are the following:

1. Smith, M. B.; March, J. *Advanced Organic Chemistry: Reactions, Mechanisms and Structures*; 5th ed.; Wiley: New York, 2001.
2. Carey, F. A.; Sundberg, R. M. *Advanced Organic Chemistry; Part A: Structure and Mechanisms; Part B: Reactions and Synthesis*; 4th ed.; Kluwer: New York, 2000/2001.
3. Lowry, T. H.; Richardson, K. S. *Mechanism and Theory in Organic Chemistry*; 3rd ed.; Addison-Wesley: New York, 1997.

## Reference Books

Reference books used frequently by organic chemists include the following.

**Handbooks**

Handbooks provide compilations of physical data, such as the boiling point of bromoethane, the melting point of the 3,5-dinitrobenzoate of 2-butanol, or the density of benzaldehyde. Some handbooks, including *The Merck Index* and the *Aldrich Catalog Handbook of Fine Chemicals*, also list references to the primary literature for preparation methods. A quick, straightforward electronic resource is chemfinder.com.

1. Lide, D. R. (Ed.) *The CRC Handbook of Chemistry and Physics*; 82nd ed.; CRC Press: Boca Raton, FL, 2001.
2. Lide, D. R.; Milne, G. W. (Eds.) *Handbook of Data on Common Organic Compounds*; CRC Press: Boca Raton, FL, 1995.
3. Dean, J. A. (Ed.) *Lange's Handbook of Chemistry*; 15th ed.; McGraw-Hill: New York, 1998.
4. *Aldrich Catalog Handbook of Fine Chemicals and Laboratory Equipment*; Aldrich Chemical Co., Milwaukee, WI, published annually.
5. Budavari, S.; O'Neill, M. J.; Smith, A.; Heckelman, P. E.; Kinneary, J. F. (Eds.) *The Merck Index: An Encyclopedia of Chemicals, Drugs and Biologicals*; 12th ed.; Chapman and Hall: New York, 1997.
6. Buckingham, J. B.; (Ed.) *Dictionary of Organic Compounds*; 6th ed.; 9 vols.; CRC Press: Boca Raton, FL, 1997.
7. Gordon, A. J.; Ford, R. A. *The Chemist's Companion: A Handbook of Practical Data, Techniques and References*; Wiley: New York, 1972.
8. *Beilstein's Handbook of Organic Chemistry*; Springer-Verlag: New York. (See discussion under Abstracts and Indexes.)

**Spectral Information**

The following reference books contain spectra of thousands of organic compounds. Additional compounds are added to the Stadtler collections (refs. 3 and 4) annually.

1. Pouchert, C. J.; Behnke, J. (Eds.) *Aldrich Library of $^{13}C$ and $^{1}H$ FT-NMR Spectra*; 3 vols.; Aldrich Chemical Co., Milwaukee, WI, 1992.
2. *The Aldrich Library of FT-IR Spectra*; 2nd ed.; 3 vols.; Aldrich Chemical Co., Milwaukee, WI, 1997.
3. *The Sadtler Collection of High Resolution (NMR) Spectra*; Sadtler Research Laboratories: Philadelphia, 1992.
4. *The Sadtler Reference (IR) Spectra*; Sadtler Research Laboratories: Philadelphia, 1992.

*Reactions, Synthetic Procedures, and Techniques*

The following list gives examples of reference books that contain descriptions of techniques used in organic synthesis (refs. 1–2) and methods, procedures and reactions used in organic synthesis (refs. 13–14). They all include extensive references to the primary literature.

1. Furniss, B. S. (Ed.) *Vogel's Textbook of Practical Organic Chemistry;* 5th ed.; Prentice Hall: Paramus, NJ, 1996.
2. Loewenthal, H. J. E. *A Guide for the Perplexed Organic Experimentalist;* 2nd ed.; Wiley: New York, 1992.
3. Larock, R. C. *Comprehensive Organic Transformations: A Guide to Functional Group Preparations;* 2nd ed.; Wiley, New York, 1999.
4. *Organic Syntheses;* Wiley: New York, 1932–present. Collective Volumes I–VI combine and index ten volumes each through Volume 59, 1988. Collective Volume VII compiles Volumes 60–64. Collective Volume VIII (1993) compiles Volumes 65–69, and Collective Volume 9 (1998) compiles Volumes 70–74. The preparations have been carefully checked in two separate research laboratories.
5. *Organic Reactions;* Wiley: New York, 1932–present.
6. Fieser, L. F.; Fieser, M. *Reagents for Organic Synthesis;* 19 vols.; Wiley: New York, 1967–2000.
7. *Handbook of Reagents for Organic Synthesis;* 4 vols.; Wiley: Chichester, UK, 1999.
8. Coffey, S. (Ed.) *Rodd's Chemistry of Carbon Compounds;* 2nd ed.; Elsevier: New York, 1964–present.
9. Katritzky, A. R.; Meth-Cohn, M.; Rees, C. W. (Eds.) *Comprehensive Organic Functional Group Transformations;* 7 vols.; Elsevier: New York, 1995.
10. Harrison, I. T.; Wade, Jr., L. G.; Smith, M. B. (Eds.) *Compendium of Organic Synthetic Methods;* 8 vols.; Wiley: New York, 1971–1995.
11. Mackie, R. D.; Smith, D. M.; Aitken, R. A. *Guidebook to Organic Synthesis;* 2nd ed.; Halsted: New York, 1990.
12. Paquette, L. A. (Ed.) *Encyclopedia of Reagents for Organic Synthesis;* Wiley: New York, 1995.
13. Sandler, S. R.; Karo, W. *Sourcebook of Advanced Laboratory Preparations;* Academic Press: San Diego, CA, 1992.
14. Trost, B. M.; Fleming, I. (Eds.) *Comprehensive Organic Synthesis;* 8 vols.; Elsevier: New York, 1992.

## Chemistry Journals

Primary journals publish original research results and as such, they are the fundamental source of most information about organic chemistry. References to primary chemistry journals are found in the chapters of general reference books such as those listed in the previous section. For example, you might find the boiling point of vinyl benzoate in *The Dictionary of Organic Compounds*, but a detailed synthetic method for it will be found in a journal article written by an organic chemist who has prepared the compound. Important current journals that publish original papers in organic chemistry include the following: *Angewandte Chemie* (International Edition in English); *Bulletin of the Chemical Society of Japan* (English);

*Canadian Journal of Chemistry; Chemische Berichte* (German); *Journal of the American Chemical Society; Journal of Bioorganic Chemistry; Journal of the Chemical Society* (London), *Chemical Communications; Journal of the Chemical Society* (London), *Perkin Transactions 1; Journal of the Chemical Society* (London), *Perkin Transactions 2; Journal of Heterocyclic Chemistry; Journal of Medicinal Chemistry; Journal of Organic Chemistry; Journal of Physical Organic Chemistry; Organic Letters; Organometallics; Nouveau Journal de Chimie* (French); *Synthesis; Synthetic Communications; Tetrahedron; Tetrahedron Letters.* Many are available online.

## Abstracts and Indexes

Because the literature of chemistry is so vast, finding specific information, such as the preparation of a particular compound or reactions where a specific type of catalyst has been used, would be extremely difficult and time consuming without the existence of a survey of the entire literature of chemistry. *Chemical Abstracts* (CA), published by the American Chemical Society, provides such a survey and is the most complete source of information on chemistry in the world.

*Chemical Abstracts* condenses the content of journal articles into abstracts and publishes indexes of the abstracts by the following categories: subject, author's name, chemical substance, molecular formula, and patent numbers. To locate a compound in *Chemical Abstracts*, it is necessary to understand the nomenclature systems used both currently and in the past; the *Index Guide* (1991) provides a list of indexing terms, rules, and cross references and is the place to begin a search. Each chemical compound has been assigned a number, called a *registry number*, which greatly facilitates finding references to that compound. Registry numbers can be located in the chemical substance index. In evaluating an abstract you need to keep in mind that it gives only a brief summary of an article; you should always consult the original journal article as the final source.

Chemical Abstract Services (CAS), the publishers of *Chemical Abstracts*, also provides a number of databases. The newest of these databases, called *SciFinder Scholar 2001*, is an excellent search engine. If it is available on your campus, you will find it to be invaluable. In addition to *SciFinder Scholar*, CAS provides STN, a more limited but nonetheless helpful database for *Chemical Abstracts*. Today, most college and university libraries are equipped to search *Chemical Abstracts* using these computerized databases. There is a fee for these services, so it is necessary for you to obtain assistance and training before undertaking an online search. Consult the library at your college or university.

*Science Citation Index* lists all articles published in the more prominent journals and also lists all the articles that were cited or referred to in current articles. Each year a *Source Index, Citation Index*, and *Permuterm Subject Index* are published as part of *Science Citation Index*. *Science Citation Index* is also available in a format called *The Web of Science* that allows online computer searching.

*Beilstein's Handbook of Organic Chemistry*, while not an abstract journal or index, serves a similar purpose in locating information about organic compounds. This unique compendium covers all known organic compounds on which information was published through 1959. For each compound, the handbook gives the name (or names), the formula, all known physical properties, methods of synthesis, biological properties, derivatives, and any other information reported about it. Organic compounds are organized into three categories: acyclic (Volumes 1–4), isocyclic (carbocyclic; Volumes 5–16), and heterocyclic (Volumes 17–27). The original volumes (Hauptwerk, H) covered the literature through 1909. Four supplements, (Ergänzungswerke, EI–EIV), cover the years 1910–1959 in ten-year increments. A fifth supplement (EV), in English instead of German, is still in preparation and covers the literature from 1960 to 1979. Every piece of information has a reference to the primary literature and thus data may be checked. Moreover, information is not incorporated into *Beilstein* until it has been evaluated by scientists. The more recent supplements continue to report information on many compounds reported in the original work or earlier supplements. Thus, corrections and updating continue.

Because the system used to classify compounds is complex, several guides to using *Beilstein* have been prepared, including *How to Use Beilstein*, Beilstein Institute, Frankfurt, 1978, and a videocassette by M. M. Vestling, *A Guide to Beilstein's Handbook*, J. Hurley Associates, Boca Raton, FL, 1988. Even with little or no knowledge of German, it is possible to find physical properties and references to the primary literature; the Beilstein Institute also publishes a booklet entitled *Beilstein Dictionary, German-English*. A search for information on a specific compound is most easily started by use of the formula index *(Formelregister)* rather than the subject index *(Sachregister)*. The second supplement *(Zweites Ergänzungswerk)* has a complete formula index in separate volumes, whereas the fourth supplement has a cumulative formula index and a subject index in each volume (Band). After carbon and hydrogen, elements are listed alphabetically. For example, the compound 4-chlorobromobenzene would be listed at $C_6H_4BrCl$ in the formula index. The boldface number immediately following the name of the compound is the volume number, while the page number in the initial edition follows next. The Roman numeral is the supplement (E) number, and it is followed by the page number in that supplement in the same volume given in boldface after the name of the compound. The entry for an organic compound in the *CRC Handbook of Chemistry and Physics* and the *Aldrich Catalog* also gives the location of the compound in Beilstein.

## More Information About the Chemistry Library

We urge you to consult the library at your college or university for assistance in conducting a search for information both in the books and journals found in your library and as online information. The following books and journal articles contain more information about chemistry information sources, how to use them, and how to plan and carry out an online search.

1.  Maizell, R. E. *How to Find Chemical Information: A Guide for Practicing Chemists, Educators, and Students*; 3rd ed.; Wiley: New York, 1998.
2.  Bottle, R. T.; Rowland, J. F. B. *Information Sources in Chemistry*; 4th ed.; Bowker: London, 1993.
3.  Smith, M. B.; March, J. *Advanced Organic Chemistry: Reactions, Mechanisms and Structures*; 5th ed.; Wiley: New York, 2001, Appendix A.
4.  Loewenthal, H. J. E. *A Guide for the Perplexed Organic Experimentalist*; 2nd ed.; Wiley: New York, 1990, pp. 1–81.
5.  Wienbroer, D. R. *Guide to Electronic Research and Documentation*; McGraw-Hill: New York, 1997.
6.  Somerville, A. N. *J. Chem. Educ.* **1991,** *68,* 553–561; 842–853; **1992,** *69,* 379–386.
7.  *Using CAS Databases on STN: Student Manual*; American Chemical Society: Washington, 1995.

# INDEX

Note: Page numbers followed by f indicate figures; those followed by t indicate tables.

# PERIODIC TABLE OF THE ELEMENTS

| 1 IA | 2 IIA | 3 IIIB | 4 IVB | 5 VB | 6 VIB | 7 VIIB | 8 VIIIB | 9 VIIIB | 10 VIIIB | 11 IB | 12 IIB | 13 IIIA | 14 IVA | 15 VA | 16 VIA | 17 VIIA | 18 VIIIA |
|---|---|---|---|---|---|---|---|---|---|---|---|---|---|---|---|---|---|
| 1 **H** 1.0079 | | | | | | | | | | | | | | | | | 2 **He** 4.00 |
| 3 **Li** 6.94 | 4 **Be** 9.01 | | | | | | | | | | | 5 **B** 10.81 | 6 **C** 12.01 | 7 **N** 14.01 | 8 **O** 16.00 | 9 **F** 19.00 | 10 **Ne** 20.18 |
| 11 **Na** 22.99 | 12 **Mg** 24.31 | | | | | | | | | | | 13 **Al** 26.98 | 14 **Si** 28.09 | 15 **P** 30.97 | 16 **S** 32.06 | 17 **Cl** 35.45 | 18 **Ar** 39.95 |
| 19 **K** 39.10 | 20 **Ca** 40.08 | 21 **Sc** 44.96 | 22 **Ti** 47.88 | 23 **V** 50.94 | 24 **Cr** 52.00 | 25 **Mn** 54.94 | 26 **Fe** 55.85 | 27 **Co** 58.93 | 28 **Ni** 58.71 | 29 **Cu** 63.54 | 30 **Zn** 65.37 | 31 **Ga** 69.72 | 32 **Ge** 72.59 | 33 **As** 74.92 | 34 **Se** 78.96 | 35 **Br** 79.91 | 36 **Kr** 83.80 |
| 37 **Rb** 85.47 | 38 **Sr** 87.62 | 39 **Y** 88.91 | 40 **Zr** 91.22 | 41 **Nb** 92.91 | 42 **Mo** 95.94 | 43 **Tc** 98.91 | 44 **Ru** 101.07 | 45 **Rh** 102.91 | 46 **Pd** 106.4 | 47 **Ag** 107.87 | 48 **Cd** 112.40 | 49 **In** 114.82 | 50 **Sn** 118.69 | 51 **Sb** 121.75 | 52 **Te** 127.60 | 53 **I** 126.90 | 54 **Xe** 131.30 |
| 55 **Cs** 132.91 | 56 **Ba** 137.34 | 71 **Lu** 174.97 | 72 **Hf** 178.49 | 73 **Ta** 180.95 | 74 **W** 183.85 | 75 **Re** 186.2 | 76 **Os** 190.2 | 77 **Ir** 192.2 | 78 **Pt** 195.09 | 79 **Au** 196.97 | 80 **Hg** 200.59 | 81 **Tl** 204.37 | 82 **Pb** 207.19 | 83 **Bi** 208.98 | 84 **Po** 210 | 85 **At** 210 | 86 **Rn** 222 |
| 87 **Fr** 223 | 88 **Ra** 226.03 | 103 **Lr** 262.1 | 104 **Rf** | 105 **Db** | 106 **Sg** | 107 **Bh** | 108 **Hs** | 109 **Mt** | 110 **Uun** | 111 **Uuu** | 112 **Uub** | 113 **Uut** | | | | | |

**Lanthanides**

| | | | | | | | | | | | | | |
|---|---|---|---|---|---|---|---|---|---|---|---|---|---|
| 57 **La** 138.91 | 58 **Ce** 140.12 | 59 **Pr** 140.91 | 60 **Nd** 144.24 | 61 **Pm** 146.92 | 62 **Sm** 150.35 | 63 **Eu** 151.96 | 64 **Gd** 157.25 | 65 **Tb** 158.92 | 66 **Dy** 162.50 | 67 **Ho** 164.93 | 68 **Er** 167.26 | 69 **Tm** 168.93 | 70 **Yb** 173.04 |

**Actinides**

| | | | | | | | | | | | | | |
|---|---|---|---|---|---|---|---|---|---|---|---|---|---|
| 89 **Ac** 227.03 | 90 **Th** 232.04 | 91 **Pa** 231.04 | 92 **U** 238.03 | 93 **Np** 237.05 | 94 **Pu** 239.05 | 95 **Am** 241.06 | 96 **Cm** 247.07 | 97 **Bk** 249.08 | 98 **Cf** 251.08 | 99 **Es** 254.09 | 100 **Fm** 257.10 | 101 **Md** 258.10 | 102 **No** 255 |

Nonmetals

Metalloids

Metals

*Molar masses quoted to the number of significant figures given here can be regarded as typical of most naturally occurring samples.